4 0000 001 427 121

W9-BFM-765

DISCARDED

Absolute Measurable Spaces

Absolute measurable space and absolute null space are very old topological notions, developed from descriptive set theory, topology, Borel measure theory and analysis. This monograph systematically develops and returns to the topological and geometrical origins of these notions. Motivating the development of the exposition are the action of the group of homeomorphisms of a space on Borel measures, the Oxtoby–Ulam theorem on Lebesgue-like measures on the unit cube, and the extensions of this theorem to many other topological spaces. Existence of uncountable absolute null space, extension of the Purves theorem, and recent advances on homeomorphic Borel probability measures on the Cantor space are among the many topics discussed. A brief discussion of set-theoretic results on absolute null space is also given.

A four-part appendix aids the reader with topological dimension theory, Hausdorff measure and Hausdorff dimension, and geometric measure theory. The exposition will suit researchers and graduate students of real analysis, set theory and measure theory.

Togo Nishiura is Professor Emeritus at Wayne State University, Detroit, and Associate Fellow in Mathematics at Dickinson College, Pennsylvania.

ENCYCLOPEDIA OF MATHEMATICS AND ITS APPLICATIONS

All the titles listed below can be obtained from good booksellers or from Cambridge University Press. For a complete series listing

visit http://www.cambridge.org/uk/series/sSeries.asp?code=EOM

63. A. C. Thompson *Minkowski Geometry*
64 R. B. Bapat and T. E. S. Raghavan *Nonnegative Matrices with Applications*
65. K. Engel *Sperner Theory*
66. D. Cvetkovic, P. Rowlinson, and S. Simic *Eigenspaces of Graphs*
67. F. Bergeron, G. Labelle, and P. Leroux *Combinational Species and Tree-Like Structures*
68. R. Goodman and N. Wallach *Representations and Invariants of the Classical Groups*
69. T. Beth, D. Jungnickel, and H. Lenz *Design Theory 1, 2nd edn*
70. A. Pietsch and J. Wenzel *Orthonormal Systems for Banach Space Geometry*
71. G. E. Andrews, R. Askey, and R. Roy *Special Functions*
72. R. Ticciati *Quantum Field Theory for Mathematicians*
73. M. Stern *Semimodular Lattices*
74. I. Lasiecka and R. Triggiani *Control Theory for Partial Differential Equations I*
75. I. Lasiecka and R. Triggiani *Control Theory for Partial Differential Equations II*
76. A. A. Ivanov *Geometry of Sporadic Groups I*
77. A. Schinzel *Polynomials with Special Regard to Reducibility*
78. H. Lenz, T. Beth, and D. Jungnickel *Design Theory II, 2nd edn*
79. T. Palmer *Banach Algebras and the General Theory of *-Algebras II*
80. O. Stormark *Lie's Structural Approach to PDE Systems*
81. C. F. Dunkl and Y. Xu *Orthogonal Polynomials of Several Variables*
82. J. P. Mayberry *The Foundations of Mathematics in the Theory of Sets*
83. C. Foias, O. Manley, R. Rosa, and R. Temam *Navier–Stokes Equations and Turbulence*
84. B. Polster and G. Steinke *Geometries on Surfaces*
85. R. B. Paris and D. Kaminski *Asymptotics and Mellin–Barnes Integrals*
86 R. McEliece *The Theory of Information and Coding, 2nd edn*
87. B. Magurn *Algebraic Introduction to K-Theory*
88. T. Mora *Solving Polynomial Equation Systems I*
89. K. Bichteler *Stochastic Integration with Jumps*
90. M. Lothaire *Algebraic Combinatorics on Words*
91. A. A. Ivanov and S. V. Shpectorov *Geometry of Sporadic Groups II*
92 P. McMullen and E. Schulte *Abstract Regular Polytopes*
93. G. Gierz et al. *Continuous Lattices and Domains*
94. S. Finch *Mathematical Constants*
95. Y. Jabri *The Mountain Pass Theorem*
96. G. Gasper and M. Rahman *Basic Hypergeometric Series, 2nd edn*
97. M. C. Pedicchio and W. Tholen (eds.) *Categorical Foundations*
98. M. E. H. Ismail *Classical and Quantum Orthogonal Polynomials in One Variable*
99. T. Mora *Solving Polynomial Equation Systems II*
100. E. Olivier and M. Eulália Vares *Large Deviations and Metastability*
101. A. Kushner, V. Lychagin, and V. Rubtsov *Contact Geometry and Nonlinear Differential Equations*
102. L. W. Beineke, R. J. Wilson, and P. J. Cameron. (eds.) *Topics in Algebraic Graph Theory*
103. O. Staffans *Well-Posed Linear Systems*
104. J. M. Lewis, S. Lakshmivarahan, and S. Dhall *Dynamic Data Assimilation*
105. M. Lothaire *Applied Combinatorics on Words*
106. A. Markoe *Analytic Tomography*
107. P. A. Martin *Multiple Scattering*
108. R. A. Brualdi *Combinatorial Matrix Classes*
110. M.-J. Lai and L. L. Schumaker *Spline Functions on Triangulations*
111. R. T. Curtis *Symmetric Generation of Groups*
112. H. Salzmann, T. Grundhöfer, H. Hähl, and R. Löwen *The Classical Fields*
113. S. Peszat and J. Zabczyk *Stochastic Partial Differential Equations with Lévy Noise*
114. J. Beck *Combinatorial Games*
115. L. Barreira and Y. Pesin *Nonuniform Hyperbolicity*
116. D. Z. Arov and H. Dym J-*Contractive Matrix Valued Functions and Related Topics*
117. R. Glowinski, J.-L. Lions, and J. He *Exact and Approximate Controllability for Distributed Parameter Systems*
118. A. A. Borovkov and K. A. Borovkov *Asympotic Analysis of Random Walks*
119. M. Deza and M. Dutour Sikirić *Geometry of Chemical Graphs*
120. T. Nishiura *Absolute Measurable Spaces*

ENCYCLOPEDIA OF MATHEMATICS AND ITS APPLICATIONS

Absolute Measurable Spaces

TOGO NISHIURA

CAMBRIDGE UNIVERSITY PRESS
Cambridge, New York, Melbourne, Madrid, Cape Town, Singapore, São Paulo, Delhi
Cambridge University Press
The Edinburgh Building, Cambridge CB2 8RU, UK

Published in the United States of America by Cambridge University Press, New York

www.cambridge.org
Information on this title: www.cambridge.org/9780521875561

First published 2008

Printed in the United Kingdom at the University Press, Cambridge

A catalog record for this publication is available from the British Library

ISBN 978-0-521-87556-1 hardback

Contents

Preface

This book is about absolute measurable spaces. What is an absolute measurable space and why study them?

To answer the first question, an absolute measurable space, simply put, is a separable metrizable space X with the property that every topological embedding of X into any separable metrizable space Y results in a set that is *μ-measurable for every continuous, complete, finite Borel measure μ on Y*. Of course, only Borel measures are considered since the topology of Y must play a role in the definition.

For an answer to the second question, observe that the notion of absolute measurable space is a topological one in the spirit of many other notions of "absolute" such as absolute Borel space, absolute G_δ space, absolute retract and many more. As the definition is topological, one is led to many topological questions about such spaces. Even more there are many possible geometric questions about such spaces upon assigning a metric to the space. Obviously, there is also a notion of "absolute null space"; these spaces are those absolute measurable spaces for which all topological copies have μ measure equal to 0. Absolute null spaces are often called "universal measure zero sets" and have been extensively studied. The same topological and geometric questions can be investigated for absolute null spaces. It is well-known that absolute Borel spaces are absolute measurable spaces. More generally, so are analytic and co-analytic spaces. Many topological and geometric questions have already been investigated in the literature for absolute Borel spaces and analytic spaces. The challenge is to prove or disprove analogues of these known results in the context of absolute measurable spaces.

It is clear that absolute measurable spaces are invariant under Borel isomorphism (Borel measurable bijection whose inverse is also Borel measurable). Consequently, each absolute measurable space will correspond to an absolute measurable subspace of the real line $\mathbb{R}$. It would be tempting to investigate only absolute measurable spaces contained in $\mathbb{R}$, which has been extensively done. This would be fine if one is interested only in, say, measure theoretic or set theoretic properties of absolute measurable spaces, but clearly inadequate if one is interested in topological or geometric structures since they may not be preserved by Borel isomorphisms. The emphasis of the book is on topological and geometric properties associated with absolute measurable spaces. Homeomorphisms will be emphasized for topological structures. For geometric structures, one must have a metric assigned to the separable metrizable space – bi-Lipschitzian maps will replace homeomorphisms.

There is a second notion called "universally measurable sets." This notion fixes a space X and considers the collection of all subsets of X that are μ-measurable for every continuous, complete, finite Borel measure μ on X. Obviously a subset of X that is an absolute measurable space is a universally measurable set in X. But a universally measurable set in a space X need not be an absolute measurable space – indeed, for a non-Lebesgue measurable set X of $\mathbb{R}$, the set X itself is a universally measurable set in X that is not an absolute measurable space. It is easily seen that X is an absolute measurable space if and only if every universally measurable set of X is an absolute measurable space.

An extensive literature exists concerning the notions of absolute measurable space and universally measurable set. The 1982 survey article [**18**], written by J. B. Brown and G. V. Cox, is devoted to a large number of classes of "singular" spaces among which is the class of absolute null spaces. Their article is essentially a broad ranging summary of the results up to that time and its coverage is so ambitious that a systematic development from the basics of real analysis and topology has not been presented. There are two other survey articles that are devoted to set theoretic results on certain singular sets. From the set theoretic point of view only subsets of the real line needed to be considered. The first article is a 1984 survey about such subsets by A. W. Miller [**110**] and the second is his 1991 update [**111**]. Absolute measurable spaces and absolute null spaces have appeared also in probability theory – that is, probability theory based on abstract measurable spaces $(X,\mathfrak{A})$ in which metrics are induced on X by imposing conditions on the σ-algebra $\mathfrak{A}$ of measurable sets. Obviously this approach to the notion of absolute measurable space concentrates on probability concepts and does not investigate topological and geometric properties. In 1984, R. M. Shortt investigated metric properties from the probability approach in [**139**] (announced in 1982 [**138**]). Also in non-book form are two articles that appeared much earlier in 1937; one is a commentary by S. Braun and E. Szpilrajn in collaboration with K. Kuratowski that appeared in the "Annexe" [**15**] to the new series of the *Fundamenta Mathematicae* and the other is a fundamental one by Szpilrajn-Marczewski [**152**] that contains a development of the notions of absolute measurable space and universally measurable subsets of a metric space with applications to singular sets. Years have passed since the two articles were written.

The book sets aside many singular sets whose definitions depend on a chosen metric; fortunately, the definition of the Lebesgue measure on the real line depends only on the arithmetic structure of the real number system and is metric independent. This setting aside of metric-dependent singular set theory permits a systematic development, beginning with the basics of topology and analysis, of absolute measurable space and universally measurable sets in a separable metrizable space. Two themes will appear. One deals with the question of the possibility of strengthening theorems by replacing absolute Borel spaces in the hypothesis of known theorems with absolute measurable spaces. The other is an investigation of the possibility of extending topological properties or geometric properties of universally measurable sets in $\mathbb{R}$ to absolute measurable spaces X other than $\mathbb{R}$. The first question is complicated by the following unresolved set theoretic question [**110**] due to R. D. Mauldin. Note that there are $\mathfrak{c}$ Borel sets in $\mathbb{R}$.

(Mauldin) *What is the cardinality of the collection of all absolute measurable subspaces of the real line* $\mathbb{R}$*?* In particular, *are there always more than* $\mathfrak{c}$ *absolute measurable subspaces of* $\mathbb{R}$*?*

The cardinality of absolute null spaces plays a role in Mauldin's question since an absolute measurable space is not necessarily the symmetric difference of an absolute Borel space and an absolute null space.

There are six chapters plus a four-part appendix. The first chapter is a systematic development of the notions of absolute measurable space and absolute null space. Clearly countable separable metrizable spaces are always absolute null spaces. Solutions of the question of the existence, under the usual axioms of set theory, of uncountable absolute null spaces are presented.

The second chapter is a systematic development of the notion of universally measurable sets in a separable metrizable space X. The concept of positive measures (loosely speaking, $\mu(U) > 0$ whenever U is a nonempty open set) is introduced. This concept leads naturally to the operation called *positive closure* which is a topological invariant. Of particular interest is the example $[0, 1]$ and $\mathsf{HOMEO}([0, 1])$, the group of all homeomorphisms of $[0, 1]$. It is a classical result that the collection of all universally measurable sets in $[0, 1]$ is generated by the Lebesgue measure λ on $[0, 1]$ and $\mathsf{HOMEO}([0, 1])$. Even more, it is known that the collection of all positive, continuous, complete, finite Borel measures on $[0, 1]$ is generated by λ and $\mathsf{HOMEO}([0, 1])$.

The topological project of replacing the space $[0, 1]$ with other absolute measurable spaces is the focus of the third chapter. This project, which addresses the second of the two above mentioned classical results, leads naturally to the Oxtoby–Ulam theorem and its many generalizations. The Oxtoby–Ulam theorem does not generalize to the Cantor space $\{0, 1\}^{\mathbb{N}}$. Fortunately there is a Radon–Nikodym derivative version of the Oxtoby–Ulam theorem which includes the Cantor space and allows the introduction of analysis into the book.

There are many results in analysis on functions $f : \mathbb{R} \to \mathbb{R}$ in the context of universally measurable sets in $\mathbb{R}$. Chapter 4 is devoted to the question of the replacement of the domain or the range of f by absolute measurable spaces. The usual approach of using Borel isomorphisms does not necessarily apply to the task at hand. But the results of Chapter 3 can be applied.

Chapter 5 is devoted to geometric properties of universally measurable sets in $\mathbb{R}^n$ – in particular, the Hausdorff measure and Hausdorff dimension of absolute null spaces. Results, due to O. Zindulka, that sharpen the classical inequalities between Hausdorff dimension and topological dimension form the main part of the chapter.

Finally, Chapter 6 is a short discussion of the set theoretic aspect of absolute measurable spaces. The literature on this aspect is quite extensive. Only a brief survey is given of the use of the continuum hypothesis and the Martin axiom in the book. Of particular interest is the topological dimension of absolute null spaces. Surprisingly, the result, due to Zindulka, depends on set axioms.

Appendix A collects together the needed descriptive set theoretic results and measure theoretic results that are used in the book. Developing notational consistency is also an objective of this part. A proof of the Purves theorem is also presented since it is extended to include universally measurable sets and universally null sets in Chapter 2.

Appendix B is a brief development of universally measurable sets and universally null sets from the measure theoretic and probability theoretic point of view, which reverses our "Borel sets lead to probability measures" to "probability measures lead to Borel sets." This reversal places emphasis on Borel isomorphism and not on homeomorphism; consequently, topological and geometrical questions are not of interest here.

Appendix C concerns Cantor spaces (metrizable spaces that are nonempty, compact, perfect and totally disconnected). Cantor spaces have many realizations, for example, k^{ω}, where k is a finite set with $\operatorname{card}(k) > 1$. The homeomorphism equivalence classes of positive, continuous, complete Borel probability measures on a topological Cantor space are not very well understood. Even the Bernoulli measures on k^{ω} are not completely understood. Extensive investigations by many authors have been made for $\operatorname{card}(k) = 2$. In this case a weaker equivalence relation introduces a connection to polynomials with coefficients in $\mathbb{Z}$. These polynomials are special Bernstein polynomials found in classical approximation theory. Recent results of R. Dougherty, R. D. Mauldin and A. Yingst [**47**] and T. D. Austin [**6**] are discussed and several examples from the earlier literature are given. The E. Akin approach of introducing topological linear order into the discussion of Cantor spaces is also included.

Finally, Appendix D is a brief survey of Hausdorff measure, Hausdorff dimension, and topological dimension. These concepts are very important ones in the book. Zindulka's new proof of the classical relationship between the Hausdorff and topological dimensions is given.

The book is somewhat self-contained; many complete proofs are provided to encourage further investigation of absolute measurable spaces.

Acknowledgements

First I want to thank Jack Brown for providing many of the references in his possession and for the help he gave me in the initial phase of the writing of the book. Next, thanks go to R. D. Mauldin for the many conversations we had on various major topics presented in the book; he has written widely on these topics and his knowledge of the literature surrounding the Oxtoby-motivated theorems greatly improved the exposition. Also, I wish to acknowledge, with thanks, several helpful conversations with A. W. Miller. I want to thank K. P. Hart for his help with certain aspects of the development of the material on Martin's axiom. Also, thanks go to P. Mattila for help with geometric measure theory. Two authors, E. Akin and O. Zindulka, were very kind in providing me with their works in prepublication form. Akin's works on the homeomorphism group of the Cantor space and his Oxtoby-motivated theorems were very influential, especially his construction of the bi-Lipschitzian equivalence of Radon–Nikodym derivatives associated with probability measures on the Cantor space; Zindulka's dimension theoretic results motivated the final two chapters and the appendix. Finally I wish to acknowledge the help and support provided by the Library and the Department of Mathematics and Computer Science of Dickinson College.

1

The absolute property

A measure space $\mathrm{M}(X, \mu)$ is a triple $(X, \mu, \mathfrak{M}(X, \mu))$, where μ is a countably additive, nonnegative, extended real-valued function whose domain is the σ-algebra $\mathfrak{M}(X, \mu)$ of subsets of a set X and satisfies the usual requirements. A subset M of X is said to be μ-measurable if M is a member of the σ-algebra $\mathfrak{M}(X, \mu)$.

For a separable metrizable space X, denote the collection of all Borel sets of X by $\mathfrak{B}(X)$. A measure space $\mathrm{M}(X, \mu)$ is said to be *Borel* if $\mathfrak{B}(X) \subset \mathfrak{M}(X, \mu)$, and if $M \in \mathfrak{M}(X, \mu)$ then there is a Borel set B of X such that $M \subset B$ and $\mu(B) = \mu(M)$.[1] Note that if $\mu(M) < \infty$, then there are Borel sets A and B of X such that $A \subset M \subset B$ and $\mu(B \setminus A) = 0$.

Certain collections of measure spaces will be referred to often – for convenience, two of them will be defined now.

NOTATION 1.1 (MEAS ; $\mathrm{MEAS}^{\mathrm{finite}}$). *The collection of all complete, σ-finite Borel measure spaces* $\mathrm{M}(X, \mu)$ *on all separable metrizable spaces X will be denoted by* MEAS. *The subcollection of* MEAS *consisting of all such measures that are finite will be denoted by* $\mathrm{MEAS}^{\mathrm{finite}}$.[2]

In the spirit of absolute Borel space, the notion of absolute measurable space will be defined in terms of μ-measurability with respect to all Borel measure spaces $\mathrm{M}(Y, \mu)$ in the collection MEAS. After the notion of absolute measurable space has been developed, the notion of absolute 0-measure space – more commonly known as absolute null space – is defined and developed. Two early solutions to the question of the existence of uncountable absolute null spaces are presented. They use the notion of *m*-convergence introduced by F. Hausdorff [**73**]. A more recent example, due to E. Grzegorek [**68**], that has other properties is also developed. The theorems due to S. Plewik [**127**, Lemma] and to I. Recław [**130**] will conclude the discussion of existence.

1.1. Absolute measurable spaces

DEFINITION 1.2. *Let X be a separable metrizable space. Then X is called an* absolute measurable space *if, for every Borel measure space* $\mathrm{M}(Y, \mu)$ *in* MEAS, *it is*

[1] Such measures are often called *regular* Borel measures. We have dropped the modifier regular for convenience.

[2] See also equations (A.4) and (A.5) on page 187 of Appendix A.

true that every topological copy M of X that is contained in Y is a member of the σ-algebra $\mathfrak{M}(Y,\mu)$. The collection of all absolute measurable spaces will be denoted by abMEAS.

Obviously, the notion of absolute measurable space is invariant under homeomorphisms. Hence it would be appropriate to define the notion of topological equivalence for Borel measure spaces on separable metrizable spaces. In order to do this we need the following definition of measures $f_{\#}\mu$ induced by measurable maps f.

DEFINITION 1.3 ($f_{\#}\mu$). *Let X and Y be separable metrizable spaces, let $\mathrm{M}(X,\mu)$ be a σ-finite Borel measure space, and let $f: X \to Y$ be a μ-measurable map. A subset M of Y is said to be* ($f_{\#}\mu$)-measurable *if there exist Borel sets A and B in Y such that $A \subset M \subset B$ and $\mu(f^{-1}[B \setminus A]) = 0$.*

It is clear that $\mathrm{M}(f_{\#}\mu, Y)$ is a complete, finite Borel measure on Y whenever $\mu(X) < \infty$, and that $\mathrm{M}(f_{\#}\mu, Y)$ is complete and σ-finite whenever f is a homeomorphism of X into Y and μ is σ-finite.[3]

DEFINITION 1.4. *σ-finite Borel measure spaces $\mathrm{M}(X,\mu)$ and $\mathrm{M}(Y,\nu)$ are said to be* topologically equivalent *if there is a homeomorphism h of X onto Y such that $h_{\#}\mu(B) = \nu(B)$ whenever $B \in \mathfrak{B}(Y)$.*

The last definition does not require that the Borel measure spaces be complete – but $h_{\#}$ does induce complete measure spaces. Hence the identity homeomorphism id_X of a space X yields a complete Borel measure space $\mathrm{M}(\mathrm{id}_{X\#}\,\mu, X)$, indeed, the measure completion of $\mathrm{M}(\mu, X)$.

It is now evident that there is no loss in assuming that the absolute measurable space X is contained in the Hilbert cube $[0,1]^{\mathbb{N}}$ for topological discussions of the notion of absolute measurable space.

1.1.1. Finite Borel measures. Often it will be convenient in discussions of absolute measurable spaces to deal only with finite Borel measure spaces rather than the more general σ-finite ones – that is, the collection $\mathrm{MEAS}^{\mathrm{finite}}$ rather than MEAS. The following characterization will permit us to do this.

THEOREM 1.5. *A separable metrizable space X is an absolute measurable space if and only if, for every Borel measure space $\mathrm{M}(Y,\mu)$ in $\mathrm{MEAS}^{\mathrm{finite}}$, it is true that every topological copy M of X that is contained in Y is a member of $\mathfrak{M}(Y,\mu)$.*

PROOF. Clearly, if a space X is an absolute measurable space, then it satisfies the condition given in the theorem. So suppose that X satisfies the condition of the theorem. Let $\mathrm{M}(Y,\mu)$ be a σ-finite Borel measure space. There is a finite Borel measure space $\mathrm{M}(Y,\nu)$ such that the σ-algebra equality $\mathfrak{M}(Y,\mu) = \mathfrak{M}(Y,\nu)$ holds (see Section A.5 of Appendix A). So $M \in \mathfrak{M}(Y,\mu)$, hence X is an absolute measurable space. □

[3] See Appendix A for more on the operator $f_{\#}$.

1.1.2. Continuous Borel measure spaces. Later it will be necessary to consider the smaller collection of all continuous Borel measure spaces.[4] If this smaller collection is used in Definition 1.2 above, it may happen that more spaces become absolute measurable spaces. Fortunately, this will not be the case because of our assumption that all measure spaces in MEAS are σ-finite. Under this assumption, for a measure μ, the set of points x for which $\mu(\{x\})$ is positive is a countable set. As continuous Borel measures have measure zero for every countable set, the collection of absolute measure spaces will be the same when one considers the smaller collection of all continuous, complete, σ-finite Borel measure spaces. The following notation will be used.

NOTATION 1.6 ($\mathrm{MEAS}^{\mathrm{cont}}$). *The collection of all continuous, complete, σ-finite Borel measure spaces* $\mathrm{M}(X,\mu)$ *on all separable metrizable spaces X is denoted by* $\mathrm{MEAS}^{\mathrm{cont}}$. *That is,*

$$\mathrm{MEAS}^{\mathrm{cont}} = \{\,\mathrm{M}(X,\mu) \in \mathrm{MEAS}\colon \mathrm{M}(X,\mu) \textit{ is continuous}\,\}. \tag{1.1}$$

1.1.3. Elementary properties. Let us describe some properties of absolute measurable spaces. Clearly, each absolute Borel space is an absolute measurable space. The M. Lavrentieff theorem (Theorem A.2) leads to a characterization of absolute Borel spaces. This characterization yields the following useful characterization of absolute measurable spaces.

THEOREM 1.7. *Let X be a separable metrizable space. The following statements are equivalent.*

(1) *X is an absolute measurable space.*
(2) *There exists a completely metrizable space Y and there exists a topological copy M of X contained in Y such that $M \in \mathfrak{M}(Y,\nu)$ for every complete, finite Borel measure space* $\mathrm{M}(Y,\nu)$.
(3) *For each complete, finite Borel measure space* $\mathrm{M}(X,\mu)$ *there is an absolute Borel space A contained in X with $\mu(X \setminus A) = 0$.*

PROOF. It is clear that the first statement implies the second.

Assume that the second statement is true and let $h\colon X \to M$ be a homeomorphism. Then $\mathrm{M}(Y, h_{\#}\mu)$ is a complete Borel measure space in $\mathrm{MEAS}^{\mathrm{finite}}$. There exists a Borel set A' such that $A' \subset M$ and $h_{\#}\mu(M \setminus A') = 0$. As Y is a completely metrizable space, the space A' is an absolute Borel space. The restricted measure space $\mathrm{M}(M, (h_{\#}\mu)|M)$ is complete and is topologically equivalent to $\mathrm{M}(X,\mu)$. So $\mu(X \setminus A) = 0$, where $A = h^{-1}[A']$. As A' is an absolute Borel space, we have A is an absolute Borel space; hence statement (3) follows.

Finally let us show statement (3) implies statement (1). Let Y be a space and let M be a topological copy of X contained in Y. Suppose that $\mathrm{M}(Y,\mu)$ is complete and finite. Then $\mathrm{M}(M, \mu|M)$ is also complete and finite. It is easily seen that statement (3) is invariant under topological equivalence of Borel measure spaces. Hence $\mathrm{M}(M, \mu|M)$

[4] See Appendix A, page 187, for the definition of continuous Borel measure space.

also satisfies statement (3). There is an absolute Borel space A such that $A \subset M$ and $(\mu|M)(M \setminus A) = 0$. As $\mu^*(M \setminus A) = (\mu|M)(M \setminus A) = 0$, we have $M \setminus A \in \mathfrak{M}(Y, \mu)$, whence $M = (M \setminus A) \cup A$ is in $\mathfrak{M}(Y, \mu)$. □

1.1.4. σ-ring properties. As an application of the above theorem, let us investigate a σ-ring property of the collection abMEAS of all absolute measurable spaces. We begin with closure under countable unions and countable intersections.

PROPOSITION 1.8. *If $X = \bigcup_{i=1}^{\infty} X_i$ is a separable metrizable space such that each X_i is an absolute measurable space, then X and $\bigcap_{i=1}^{\infty} X_i$ are absolute measurable spaces.*

PROOF. Let Y be a completely metrizable extension of X and ν be a complete, finite Borel measure on Y. Then $X_i \in \mathfrak{M}(Y, \nu)$ for every i. Hence $X \in \mathfrak{M}(Y, \nu)$ and $\bigcap_{i=1}^{\infty} X_i \in \mathfrak{M}(Y, \nu)$. Theorem 1.7 completes the proof. □

PROPOSITION 1.9. *If $X = X_1 \cup X_2$ is a separable metrizable space such that X_1 and X_2 are absolute measurable spaces, then $X_1 \setminus X_2$ is an absolute measurable space.*

PROOF. Let Y be a completely metrizable extension of X and ν be a complete, finite Borel measure on Y. Then $X_i \in \mathfrak{M}(Y, \nu)$ for $i = 1, 2$. Hence $X_1 \setminus X_2$ is in $\mathfrak{M}(Y, \nu)$. Theorem 1.7 completes the proof. □

The σ-ring property of the collection abMEAS has been established. The next proposition follows from the ring properties.

PROPOSITION 1.10. *If X is a Borel subspace of an absolute measurable space, then X is an absolute measurable space.*

PROOF. Let Y be an absolute measurable space that contains X as a Borel subspace. Let Y_0 be a completely metrizable extension of Y. There exists a Borel subset B of Y_0 such that $X = Y \cap B$. As Y_0 is a completely metrizable space, we have that B is an absolute Borel space, whence an absolute measurable space. The intersection of the spaces Y and B is an absolute measurable space. □

1.1.5. Product properties. A finite product theorem for absolute measurable spaces is easily shown.

THEOREM 1.11. *A nonempty, separable, metrizable product space $X_1 \times X_2$ is also an absolute measurable space if and only if X_1 and X_2 are nonempty absolute measurable spaces.*

The proof is a consequence of the following proposition whose proof is left to the reader as it follows easily from Lemma A.34 in Appendix A.

PROPOSITION 1.12. *Let $\mathrm{M}(Y_1 \times Y_2, \mu)$ be a complete, σ-finite Borel measure space. If X_1 is an absolute measurable subspace of Y_1, then $X_1 \times Y_2$ is μ-measurable.*

PROOF OF THEOREM. Suppose that X_1 and X_2 are nonempty absolute measurable spaces and let Y_1 and Y_2 be completely metrizable extensions of X_1 and X_2, respectively. Consider any complete, finite Borel measure space $\mathrm{M}(Y_1 \times Y_2, \mu)$. By the proposition, $X_1 \times Y_2$ is a μ-measurable subset of $Y_1 \times Y_2$. As $Y_1 \times Y_2$ is a completely metrizable space, we have by the characterization theorem that $X_1 \times Y_2$ is an absolute measurable space. Analogously, $Y_1 \times X_2$ is an absolute measurable space. By the σ-ring properties of absolute measurable spaces, we have that $X_1 \times X_2$ is an absolute measurable space.

For the converse, assume $X_1 \times X_2$ is a nonempty absolute measurable space. Then X_1 and X_2 are nonempty. By Proposition 1.10, we have that $X_1 \times \{x_2\}$ and $\{x_1\} \times X_2$ are absolute measurable spaces. Consequently, X_1 and X_2 are absolute measurable spaces. □

Let us turn to a countable product theorem. To this end, we may assume that X_i, $i \in \mathbb{N}$, is a countable collection of absolute measurable spaces contained in the Hilbert cube $[0,1]^{\mathbb{N}}$. Clearly, $\times_{i\in\mathbb{N}} X_i \subset \left([0,1]^{\mathbb{N}}\right)^{\mathbb{N}}$. A simple application of the finite product theorem and the σ-ring property gives

THEOREM 1.13. *If X_i, $i \in \mathbb{N}$, is a sequence of absolute measurable spaces, then $\times_{i\in\mathbb{N}} X_i$ is an absolute measurable space.*

1.1.6. Inclusion properties.

There are several subclasses of the class of all separable metrizable spaces that are naturally associated with the notion of absolute measurable space. Let us define them now. In order to do this we will need the definitions of analytic and co-analytic spaces.

DEFINITION 1.14. *A separable metrizable space is said to be* analytic *if it is the image of a continuous map on $\mathcal{N}$, where $\mathcal{N}$ is the space $\{x \in [0,1]\colon x \text{ is irrational}\}$. A separable metrizable space is said to be* co-analytic *if it is homeomorphic to the complement of an analytic subspace of some completely metrizable space.*

The space $\mathcal{N}$ is topologically the same as the product space $\mathbb{N}^{\mathbb{N}}$.

It is known that analytic spaces and co-analytic spaces are topologically invariant. Moreover, we have that a subset of a separable metrizable space Y that is an analytic space or a co-analytic space is μ-measurable for every measure space $\mathrm{M}(Y, \mu)$ in MEAS. Hence these spaces are also absolute measurable spaces. It is known that a space is both analytic and co-analytic if and only if it is an absolute Borel space. Also, a separable metrizable space is completely metrizable if and only if it is an absolute G_δ space.[5]

Consider the following classes of spaces.

MET : the class of separable metrizable spaces.
abMEAS : the class of absolute measurable spaces.
ANALYTIC : the class of analytic spaces.
CO-ANALYTIC : the class of co-analytic spaces.

[5] See Appendix A for the assertions made in the paragraph.

abBOR : the class of absolute Borel spaces.
$\mathsf{MET}_{\mathrm{comp}}$: the class of completely metrizable spaces.
ab G_δ : the class of absolute G_δ spaces.

We have the following inclusions.[6]

$$\mathsf{MET} \supset \mathsf{abMEAS} \supset \mathsf{ANALYTIC} \cup \mathsf{CO\text{-}ANALYTIC},$$
$$\mathsf{ANALYTIC} \cap \mathsf{CO\text{-}ANALYTIC} = \mathsf{abBOR} \supset \mathsf{ab\,G}_\delta = \mathsf{MET}_{\mathrm{comp}}.$$

As there are non-Lebesgue measurable subsets of $\mathbb{R}$, the first inclusion of the first line is a proper one. That the second inclusion of the first line is proper will be illustrated by totally imperfect spaces[7] that are also uncountable absolute measure spaces. Such spaces will be shown to exist later in the chapter; they are the uncountable absolute null spaces. To make sure that the space is not co-analytic as well, take a disjoint topological sum of the space with the space that is analytic but not co-analytic.

1.1.7. Invariance of absolute measurable spaces. The collection ANALYTIC is invariant under Borel measurable maps (Theorem A.13) and the collection abBOR is invariant under injective Borel measurable maps (Theorem A.15). There are images of absolute Borel spaces under Borel measurable maps that are not absolute Borel spaces – of course, the images are analytic spaces. This cannot happen if the Borel measurable maps are restricted further to be $\mathfrak{B}$-maps, whose definition (Definition A.18) is repeated next.

DEFINITION 1.15. *Let X and Y be separable metrizable spaces. A Borel measurable mapping $f: X \to Y$ is a* $\mathfrak{B}$-map[8] *if $f[B] \in \mathfrak{B}(Y)$ whenever $B \in \mathfrak{B}(X)$.*

By the R. Purves Theorem A.43, abBOR is invariant under $\mathfrak{B}$-maps. Indeed, let $f: X \to Y$ be a surjective $\mathfrak{B}$-map and let X be in abBOR. Then $\mathrm{card}(U(f)) \leq \aleph_0$ by Purves's theorem, where $U(f)$ is the set of uncountable order of f. Hence, by Theorem A.22, Y is an absolute Borel space. As $B \in \mathfrak{B}(Y)$ implies $B \in \mathsf{abBOR}$, the assertion follows.

An invariance property also holds for the collection abMEAS under Borel isomorphisms. In our book we shall rename Borel isomorphism to be $\mathfrak{B}$-*homeomorphism*.

THEOREM 1.16. *Let $f: X \to Y$ be a surjective $\mathfrak{B}$-homeomorphism of separable metrizable spaces X and Y. Then X is an absolute measurable space if and only if Y is an absolute measurable space.*

PROOF. Suppose $X \in \mathsf{abMEAS}$ and let $\mathrm{M}(Y, \mu)$ be a complete, finite Borel measure space. Then the Borel measure space $\mathrm{M}(X, f^{-1}{}_{\#}\mu)$ is complete and finite. There is an absolute Borel space A' with $A' \subset X$ and $\big(f^{-1}{}_{\#}\mu\big)(X \setminus A') = 0$. Observe that

[6] See also Section A.3.1 of Appendix A.
[7] Totally imperfect spaces are those nonempty separable metrizable spaces that contain no topological copies of the space $\{0, 1\}^{\mathbb{N}}$.
[8] As mentioned in the footnote to Definition A.18, in **[129]** R. Purves calls such maps *bimeasurable* with the extra requirement that X and $f[X]$ be absolute Borel spaces. See that footnote for further comments.

$A = f[A']$ is an absolute Borel space that is contained in Y. Since $\mu(Y \setminus A) = (f^{-1}{}_{\#}\mu)(X \setminus A') = 0$, it follows that Y is an absolute measurable space by Theorem 1.7. □

It was observed that each Borel measure space $\mathrm{M}(X, \mu)$ in MEAS is, in a topological sense, determined by some complete, σ-finite Borel measure on the Hilbert cube $[0, 1]^{\mathbb{N}}$ since X can be topologically embedded into the Hilbert cube. The above theorem shows that, in a measure theoretic sense, each Borel measure space in MEAS is determined by some σ-finite Borel measure on the topological space $\{0, 1\}^{\mathbb{N}}$. Indeed, there is a $\mathfrak{B}$-homeomorphism of the Hilbert cube $[0, 1]^{\mathbb{N}}$ onto $\{0, 1\}^{\mathbb{N}}$, which is topologically embeddable in the interval $[0, 1]$. Hence measure theoretical properties can be studied by concentrating on the topological spaces $\{0, 1\}^{\mathbb{N}}$ and $[0, 1]$. Of course, topological properties are not preserved by $\mathfrak{B}$-homeomorphisms. Consequently, homeomorphisms are still important for our considerations.

1.1.8. More characterizations. Let us conclude this section with one more characterization theorem.

THEOREM 1.17. *Let X be a separable metrizable space. The following statements are equivalent.*

(1) *X is an absolute measurable space.*
(2) *There is a topological copy M of X contained in some absolute measurable space Y such that $M \in \mathfrak{M}(Y, \nu)$ for every complete, finite Borel measure space* $\mathrm{M}(Y, \nu)$.

PROOF. That the first statement implies the second is trivial.

Let us show that the second statement implies the first. To this end let Y be an absolute measurable space and let M be a topological copy of X contained in Y such that $M \in \mathfrak{M}(Y, \mu)$ for every complete, finite Borel measure μ. Let $\mathrm{M}(X, \nu)$ be a complete, finite Borel measure space. Then $\mu = f_{\#}\nu$ is a complete, finite Borel measure on Y, where f yields the embedding. As Y is an absolute measurable space there is an absolute Borel subspace B of Y such that $\mu(Y \setminus B) = 0$. Since $M \in \mathfrak{M}(Y, \mu)$, there exists a set A in $\mathfrak{B}(Y)$ such that $A \subset M$ and $\mu(M \setminus A) = 0$. We may assume $A \subset B$. As $\mu(M) = \mu(Y)$, we have $0 = \mu(M \setminus A) = \mu(Y \setminus A) = \nu(X \setminus f^{-1}[A])$. Since $f^{-1}[A]$ is an absolute Borel space, X is an absolute measurable space. □

1.2. Absolute null spaces

A natural collection of separable metrizable spaces is the one consisting of those spaces X whose topological copies in Y are null sets for every $\mathrm{M}(Y, \mu)$ in $\mathrm{MEAS}^{\mathrm{cont}}$ (see equation (1.1) above), that is, absolute 0-measure spaces. The present day convention is to call these spaces absolute null spaces.

DEFINITION 1.18. *Let X be a separable metrizable space. Then X is called an* absolute null space *if, for every* $\mathrm{M}(Y, \mu)$ *in* $\mathrm{MEAS}^{\mathrm{cont}}$, *it is true that every topological copy M of X that is contained in Y is a member of $\mathfrak{N}(Y, \mu)$, that is, $\mu(M) = 0$. The collection of all absolute null spaces will be denoted by* abNULL.

The reader is reminded that, unlike the definition of absolute measurable spaces, the complete Borel measure spaces $\mathrm{M}(Y, \mu)$ are required to be continuous; for, without this additional condition, only the empty space would be a member of abNULL.

There is the following analogue of Theorem 1.5.

THEOREM 1.19. *A separable metrizable space X is an absolute null space if and only if, for every continuous, complete, finite Borel measure space $\mathrm{M}(Y, \mu)$, it is true that every topological copy M of X that is contained in Y is a member of $\mathfrak{N}(Y, \mu)$.*

The proof is analogous to that of Theorem 1.5 and is left to the reader.

1.2.1. Characterization. We have the following characterizations of absolute null spaces.

THEOREM 1.20. *Let X be a separable metrizable space. Then the statement*

(o) *X is an absolute null space*

is equivalent to each of the following four statements.

(α) *If X is a subspace of a separable completely metrizable space Y and if $\mathrm{M}(Y, \mu)$ is a continuous, complete, σ-finite Borel measure space, then $\mu(X) = 0$.*
(β) *If $\mathrm{M}(X, \mu)$ is a continuous, complete, σ-finite Borel measure space, then $\mu(X) = 0$.*
(γ) *Every subspace of X is an absolute measurable space.*
(δ) *X is both a totally imperfect space and an absolute measurable space.*

PROOF. Clearly the statement (o) implies statement (α). That (α) implies (β) is equally clear since, with the aid of the inclusion map, any continuous, complete, σ-finite Borel measure space $\mathrm{M}(X, \mu)$ can be extended to a continuous, complete, σ-finite Borel measure space $\mathrm{M}(Y, \nu)$ for any completely metrizable extension Y of X.

To prove that (β) implies (o) let Y be a separable metrizable space that contains X and let $\mathrm{M}(Y, \nu)$ be a continuous, complete, finite Borel measure space. Then $\mu = \nu|X$ is a continuous, complete, finite Borel measure on X. If X satisfies (β), then $0 = \mu(X) = \nu^*(X)$, whence $\nu(X) = 0$. Thereby (β) implies (o).

Obviously (o) implies (γ).

To show that (γ) implies (δ), assume that X satisfies (γ). Then X is an absolute measurable space. Suppose that X is not totally imperfect. Then X contains a topological copy of $\{0, 1\}^{\mathbb{N}}$. As $\{0, 1\}^{\mathbb{N}}$ contains a nonabsolute measurable space, (γ) is not satisfied by X. Hence X satisfies (δ).

Finally, suppose that X satisfies (δ) and let $\mathrm{M}(X, \mu)$ be a continuous, complete, σ-finite Borel measure space. As X is an absolute measurable space, by Theorem 1.7, there is an absolute Borel subspace A of X such that $\mu(X \setminus A) = 0$. As A is also totally imperfect, we have $\operatorname{card}(A) \leq \aleph_0$ whence $X \in \mathfrak{N}(Y, \mu)$. Hence ($\delta$) implies ($\beta$). □

1.2.2. Cardinal number consequences. At this point let us consider cardinal numbers of sets which have not been assigned topological structures. We shall consider nonnegative real-valued, continuous measures μ on sets X for which every subset is μ-measurable, denoted by the measure space $(X, \mu, P(X))$, where $P(X) = \{E : E \subset X\}$ is the power set of X. If one assigns the discrete topology to X, then X is a completely metrizable space (but not necessarily separable). With this topology, we have $\mathfrak{B}(X) = P(X)$, whence the continuous, complete, finite Borel measure space $\mathrm{M}(X, \mu)$ is exactly the same as $(X, \mu, P(X))$. We have the following theorem due to W. Sierpiński and E. Szpilrajn [**142**].

THEOREM 1.21 (Sierpiński–Szpilrajn). *If* $\mathfrak{n}$ *is the cardinality of an absolute null space* X_0, *then each nonnegative real-valued, continuous measure space* $(X, \mu, P(X))$ *with* $\operatorname{card}(X) = \mathfrak{n}$ *is the* 0 *measure space.*

PROOF. Let X_0 be a separable metrizable space with $\operatorname{card}(X_0) = \mathfrak{n}$ and let $f : X \to X_0$ be a bijective map. The discrete topology on X makes f continuous, whence Borel measurable. Then $\nu = f_{\#}\mu$ is a continuous, complete, finite Borel measure on X_0. (Note that $f_{\#}\mu$ is well defined since f is a bijection.) So μ is identically equal to 0 if and only if ν is identically equal to 0. The above characterization, Theorem 1.20, completes the proof. □

1.2.3. Product theorem. As an application of the above characterization theorem we will give a product theorem for absolute null spaces.

THEOREM 1.22. *A nonempty, separable metrizable product space* $X_1 \times X_2$ *is also an absolute null space if and only if* X_1 *and* X_2 *are nonempty absolute null spaces.*

PROOF. Suppose that X_1 and X_2 are nonempty absolute null spaces. By the characterization theorem, X_1 and X_2 are totally imperfect spaces and absolute measurable spaces. By Proposition A.26, $X_1 \times X_2$ is a totally imperfect space; and, by Theorem 1.11, $X_1 \times X_2$ is an absolute measurable space. Hence by the characterization theorem, the product $X_1 \times X_2$ is an absolute null space.

Conversely, suppose that $X_1 \times X_2$ is a nonempty absolute null space, whence $X_1 \times X_2$ is a totally imperfect space and an absolute measurable space. By Theorem 1.11, X_1 and X_2 are absolute measurable spaces. Also, by Proposition A.26, $X_1 \times \{x_2\}$ and $\{x_1\} \times X_2$ are totally imperfect spaces. Hence, X_1 and X_2 are nonempty, absolute null spaces by the characterization theorem. □

Observe that $\{0, 1\}^{\mathbb{N}}$ and $\mathbb{N}^{\mathbb{N}}$ are countable products of absolute null spaces. Hence, unlike absolute measurable spaces, there is no countable product theorem for absolute null spaces.

1.2.4. A mapping theorem. A proof of the converse part of the above product theorem can be achieved by means of the next mapping theorem.

THEOREM 1.23. *Let* X *and* Y *be separable metrizable spaces and let* $f : X \to Y$ *be a Borel measurable map such that* $f[X]$ *is an absolute null space. Then,* X *is an absolute null space if and only if* $f^{-1}[\{y\}]$ *is an absolute null space for each* y *in* $f[X]$.

PROOF. Suppose $X \in \mathsf{abNULL}$. Since $f^{-1}[\{y\}]$ is a subset of X, we have $f^{-1}[\{y\}] \in \mathsf{abNULL}$.

To prove the converse, let μ be a continuous, complete, finite Borel measure on X. Then $f_{\#}\mu$ is a finite, complete Borel measure on Y. Let us show that it is also continuous. As $f^{-1}[\{y\}]$ is a Borel set and $f^{-1}[\{y\}] \in \mathsf{abNULL}$, we have $\mu\big(f^{-1}[\{y\}]\big) = 0$, whence $f_{\#}\mu\big(\{y\}\big) = 0$. So, $f_{\#}\mu$ is continuous. As $f[X] \in \mathsf{abNULL}$, there is a Borel set B in Y such that $f[X] \subset B$ and $f_{\#}\mu(B) = 0$. Hence $\mu(X) = 0$. □

We leave the second proof of the product theorem to the reader.

Here is an interesting and useful lemma. Its proof is left to the reader.

LEMMA 1.24. *Let X and Y be separable metrizable spaces and let $f : X \to Y$ be an arbitrary surjection. Then there is a separable metrizable space Z and there are continuous maps $f_1 : Z \to X$ and $f_2 : Z \to Y$ such that $f = f_2 f_1^{-1}$, f_1 is bijective, and f_2 is surjective. Indeed, $Z = \operatorname{graph}(f)$ and the natural projection maps satisfy the requirements.*

The following is an application of the lemma to absolute null spaces.

THEOREM 1.25. *Let X be an absolute null space and Y be an arbitrary nonempty separable metrizable space. If $f : X \to Y$ is an arbitrary function, then the graph of f is an absolute null space. Consequently, $f[X]$ is the continuous image of some absolute null space.*

PROOF. As the natural projection map f_1 of the lemma is a continuous bijection of $\operatorname{graph}(f)$ onto X, the above Theorem 1.23 applies. The map f_2 of the lemma completes the proof. □

***1.2.5. σ-ideal property of* abNULL.** Theorem 1.20 together with σ-ring properties of abMEAS and Proposition A.26 will yield the following σ-ideal property of the collection abNULL of all absolute null spaces.

PROPOSITION 1.26. *The collection* abNULL *possesses the properties*

(1) *if $X \subset Y \in \mathsf{abNULL}$, then $X \in \mathsf{abNULL}$,*
(2) *if a separable metrizable space X is a countable union of subspaces from* abNULL, *then $X \in \mathsf{abNULL}$.*

1.3. Existence of absolute null spaces

It is obvious that every countable space is an absolute null space. Hence the question of the existence of an uncountable absolute null space arises. Hausdorff gave a sufficient condition for the existence of such spaces in **[73]**. With the aid of this condition, two examples of uncountable absolute null spaces will be presented, the first by Hausdorff **[73]** and the second by Sierpiński and Szpilrajn **[142]**. Hausdorff's condition is shown to yield a theorem that characterizes the existence of uncountable absolute null spaces contained in separable metrizable spaces, more briefly, uncountable absolute null subspaces.

1.3.1. Hausdorff sufficient condition. To give Hausdorff's theorem on sufficiency we must define the following.

DEFINITION 1.27. *Let X be an uncountable separable metrizable space. A transfinite sequence B_α, $\alpha < \omega_1$, of subsets of X is said to be* m-convergent *in X if*

(1) *B_α is a nonempty μ-measurable subset of X for every α and for every continuous, complete, σ-finite Borel measure space* $\mathrm{M}(X, \mu)$,
(2) *$B_\alpha \cap B_\beta = \emptyset$ whenever $\alpha \neq \beta$,*
(3) *$X = \bigcup_{\alpha<\omega_1} B_\alpha$,*
(4) *for each continuous, complete, σ-finite Borel measure space* $\mathrm{M}(X, \mu)$ *there exists an ordinal number β such that $\beta < \omega_1$ and $\mu\big(\bigcup_{\beta\le\alpha<\omega_1} B_\alpha\big) = 0$.*

The next result, which is implicit in [**73**], is due to Hausdorff.

THEOREM 1.28 (Hausdorff). *Assume X is an uncountable separable metrizable space. If there exists a transfinite sequence B_α, $\alpha < \omega_1$, that is m-convergent in X, then X' is an absolute null subspace of X with* $\mathrm{card}(X') = \aleph_1$ *whenever* $0 < \mathrm{card}(X' \cap B_\alpha) \le \aleph_0$ *for every α.*

The proof will follow immediately from the next lemma.

LEMMA 1.29. *Let X be a separable metrizable space and let B_α, $\alpha < \omega_1$, be a transfinite sequence that is m-convergent in X. If B'_α is a nonempty absolute Borel space contained in B_α for each α, then $X' = \bigcup_{\alpha<\omega_1} B'_\alpha$ is an absolute measurable space. Moreover, the set X' is an uncountable absolute null space if and only if* $0 < \mathrm{card}(X' \cap B_\alpha)$ *for uncountably many α, and* $\mathrm{card}(X' \cap B_\alpha) \le \aleph_0$ *for each α.*

PROOF. Let Y be a separable completely metrizable space that contains X and let $\mathrm{M}(Y, \nu)$ be a complete, finite Borel measure space. Let $\mu = \nu|X$ be the restriction measure. Then there is an ordinal number β with $\beta < \omega_1$ and $\mu\big(\bigcup_{\beta\le\alpha<\omega_1} B_\alpha\big) = 0$. So, $\nu^*\big(\bigcup_{\beta\le\alpha<\omega_1} B'_\alpha\big) = \mu\big(\bigcup_{\beta\le\alpha<\omega_1} B'_\alpha\big) = 0$ because $\bigcup_{\beta\le\alpha<\omega_1} B'_\alpha \subset X$ and $\mu = \nu|X$. From the completeness of the measure ν we have $\nu\big(\bigcup_{\beta\le\alpha<\omega_1} B'_\alpha\big) = 0$. Since B'_α, $\alpha < \beta$, is a countable collection of absolute Borel spaces, we see that X' is ν-measurable, whence X' is an absolute measurable space by Theorem 1.7. Moreover, $\bigcup_{\beta\le\alpha<\omega_1} B'_\alpha \in \mathsf{abMEAS}$ whenever $\beta < \omega_1$.

To prove the second statement, assume that X' is an uncountable absolute null space. Then $X' \cap B_\alpha = B'_\alpha$ is an absolute null space as well as an absolute Borel space, whence $\mathrm{card}(X' \cap B_\alpha) \le \aleph_0$ for each α. As X' is uncountable, we have $0 < \mathrm{card}(X' \cap B_\alpha)$ for uncountably many α. Conversely, for each α, suppose that $\mathrm{card}(X' \cap B_\alpha) \le \aleph_0$, and let $\mathrm{M}(X', \nu)$ be a continuous, complete, σ-finite Borel measure space. With $\mu = f_\#\nu$, where f is the inclusion map $X' \subset X$, we have $\mathrm{M}(X, \mu)$ is also a continuous, complete, σ-finite Borel measure space. Hence there is a β with $\mu\big(\bigcup_{\beta\le\alpha<\omega_1} B_\alpha\big) = 0$. Consequently we have $\nu\big(\bigcup_{\beta\le\alpha<\omega_1} B'_\alpha\big) = 0$. As $\mathrm{card}\big(\bigcup_{\alpha<\beta} B'_\alpha\big) \le \aleph_0$ we have $\nu(X') = 0$. Hence $X' \in \mathsf{abNULL}$ by statement (β) of Theorem 1.7. We have that X' is uncountable since $0 < \mathrm{card}(X' \cap B_\alpha)$ for uncountably many α. □

1.3.2. The Hausdorff example. Hausdorff proved in [**73**] the existence of a transfinite sequence B_α, $\alpha < \omega_1$, that is *m*-convergent in $\{0,1\}^{\mathbb{N}}$. As the proof is quite elementary we shall present it. We follow the proof found in R. Laver [**88**].

Although the proof can be carried out in the space $\{0,1\}^{\mathbb{N}}$, the description of the constructions is easier made in the topologically equivalent space ${}^{\mathbb{N}}\{0,1\}$, the collection of all functions from $\mathbb{N}$ into $\{0,1\}$. Hence we shall work in the space ${}^{\mathbb{N}}\{0,1\}$.

Let f and g be members of ${}^{\mathbb{N}}\{0,1\}$. Then f is said to be *eventually* less than or equal to g, denoted $f \leq^* g$, if there is an m in $\mathbb{N}$ such that $f(n) \leq g(n)$ whenever $n \geq m$. We write $f <^* g$ if $f \leq^* g$ holds and $g \leq^* f$ fails (that is, $f(n) < g(n)$ for infinitely many n). It is not difficult to see that $\leq^*$ and $<^*$ are transitive. We shall use $[f,g]$ to denote the set $\{x : f \leq^* x \leq^* g\}$. Observe that $[f,f]$ is a countably infinite set and that $[0,1] = {}^{\mathbb{N}}\{0,1\}$, where 0 and 1 are respectively constantly 0 and 1 on $\mathbb{N}$. Note that $\{x : 0 \leq x(n) \leq f(n)\}$ is compact for each n. Since $[0,f] = \bigcup_{m=1}^{\infty} \bigcap_{m \leq n} \{x : 0 \leq x(n) \leq f(n)\}$ we have that $[0,f]$ is a σ-compact set. Moreover, the above Hausdorff's binary relation $\leq^*$, as a subset of ${}^{\mathbb{N}}\{0,1\} \times {}^{\mathbb{N}}\{0,1\}$, is equal to the set $\{(f,g) : f \leq^* g\} = \bigcup_{m=1}^{\infty} \bigcap_{m \leq n} \{(f,g) : f(n) \leq g(n)\}$, which is a σ-compact subset of ${}^{\mathbb{N}}\{0,1\} \times {}^{\mathbb{N}}\{0,1\}$.

Hausdorff asserted: *If* $[f_\alpha, g_\alpha]$, $\alpha < \omega_1$, *is a nested collection whose intersection* $\bigcap_{\alpha < \omega_1} [f_\alpha, g_\alpha]$ *is empty, then the nonempty sets among*

$$B_0 = [0,1] \setminus [f_0, g_0], \qquad B_\alpha = [f_\alpha, g_\alpha] \setminus [f_{\alpha+1}, g_{\alpha+1}], \quad 0 < \alpha < \omega_1,$$

is m-convergent in ${}^{\mathbb{N}}\{0,1\}$. Such a nested collection of sets was called an Ω-Ω^* *gap* by Hausdorff. His assertion will follow if, for each continuous, complete, finite Borel measure μ on ${}^{\mathbb{N}}\{0,1\}$, there is an α with $\alpha < \omega_1$ such that $\mu([f_\alpha, g_\alpha]) = 0$. Let us show that this is so.

Let $T_\alpha = [f_\alpha, g_\alpha]$, $\alpha < \omega_1$, be an Ω-Ω^* gap. As $\mu(T_\alpha)$, $\alpha < \omega_1$, is a non-increasing transfinite sequence of real numbers, there is an η with $\eta < \omega_1$ such that $\mu(T_\alpha) = \mu(T_\eta)$ whenever $\eta < \alpha$. Let us suppose $\varepsilon = \mu(T_\eta) > 0$ and derive a contradiction. As $\{f_\eta(n), g_\eta(n)\} \subset \{0,1\}$ for each n, there is a j in T_η such that $\mu(\{f \in T_\eta : f(n) = j(n)\}) \geq \varepsilon/2$ for every n. Let α be such that $\alpha > \eta$ and $\{f_\eta, g_\eta, j\} \cap T_\alpha = \emptyset$. As $f_\eta <^* f_\alpha <^* g_\alpha <^* g_\eta$, the set of n for which $0 \leq f_\eta(n) < g_\eta(n) \leq 1$ holds is infinite and among them are infinitely many with $f_\alpha(n) = g_\alpha(n)$. Let n_i be an enumeration of those n such that $0 = f_\eta(n) \leq f_\alpha(n) = g_\alpha(n) \leq g_\eta(n) = 1$. Then $T_\alpha = \bigcup_{m=1}^{\infty} T_{\alpha,m}$, where

$$T_{\alpha,m} = \{f \in T_\alpha : f(n_i) = f_\alpha(n_i) \text{ whenever } i > m\}.$$

Pick an m with $\mu(T_{\alpha,m}) > \varepsilon/2$. Since $j \notin T_\alpha$ there is an n_i with $i > m$ such that $j(n_i) \neq f_\alpha(n_i)$. Hence, for f in T_η, we have $f(n_i) = j(n_i)$ if and only if $f(n_i) \neq f_\alpha(n_i)$. So,

$$T_\eta \supset \{f \in T_\alpha : f(n_i) \neq j(n_i)\} = \{f \in T_\alpha : f(n_i) = f_\alpha(n_i)\} \supset T_{\alpha,m}.$$

We now have $\varepsilon = \mu(T_\eta) \geq \mu\big(\{f \in T_\eta : f(n_i) = j(n_i)\}\big) + \mu(T_{\alpha,m}) > \varepsilon/2 + \varepsilon/2$ and a contradiction has appeared. Hausdorff's assertion is proved.

It remains to prove the existence of an Ω-Ω^* gap. To this end, define

$$\delta(f,g) = \min\{\, m : f(n) \leq g(n) \text{ whenever } m \leq n \,\}.$$

Of course, $\delta(f,g) = \infty$ if $f \leq^* g$ fails. The following statement is easily proved:

$$\delta(f,g) \leq \max\{\, \delta(f,h), \delta(h,g) \,\} \text{ whenever } f \leq^* h \leq^* g.$$

For subsets F of ${}^{\mathbb{N}}\{0,1\}$ and for h in ${}^{\mathbb{N}}\{0,1\}$ such that $f <^* h$ for every f in F, Hausdorff defined in [**73**] the *property* $P(F,h)$:

$$\operatorname{card}(\{f \in F : \delta(f,h) = m\}) < \aleph_0 \text{ for every } m \text{ in } \mathbb{N}.$$

He observed the following easily proved fact about this property: *If F and G are nonempty subsets of ${}^{\mathbb{N}}\{0,1\}$ such that $f <^* g$ for every f in F and every g in G, and such that $P(F,g)$ holds for every g in G, then $P(F,h)$ holds whenever h satisfies $f <^* h <^* g$ for every f in F and every g in G.*

For an h in ${}^{\mathbb{N}}\{0,1\}$ and a nonempty subset F of ${}^{\mathbb{N}}\{0,1\}$ define

$$F_{<h} = \{f \in F : f <^* h\}.$$

Hausdorff also proved the next three key lemmas.

Lemma 1.30. *Let $\alpha < \omega_1$. If*

$$F = \{f_\beta : \beta < \alpha\} \quad \textit{and} \quad G = \{g_\beta : \beta < \alpha\}$$

are such that $f_\beta \leq^* f_{\beta+1} <^* g_{\beta+1} \leq^* g_\beta$ for every β, then there exists an h such that $f_\beta <^* h <^* g_\beta$ for every β. Moreover, if $P(F_{<f_\beta}, g_\beta)$ holds, then $P(F_{<f_\beta}, h)$ holds.

Proof. Let β_k, $k = 1,2,\ldots$, be a sequence such that $\beta_k \leq \beta_{k+1}$ for all k and $\lim_{k\to\infty} \beta_k = \alpha$. One can easily construct a sequence $N_0, N_1, N_2, \ldots$ of disjoint intervals of $\mathbb{N}$ with $\mathbb{N} = \bigcup_{k=0}^{\infty} N_k$ such that, for each k,

$$f_{\beta_1}(n) \leq \cdots \leq f_{\beta_{k-1}}(n) \leq f_{\beta_k}(n) \leq g_{\beta_k}(n) \leq g_{\beta_{k-1}}(n) \leq \cdots \leq g_{\beta_1}(n),$$

for every n in $N_{2k-1} \cup N_{2k}$, and $f_{\beta_k}(n) < g_{\beta_k}(n)$ for some n in N_{2k-1} and also for some n in N_{2k}. For $k = 0$, let $h(n) = f_{\beta_1}(n)$ for every n in N_0; and, for $k > 0$, let $h(n) = f_{\beta_k}(n)$ for every n in N_{2k-1}, and $h(n) = g_{\beta_k}(n)$ for every n in N_{2k}. For every k, it follows easily that $f_{\beta_k}(n) \leq h(n) \leq g_{\beta_k}(n)$ for every n in $\bigcup_{j\geq k} N_j$, that $h(n) < g_{\beta_k}(n)$ infinitely often in $\bigcup_{j\geq k} N_{2j-1}$, and that $f_{\beta_k}(n) < h(n)$ infinitely often in $\bigcup_{j\geq k} N_{2j}$. Hence $f_{\beta_k} <^* h <^* g_{\beta_k}$ for every k. The lemma follows from the above Hausdorff observation. □

Lemma 1.31. *Let $\{f_i : i = 1,2,\ldots\} \cup \{h, h'\} \subset {}^{\mathbb{N}}\{0,1\}$ be such that $f_i <^* f_{i+1} <^* h' <^* h$ for every i. For each i let H_i be a finite subset of ${}^{\mathbb{N}}\{0,1\}$ such that $f_i \in H_i$ and* *such that* $f_i <^* f <^* f_{i+1}$ *whenever* $f \in H_i$ *and* $f \neq f_i$. *Then there is a g in* ${}^{\mathbb{N}}\{0,1\}$ *such that, for each i,* $f <^* g <^* h'$ *and* $\delta(f,g) \geq i-1$ *whenever* $f \in H_i$.

PROOF. Let $m_{k,0} = \max\{\delta(f_i,f) : f \in H_i, i \leq k\}$. Since it is given that $f_i <^* f_{i+1} <^* h' <^* h$ for every i we can inductively select a sequence m_k, $k = 1, 2, \ldots$, such that m_k and $N_k = [m_k, m_{k+1})$ satisfy

(1) $m_k \geq m_{k,0} + k$,
(2) $f(n) \leq f_{j+1}(n) \leq h'(n) \leq h(n)$ for n in N_k and f in H_j whenever $j \leq k$,
(3) there is an n_0 in N_k such that $f(n_0) = f_{k+1}(n_0) < h'(n_0)$ whenever $f \in H_k$,
(4) there is an n_1 in N_k such that $f_k(n_1) < f_{k+1}(n_1) = h'(n_1) = h(n_1)$,
(5) for each f in H_k that is not f_k there is an n_2 in N_k such that $f_k(n_2) < f(n_2)$.

Define $N_0 = [0, m_1)$, and define g to be

$$g(n) = \begin{cases} f_1(n), & \text{if } n \in N_0 \\ f_k(n), & \text{if } n \in N_k \text{ whenever } k > 0. \end{cases}$$

It is easily seen that $\mathbb{N} = \bigcup_{k=0}^{\infty} N_k$ and $f <^* g <^* h'$ whenever $f \in H_k$. For $k > 1$ let us compute a lower bound for $\delta(f,g)$ for each f in H_k. In N_{k-1} there is an n_1 such that $g(n_1) = f_{k-1}(n_1) < f_k(n_1)$, and $f_k(n) \leq g(n)$ whenever $n \geq m_k$. Hence $\delta(f_k, g) \geq m_{k-1} \geq k-1$. For each f in H_k that is not f_k, there is an n_2 in N_k such that $g(n_2) = f_k(n_2) < f(n_2)$, and $f(n) \leq f_{k+1}(n) \leq g(n)$ whenever $n \geq m_{k+1}$. Hence $\delta(f,g) \geq m_k \geq k$. Consequently, $\delta(f,g) \geq k-1$ whenever $f \in H_k$. □

The proof of the next lemma is taken from [73].

LEMMA 1.32. *Let* α *be a countable limit ordinal number. If h and* $F = \{f_\beta : \beta < \alpha\}$ *are such that* $f_0 <^* f_1 <^* \ldots <^* f_\beta <^* h$ *and such that* $P(F_{<f_\beta}, h)$ *holds whenever* $\beta < \alpha$, *then there exists a g such that* $f_\beta <^* g <^* h$ *whenever* $\beta < \alpha$ *and such that* $P(F,g)$ *holds.*

PROOF. As $<^*$ orders F we shall write $F_{<\beta}$ for $F_{<f_\beta}$. There is an increasing sequence β_k, $k = 1, 2, \ldots$, of ordinal numbers such that $\lim_{k\to\infty} \beta_k = \alpha$ and $\beta_1 = 0$. For each f and each m, define $B_m(f) = \{f_\beta \in F : \delta(f_\beta, f) = m\}$.

We follow Hausdorff's argument. Define the possibly infinite sets

$$A^k = \bigcup_{m \leq k} \{f_\beta : f_\beta \in B_m(h)\}, \quad k = 1, 2, \ldots,$$

and, for each j, define the finite set (since $P(F_{<\beta}, h)$ holds)

$$H_j^k = \{f_{\beta_j}\} \cup \left(A^k \cap (F_{<\beta_{j+1}} \backslash F_{<\beta_j})\right).$$

Observe that $F = \bigcup_{k=1}^{\infty} A^k$ and $A^k = \bigcup_{j=0}^{\infty} H_j^k$.

With the aid of Lemmas 1.30 and 1.31, we can find a g^1 such that $f_\beta <^* g^1 <^* h$ for every β and such that $\delta(f_\beta, g^1) \geq i-1$ whenever $f_\beta \in H_i^1$. Let $f_\beta \in B_m(g^1) \cap A^1$. There is an i such that $f_{\beta_i} \leq^* f_\beta <^* f_{\beta_{i+1}}$. Hence, $i - 1 \leq \delta(f_\beta, g^1) = m$, or

$i \le m+1$. That is, $B_m(g^1) \cap A^1 \subset F_{<\beta_{m+2}} \cap A^1 \subset \bigcup_{j=0}^{m+2} H_j^1$. As the right-hand side of the inclusion is finite we have that $P(A^1, g^1)$ holds. Inductively employing Lemma 1.30, we can construct a sequence g^k, $k = 1, 2, \ldots$, such that $f_\beta <^* \ldots <^* g^{k+1} < g^k <^* \ldots <^* g^1 <^* h$ whenever $\beta < \alpha$ and such that $P(A^k, g^k)$ holds for every k. By Lemma 1.30 there is a g such that $f_\beta <^* g <^* g^k$ for every β and every k. We prove that $P(F, g)$ holds by establishing a contradiction. Suppose that $F' = \{f_\gamma : \delta(f_\gamma, g) = m\}$ is infinite for some m. From $\delta(f_\gamma, h) \le \max\{\delta(f_\gamma, g), \delta(g, h)\}$ we infer $F' \subset A^k$ for some k. So $P(A^k, g)$ also fails for this k. But, $P(A^k, g^k)$ holds by the construction. By Hausdorff's observation, we have that $P(A^k, g)$ holds. Thereby a contradiction has occurred. □

The last lemma is the inductive step for a transfinite construction in the Hausdorff existence theorem.

THEOREM 1.33 (Hausdorff). *There exists an Ω-Ω^* gap.*

PROOF. Let $f_0 = 0$ and $g_0 = 1$. Suppose for $\alpha < \omega_1$ that

$$f_0 <^* f_1 <^* \ldots <^* f_\beta <^* f_{\beta+1} <^* \ldots <^* g_{\beta+1} <^* g_\beta <^* \ldots g_1 <^* g_0$$

is such that $P(F_{<\beta}, g_\beta)$ holds whenever $\beta < \alpha$. We must find f_α and g_α such that $f_\beta <^* f_\alpha <^* g_\alpha <^* g_\beta$ whenever $\beta < \alpha$ and such that $P(F_{<\alpha}, g_\alpha)$ holds. This is very easy if α is not a limit ordinal. For the limit ordinal case, there are h and h' such that $f_\beta <^* h' <^* h <^* g_\beta$ whenever $\beta < \alpha$. An application of Lemma 1.32 provides a g_α such that $P(F_{<\alpha}, g_\alpha)$ holds. Let f_α be such that $f_\beta < f_\alpha < g_\alpha$ whenever $\beta < \alpha$. The α-th step of the transfinite construction is now completed. Suppose that there is an h such that $f_\alpha <^* h <^* g_\alpha$ for every α. Then for some m there will be uncountably many f_β such that $\delta(f_\beta, h) = m$. Hence $P(F_{<\alpha}, h)$ fails for some α, which contradicts the Hausdorff observation that $P(F_{<\alpha}, h)$ holds whenever $f_\alpha < h < g_\alpha$. □

1.3.3. The Sierpiński and Szpilrajn example. Sierpiński and Szpilrajn gave this example in [**142**]. It uses the constituent decomposition of co-analytic spaces (see Appendix A, page 181).

THEOREM 1.34 (Sierpiński–Szpilrajn). *Every co-analytic space X that is not an analytic space has a transfinite sequence B_α, $\alpha < \omega_1$, in* abBOR *that is m-convergent in X.*

PROOF. From equation (A.1) on page 181 of Appendix A, we have

$$X = Y \setminus A = \bigcup_{\alpha<\omega_1} A_\alpha,$$

where Y is some separable completely metrizable space and A is an analytic space. From Theorem A.5, Corollary A.7 and Theorem A.9, we have that the constituents A_α are absolute Borel spaces and that the collection of those constituents which are nonempty is uncountable. Hence there is a natural transfinite subsequence B_α, $\alpha < \omega_1$, of Borel sets such that the first three conditions of the definition of

m-convergence in X are satisfied. To verify the fourth condition let $\mathrm{M}(X, \mu)$ be a complete, σ-finite Borel measure space. There is a σ-compact subset E of Y such that $E \subset X$ and $\mu(X \setminus E) = 0$. By equations (A.2) and (A.3) of Theorem A.6, there is an ordinal number β such that $\beta < \omega_1$ and $E \subset \bigcup_{\alpha<\beta} B_\alpha$, whence $\mu\big(\bigcup_{\beta \leq \alpha < \omega_1} B_\alpha\big) = 0$. Thereby the m-convergence is verified. □

We infer from the above theorem that $\{0,1\}^{\mathbb{N}}$ has a transfinite sequence $B_\alpha, \alpha < \omega_1$, that is m-convergent in $\{0,1\}^{\mathbb{N}}$. Indeed, select a co-analytic subset X of $\{0,1\}^{\mathbb{N}}$ that is not an analytic space and let B'_α, $\alpha < \omega_1$, be m-convergent in X as provided by the lemma. Then simply let $B_0 = B'_0 \cup (\{0,1\}^{\mathbb{N}} \setminus X)$ and $B_\alpha = B'_\alpha$ for $\alpha \neq 0$.

1.3.4. A characterization theorem. We now have the promised theorem.

THEOREM 1.35. *Let X be a separable metrizable space. Then the following three statements are equivalent.*

(1) *X contains an uncountable absolute null space.*
(2) *X contains an uncountable absolute measurable space.*
(3) *X has a transfinite sequence B_α, $\alpha < \omega_1$, that is m-convergent in X.*

PROOF. As absolute null spaces are absolute measurable spaces, we have that statement (1) implies statement (2).

To prove that statement (2) implies statement (3), suppose that X' is an absolute measurable space that is contained in X. If X' is already an uncountable absolute null space, then X' contains a subset X_0 with $\operatorname{card}(X_0) = \aleph_1$. Let $B_0 = X \setminus X_0$ and B_α be singleton subsets of X_0 for the remaining ordinal numbers α. Hence statement (3) follows for this case. So assume that X' is not an absolute null space. Then there exists a continuous, complete, σ-finite Borel measure space $\mathrm{M}(X, \mu)$ with $\mu(X') > 0$. As X' is an absolute measurable space there is an absolute Borel space B contained in X' with $\mu(X' \setminus B) = 0$. So B is uncountable. Let Y be a topological copy of $\{0,1\}^{\mathbb{N}}$ contained in B and define $B_0 = (X \setminus Y) \cup B'_0$ and $B_\alpha = B'_\alpha$ for $0 < \alpha < \omega_1$, where B'_α, $\alpha < \omega_1$ is a transfinite sequence that is m-convergent in Y. It is easily seen that $B_\alpha, \alpha < \omega_1$, is m-convergent in X.

That statement (3) implies statement (1) is Hausdorff's sufficiency theorem (Theorem 1.28). □

This characterization theorem has been proved without the aid of the continuum hypothesis. Also, as the above two examples provide the existence of uncountable absolute null spaces without the aid of the continuum hypothesis, we have

THEOREM 1.36. *There exist absolute null spaces of cardinality $\aleph_1$.*

The use of the continuum hypothesis in the early years of the subject of absolute null spaces will be commented on below. Absolute null spaces have played a special role in what is called the Ulam numbers. That is, in Section 1.2.2 it was shown that $\aleph_1$ is an Ulam number. Other remarks on Ulam numbers will be given below in the Comment section.

1.3.5. Existence under the continuum hypothesis. There is a novel presentation of the existence of absolute null spaces with the aid of the continuum hypothesis in the book *Measure and Category* by J. C. Oxtoby [**120**]. The development there uses a partition theorem that will permit the use of the Hausdorff condition of the characterization theorem (Theorem 1.35).

Let us begin with the statement of the *partition theorem*. This theorem is a purely set theoretic one; that is, there are no topological assumptions made. Also the continuum hypothesis is not required. For the reader's benefit, we shall include also the beautiful proof in [**120**].

THEOREM 1.37. *Let X be a set with* $\operatorname{card}(X) = \aleph_1$, *and let $\mathcal{K}$ be a class of subsets of X with the following properties*:

(1) *$\mathcal{K}$ is a σ-ideal,*
(2) *the union of $\mathcal{K}$ is X,*
(3) *$\mathcal{K}$ has a subclass $\mathcal{G}$ with* $\operatorname{card}(\mathcal{G}) = \aleph_1$ *and the property that each member of $\mathcal{K}$ is contained in some member of $\mathcal{G}$,*
(4) *the complement of each member of $\mathcal{K}$ contains a set with cardinality $\aleph_1$ that belongs to $\mathcal{K}$.*

Then X can be decomposed into $\aleph_1$ disjoint sets X_α, each of cardinality $\aleph_1$, such that a subset E of X belongs to $\mathcal{K}$ if and only if E is contained in a countable union of the sets X_α. Moreover, each X_α is in the σ-ring generated by $\mathcal{G}$.

PROOF. Let G_α, $\alpha < \omega_1$, be a well-ordering of $\mathcal{G}$. For each α define

$$H_\alpha = \bigcup_{\beta\le\alpha} G_\beta \quad \text{and} \quad K_\alpha = H_\alpha \setminus \bigcup_{\beta<\alpha} H_\beta .$$

Put $B = \{\alpha \colon K_\alpha \text{ is uncountable}\}$. Properties (1), (3) and (4) imply $\sup B = \omega_1$. Therefore there exists an order-preserving bijection φ of $\{\alpha \colon \alpha < \omega_1\}$ onto B. For each α, define

$$X_\alpha = H_{\varphi(\alpha)} \setminus \bigcup_{\beta<\alpha} H_{\varphi(\beta)} .$$

By construction and property (1), the sets X_α are disjoint and belong to $\mathcal{K}$. Since $X_\alpha \supset K_{\varphi(\alpha)}$, each of the sets X_α has cardinality $\aleph_1$. For each β we have $\beta < \varphi(\alpha)$ for some α ; therefore,

$$G_\beta \subset H_\beta \subset H_{\varphi(\alpha)} = \bigcup_{\gamma\le\alpha} X_\gamma .$$

Hence, by property (3), each member of $\mathcal{K}$ is contained in a countable union of the sets X_α. Using property (2), we see that

$$X = \bigcup_{K\in\mathcal{K}} K \subset \bigcup_{\alpha<\omega_1} X_\alpha .$$

Thus $\{X_\alpha : \alpha < \omega_1\}$ is a decomposition of X with the required properties. □

As an application of this decomposition, consider the σ-ring that consists of all sets of first category of Baire.

LEMMA 1.38. *Assume the continuum hypothesis. In an uncountable, separable completely metrizable space that contains no isolated points, the collection $\mathcal{K}$ of all sets of the first category of Baire and the collection $\mathcal{G}$ of all F_σ first category sets satisfy the conditions of the partition theorem.*

PROOF. Under the continuum hypothesis the cardinality of $\mathcal{G}$ is $\aleph_1$. As X is of the second category of Baire and contains no isolated points, the complement of a first category set is uncountable. Hence the conditions of the partition theorem are easily verified. Moreover, the sets X_α of the partition are absolute Borel spaces. □

We now have the following theorem.

THEOREM 1.39 (Lusin). *Assume the continuum hypothesis. In an uncountable, separable completely metrizable space that contains no isolated points, there exists a set X of cardinality $\mathfrak{c}$ $(= \aleph_1)$ such that every set K of the first category of Baire satisfies* $\operatorname{card}(K \cap X) \leq \aleph_0$.

PROOF. From the decomposition, which is assured by the above lemma, select exactly one point from each X_α to form the set X. Let K be any set of the first category of Baire. There is a β with $\beta < \omega_1$ and $K \subset \bigcup_{\alpha \leq \beta} X_\alpha$. Hence, $\operatorname{card}(K \cap X) \leq \aleph_0$. □

Finally we have need of the following lemma which will be left as an exercise.

LEMMA 1.40. *Let X be an uncountable, separable completely metrizable space and let* $\mathrm{M}(X, \mu)$ *be a continuous, complete, σ-finite Borel measure space on X. If D is a countable dense subset of X, then there exists a G_δ subset E of X that contains D such that $\mu(E) = 0$ and $X \setminus E$ is an uncountable F_σ subset of X of the first category of Baire.*

Now the above lemma and Theorem 1.37 (the partition theorem) yield a transfinite sequence X_α, $\alpha < \omega_1$, in X that is m-convergent, whence there exists an uncountable absolute null space in X. That is, the following proposition results. *Assume the continuum hypothesis. If X is an uncountable, separable completely metrizable space that contains no isolated points, then there exists an uncountable absolute null space contained in X.* Observe that every uncountable absolute Borel space contains a topological copy of the Cantor space. Hence the continuum hypothesis implies the existence of an uncountable absolute null space contained in each uncountable completely metrizable space.

We remind the reader that it was the novelty of the partition theorem that motivated this section on the continuum hypothesis.

1.4. Grzegorek's cardinal number κ_G

It is easily seen that uncountable absolute measurable spaces that are not absolute null spaces must have cardinality $\mathfrak{c} = 2^{\aleph_0}$. The question of the existence of a non absolute null space X that has the same cardinality as some uncountable absolute null space X is addressed in this section. This is a question of S. Banach [7, Problem P 21], which was proposed in a measure theoretic form and which is precisely stated in footnote 10 after all the required definitions are given. We shall present an example, due to Grzegorek [**68**], which gives an affirmative answer. The existence of the example relies on the cardinal number κ_G used by Grzegorek, which will be defined shortly.

1.4.1. An embedding. We have seen earlier that the notion of absolute null space has connections to the Ulam numbers, a purely set theoretic notion related to the existence of measures. (See also the Comment section for Ulam numbers.) It will be convenient now to consider a purely *measure theoretic setting*.[9] Let S be a set and let $\mathfrak{A}$ be a σ-algebra of subsets of S. A nonempty member C of $\mathfrak{A}$ is called an *atom* if it is a minimal element under inclusion. Denote by $\mathcal{A}(\mathfrak{A})$ the collection of all atoms that are contained in $\mathfrak{A}$. A *countably generated* σ-algebra $\mathfrak{A}$ is a pair $(\mathfrak{A}, \mathfrak{E})$ such that $\mathfrak{E} \subset \mathfrak{A}$ and $\mathfrak{A}$ is the smallest σ-algebra that contains $\mathfrak{E}$. (Often we will not display $\mathfrak{E}$.) A countably additive, finite measure μ on a countably generated σ-algebra $\mathfrak{A}$ on S is called a *nontrivial continuous measure* if $0 < \mu(S)$ and if $\mu(C) = 0$ whenever $C \in \mathcal{A}(\mathfrak{A})$. A countably generated σ-algebra $\mathfrak{A}$ on S is said to be *measurable* if there exists a nontrivial, continuous measure on $\mathfrak{A}$. Otherwise, $\mathfrak{A}$ is said to be *nonmeasurable* (hence, $\mu(S) = 0$). A σ-algebra $\mathfrak{A}$ on S is said to be *separable* if it is countably generated and $\mathcal{A}(\mathfrak{A}) = \{\{s\}: s \in S\}$.[10] Observe that the set X of a separable metrizable space has associated with it the natural σ-algebra $\mathfrak{B}(X)$ of all Borel subsets of X. Even more, this σ-algebra is separable – indeed, any collection $\mathfrak{E}$ that is a countable basis for the open sets of X generates $\mathfrak{B}(X)$. The σ-algebra $\mathfrak{B}(X)$ is measurable if and only if there is a nontrivial, finite, continuous Borel measure on the separable metrizable space.

Let us show that there is a natural injection of the set S into the product space $\{0,1\}^{\mathbb{N}}$ produced by a separable σ-algebra $\mathfrak{A}$ on a set S. Suppose that $(\mathfrak{A}, \mathfrak{E})$ is a separable σ-algebra on a set S and let $\{E_i : i \in \mathbb{N}\}$ be a well ordering of the collection $\mathfrak{E}$. For each set E_i let $f_i : S \to \{0,1\}$ be its characteristic function; that is, $f_i(s) = 1$ if $s \in E_i$, and $f_i(s) = 0$ if $s \notin E_i$. This sequence of characteristic functions defines an injection f of S into $\{0,1\}^{\mathbb{N}}$. Observe that the topology of $\{0,1\}^{\mathbb{N}}$ induces a topology on $f[S]$ and hence a natural topology τ on S is induced by $\mathfrak{E}$. Indeed, the set

$$E_i = \{s \in S : f_i(s) = 1\} = f^{-1}\big[\{x \in \{0,1\}^{\mathbb{N}} : x_i = 1\}\big]$$

is both closed and open in the topology τ. Also, $S \setminus E_i$ is both closed and open in the topology τ. Hence the natural topology τ on S induced by $\mathfrak{E}$ is separable and metrizable, the collection $\mathfrak{A}$ is the collection of all Borel sets in this topological space, and this topological space is topologically embeddable into $\{0,1\}^{\mathbb{N}}$. Consequently, if the resulting topological space is an absolute null space, then the separable σ-algebra $\mathfrak{A}$

[9] A discussion of the measure theoretic and probability theoretic approaches to absolute measurable spaces and absolute null spaces can be found in Appendix B.

[10] The above mentioned Grzegorek example solves the *Banach problem* in [7, Problem P 21] which is the following: *Does there exist two countably generated σ-algebras $\mathfrak{A}_1$ and $\mathfrak{A}_2$ on $\mathbb{R}$ that are measurable such that the σ-algebra generated by $\mathfrak{A}_1 \cup \mathfrak{A}_2$ is nonmeasurable?* Banach's problem will be discussed further in Chapter 6.

on S is nonmeasurable. And, if the separable σ-algebra $\mathfrak{A}$ on S is nonmeasurable, then every finite, complete Borel measure μ on the topological space S is trivial. Hence the separable metrizable space S is an absolute null space if and only if the separable σ-algebra $\mathfrak{A}$ on S is nonmeasurable.

Let us summarize the above discussion as follows.

THEOREM 1.41. *Suppose that X is a separable metrizable space. Then the natural σ-algebra $\mathfrak{B}(X)$ of the set X is separable and measurable if and only if there is a nontrivial, continuous, finite, Borel measure space* $\mathrm{M}(X, \mu) = \big(X, \mu, \mathfrak{M}(X, \mu)\big)$.

THEOREM 1.42. *Let $\mathfrak{A}$ be a separable σ-algebra on a set S and let $\mathfrak{E}$ be a countable subcollection of $\mathfrak{A}$ that generates $\mathfrak{A}$. Then there is a natural topology τ on the set S induced by $\mathfrak{E}$ such that $\mathfrak{A}$ is precisely the collection of all Borel subsets of S in the topology τ, and this topological space is embeddable into* $\{0, 1\}^{\mathbb{N}}$. *Moreover, the topological space is an absolute null space if and only if* $(\mathfrak{A}, \mathfrak{E})$ *is a nonmeasurable separable σ-algebra on the set S.*

1.4.2. Grzegorek's example. A set S is said to *support a measurable separable σ-algebra* if there exists a σ-algebra $\mathfrak{A}$ of subsets of S such that $\mathfrak{A}$ is measurable and separable. Following Grzegorek [**68**], we define κ_{G} to be the cardinal number

$$\kappa_{\mathrm{G}} = \min \{ \operatorname{card}(S) \colon S \text{ supports a measurable separable } \sigma\text{-algebra} \}.$$

Note that $\operatorname{card}(S) = \mathfrak{c}$ whenever S is a Borel measurable subset of $\mathbb{R}$ with positive Lebesgue measure. Hence, $\aleph_1 \leq \kappa_{\mathrm{G}} \leq \mathfrak{c}$. Also, if S is a subset of a separable metrizable space X such that $\operatorname{card}(S) < \kappa_{\mathrm{G}}$, then $\mu(S) = 0$ for every continuous, complete, finite Borel measure on X. Indeed, suppose that $\operatorname{card}(S) < \kappa_{\mathrm{G}}$ and that some continuous, complete, finite Borel measure μ on S has $\mu(S) > 0$. Then $\mathfrak{A} = \mathfrak{B}(S)$ is a measurable separable σ-algebra since the measure space $\mathrm{M}(S, \nu)$ with $\nu = \mu | \mathfrak{B}(S)$ (the function μ restricted to $\mathfrak{B}(S)$) is nontrivial. Hence $\kappa_{\mathrm{G}} \leq \operatorname{card}(S) < \kappa_{\mathrm{G}}$ and a contradiction has occurred.

Denote by $\mathsf{M}\big(\{0, 1\}^{\mathbb{N}}\big)$ the collection of all continuous, complete, finite Borel measures on $\{0, 1\}^{\mathbb{N}}$. For a subset S of $\{0, 1\}^{\mathbb{N}}$ and a μ in $\mathsf{M}\big(\{0, 1\}^{\mathbb{N}}\big)$, the outer measure $\mu^*(S)$ is a well defined nonnegative real number. Define the cardinal number κ_0 as follows:

$$\kappa_0 = \min \Big\{ \operatorname{card}(S) \colon \mu^*(S) > 0 \text{ for some } \mu \text{ in } \mathsf{M}\big(\{0, 1\}^{\mathbb{N}}\big) \Big\}.$$

We prove the following lemma due to Grzegorek [**68**].

LEMMA 1.43. *The cardinal numbers κ_{G} and κ_0 are equal.*

PROOF. Let $S \subset \{0, 1\}^{\mathbb{N}}$ and $\mu \in \mathsf{M}\big(\{0, 1\}^{\mathbb{N}}\big)$ with $\mu^*(S) > 0$. Then the restricted measure $\nu = \mu | S$ and $\mathfrak{A} = \mathfrak{B}(S)$ will result in a measurable separable σ-algebra $\mathfrak{A}$ on the set S, whence $\kappa_{\mathrm{G}} \leq \kappa_0$. To prove the other inequality, let μ be a nontrivial measure on a separable σ-algebra $\mathfrak{A}$ on a set S. A topology τ on the set S corresponding to this separable σ-algebra $\mathfrak{A}$ results in a topological embedding f of S into $\{0, 1\}^{\mathbb{N}}$. Hence

$\nu = f_{\#}\mu$ is a nontrivial, continuous, complete, finite Borel measure on $\{0,1\}^{\mathbb{N}}$ and $\nu^*(f[S]) = \mu(S) > 0$. As $\mathrm{card}(f[S]) = \mathrm{card}(S)$, we have $\kappa_0 \le \mathrm{card}(S)$. Therefore, $\kappa_0 \le \kappa_G$ and the lemma is proved. □

We now give Grzegorek's theorem.

THEOREM 1.44 (Grzegorek). *Let $S \subset \{0,1\}^{\mathbb{N}}$ and let μ be a continuous, complete, finite Borel measure on $\{0,1\}^{\mathbb{N}}$ such that* $\mathrm{card}(S) = \kappa_0$ *and* $\mu^*(S) > 0$. *Then there exists a nonmeasurable separable σ-algebra $\mathfrak{A}$ on S.*

PROOF. Let $\nu = \mu|S$ be the nontrivial continuous measure on the σ-algebra $\mathfrak{B}(S)$ of all Borel subsets of the topological space S and let $\mathcal{B} = \{\, U_i : i < \omega_0 \,\}$ be a countable base for the open subsets of S. Let s_α, $\alpha \in \kappa_G$,[11] be a well ordering of the set S. For each α let G_α be an open subset of the metrizable space S such that $\{\, s_\beta : \beta < \alpha \,\} \subset G_\alpha$ and $\nu(G_\alpha) \le \frac{1}{2}\,\nu(S)$.

Define Y to be the following subset of the product $\kappa_G \times S$.

$$Y = \bigcup_{\alpha \in \kappa_G} \big(\{\alpha\} \times G_\alpha\big).$$

For each i, let

$$E_i = \{\alpha \in \kappa_G : U_i \subset G_\alpha\,\}.$$

Let us show

$$Y = \bigcup_{i<\omega_0} E_i \times U_i.$$

Clearly, if $\alpha \in E_i$, then $\{\alpha\} \times U_i \subset \{\alpha\} \times G_\alpha$. Hence the left-hand member of the equality contains the right-hand member. Let $(\alpha, s) \in Y$. Then $s \in G_\alpha$. Hence there is an i such that $s \in U_i \subset G_\alpha$. Obviously $\alpha \in E_i$ for the same i, thereby the left-hand member is contained in the right-hand member and the equality is established. Let $(\mathfrak{A}_1, \mathfrak{E}_1)$ be any separable σ-algebra on κ_G. Then there is a separable σ-algebra $(\mathfrak{A}, \mathfrak{E})$ on the set κ_G generated by $\mathfrak{E}_1$ and the family $\{\, E_i : i \in \aleph_0 \,\}$. We claim that this separable σ-algebra on κ_G is nonmeasurable. Assume to the contrary that there is a nontrivial continuous measure λ on $\mathfrak{A}$.

For each member s_γ of the set S, the set $\{\, \beta : (\beta, s_\gamma) \in Y \,\}$ contains $\{\, \beta : \beta > \gamma \,\}$. Hence the λ-measure of every horizontal section is equal to $\lambda(\kappa_G)$. Also, every vertical section of Y has ν-measure not greater than $\frac{1}{2}\,\nu(S)$. Now observe that Y is in the σ-algebra on $\kappa_G \times S$ generated by the σ-algebras $\mathfrak{A}$ and $\mathfrak{B}(S)$. Hence the Fubini theorem may be applied to χ_Y. We get $0 < \lambda(\kappa_G)\nu(S) = \int_S (\int_{\kappa_G} \chi_Y \, d\lambda) \, d\nu = \int_Y d(\lambda \times \nu) = \int_{\kappa_G} (\int_S \chi_Y \, d\nu) \, d\lambda \le \frac{1}{2}\nu(S)\lambda(\kappa_G)$ and a contradiction has occurred. □

COROLLARY 1.45. *There exists a subspace X of $\{0,1\}^{\mathbb{N}}$ such that X is an absolute null space with* $\mathrm{card}(X) = \kappa_G$.

[11] A cardinal number κ will be identified with the minimal initial segment of ordinal numbers whose cardinal number is κ.

With the aid of Theorem 1.21 we have the following corollary, where $P(X)$ is the power set of X.

COROLLARY 1.46. *If* $(X, \mu, P(X))$ *is a nonnegative real-valued, continuous measure space with* $\operatorname{card}(X) = \kappa_G$, *then* μ *is identically equal to* 0.

1.4.3.* abNULL *is not preserved by Borel measurable maps. Let us begin with a simple proposition.

PROPOSITION 1.47. *There exists a subset* Y *of* $\{0,1\}^{\mathbb{N}}$ *and there exists a continuous, complete,* σ*-finite Borel measure space* $\mathrm{M}(\{0,1\}^{\mathbb{N}}, \mu)$ *such that* $\operatorname{card}(Y) = \kappa_G$, Y *is not* μ*-measurable and* $\mu^*(Y) > 0$. *Clearly, the subspace* Y *of* $\{0,1\}^{\mathbb{N}}$ *is not an absolute null space.*

PROOF. Let Z be a subset of $\{0,1\}^{\mathbb{N}}$ and μ be a continuous, complete, finite Borel measure on $\{0,1\}^{\mathbb{N}}$ with $\operatorname{card}(Z) = \kappa_G$ and $\mu^*(Z) > 0$. If Z is not μ-measurable, then let $Y = Z$. Suppose that Z is μ-measurable. Let X_1 and X_2 be a Bernstein decomposition of $\{0,1\}^{\mathbb{N}}$. That is, X_1 and X_2 are disjoint sets whose union is $\{0,1\}^{\mathbb{N}}$, and X_1 and X_2 are totally imperfect. Clearly, $\mu^*(X_1 \cap Z) > 0$. Indeed, suppose that $\mu^*(X_1 \cap Z) = 0$. Then $X_2 \cap Z$ would be a μ-measurable set with positive μ measure. So there would exist a nonempty perfect subset of it, whence of X_2. But X_2 cannot contain a nonempty perfect set, and a contradiction has occurred. Also $Y = X_1 \cap Z$ is not μ-measurable since every F_σ kernel of Y is a countable set. Observe that $\kappa_G = \kappa_0 \leq \operatorname{card}(Y) \leq \operatorname{card}(Z) = \kappa_G$ holds to complete the proof. □

THEOREM 1.48. *Let* Y *be a subspace of* $\{0,1\}^{\mathbb{N}}$ *with* $\operatorname{card}(Y) = \kappa_G$ *that is not an absolute null space. Then there is a continuous bijection* $f\colon N \to Y$ *of an absolute null space* N *contained in* $\{0,1\}^{\mathbb{N}}$.

PROOF. By Corollary 1.45 there is an absolute null space X contained in $\{0,1\}^{\mathbb{N}}$ with $\operatorname{card}(X) = \kappa_G$. Let $\varphi\colon X \to Y$ be any bijection. Since $\{0,1\}^{\mathbb{N}} \times \{0,1\}^{\mathbb{N}}$ is topologically the same as $\{0,1\}^{\mathbb{N}}$, Theorem 1.25 completes the proof. □

For the next theorem recall that the set of uncountable order of $f\colon X \to Y$ is the set $U(f) = \{y \in Y\colon \operatorname{card}(f^{-1}[\{y\}]) > \aleph_0\}$.

THEOREM 1.49 (Grzegorek). *Let* $f\colon X \to Y$ *be a Borel measurable map from an absolute Borel space* X *into a separable metrizable space* Y. *If* $\operatorname{card}(U(f)) > \aleph_0$, *then there is a topological copy* X^* *of* $\{0,1\}^{\mathbb{N}} \times \{0,1\}^{\mathbb{N}}$ *contained in* X *and there is a topological copy* Y^* *of* $\{0,1\}^{\mathbb{N}}$ *contained in* Y *such that* $f|X^*$ *is a continuous surjection of* X^* *to* Y^* *and such that* $f[M]$ *is not an absolute measurable space for some absolute null space* M *contained in* X^*.

PROOF. We infer from Corollary A.60 and the proof of Lemma A.61 the existence of a continuous map $f^*\colon \{0,1\}^{\mathbb{N}} \times \{0,1\}^{\mathbb{N}} \to \{0,1\}^{\mathbb{N}}$ and continuous injections $\Theta\colon \{0,1\}^{\mathbb{N}} \times \{0,1\}^{\mathbb{N}} \to X$ and $\vartheta\colon \{0,1\}^{\mathbb{N}} \to Y$ such that $f^*(x^*, y^*) = y^*$ for every

(x^*, y^*) in $\{0,1\}^{\mathbb{N}} \times \{0,1\}^{\mathbb{N}}$ and such that the diagram

$$\begin{array}{ccc} X & \xrightarrow{f} & Y \\ \uparrow \Theta & & \uparrow \vartheta \\ \{0,1\}^{\mathbb{N}} \times \{0,1\}^{\mathbb{N}} & \xrightarrow{f^*} & \{0,1\}^{\mathbb{N}} \end{array}$$

commutes, that is, $f \succ_B f^*$ (see page 195 for the definition of the B-successor relation $\succ_B$). Let $X^* = \Theta\big[\{0,1\}^{\mathbb{N}} \times \{0,1\}^{\mathbb{N}}\big]$ and $Y^* = \vartheta\big[\{0,1\}^{\mathbb{N}}\big]$. By Theorem 1.48 there is a continuous bijection $g: N \to Y_0$ of an absolute null space N contained in $\{0,1\}^{\mathbb{N}}$ onto a non-absolute measurable space Y_0 contained in $\{0,1\}^{\mathbb{N}}$. By Theorem 1.25, we have that graph(g) is an absolute null space. Let $M = \Theta\big[\text{graph}(g)\big]$. Observe that M is an absolute null space and that $\vartheta[Y_0]$ is a non absolute measurable space. From the commutative diagram, $f|X^* = \vartheta f^* \Theta^{-1}$. As $f[M] = \vartheta[Y_0]$, the proof is completed. □

THEOREM 1.50. *Let $f: X \to Y$ be a $\mathfrak{B}$-map, where X is an absolute Borel space. Then $f[M]$ is an absolute measurable space whenever M is an absolute measurable space contained in X.*

PROOF. As f is a $\mathfrak{B}$-map and X is an absolute Borel space, we have from Purves's theorem that $\text{card}(U(f)) \leq \aleph_0$. So $C = X \setminus f^{-1}[U(f)]$ is an absolute Borel space. Let us return to the proof of Theorem A.22. It was shown there that there is a continuous injection $g: \mathcal{N} \to \text{graph}(f|C)$ such that $\text{graph}(f|C) \setminus g[\mathcal{N}]$ is a countable set. With the natural projection $\pi: \text{graph}(f|C) \to Y$, the composition $h = \pi g$ provides a collection B_n, $n = 1, 2, \ldots$, of Borel subsets of $\mathcal{N}$ such that $\mathcal{N} = \bigcup_{n=1}^{\infty} B_n$ and $h|B_n$ is a homeomorphism for each n. Note that $C_n = C \cap \pi_1 g[B_n]$, where π_1 is the natural projection of graph$(f|C)$ onto C, is an absolute Borel space and $f|C_n$ is $\mathfrak{B}$-homeomorphism of C_n onto $h[B_n]$. As $M \cap C_n$ is an absolute measurable space, we have that $f[M \cap C_n]$ is an absolute measurable space for each n. Also, $C \setminus \pi g[\mathcal{N}]$ is a countable set. Thus we have shown that $f[M \cap C]$ is an absolute measurable space. Since $U(f) \supset f[M \setminus C]$, we have that $f[M]$ is an absolute measurable space. □

The proof given here is essentially that of R. B. Darst [**37**]. A "converse" statement will be investigated in the next chapter on universally measurable sets in a fixed space X, namely, we shall present the theorem due to Darst and Grzegorek.

COROLLARY 1.51. *Let $f: X \to Y$ be a $\mathfrak{B}$-map, where X is an absolute Borel space. Then $f[M]$ is an absolute null space whenever M is an absolute null space contained in X.*

PROOF. Let us prove that every subset of $f[M]$ is an absolute measurable space. To this end, let $Y_0 \subset f[M]$. Note that $M_0 = M \cap f^{-1}[Y_0]$ is an absolute measurable space since M is an absolute null space. Clearly $f[M_0] = Y_0$, whence an absolute measurable space. We have that $f[M]$ is an absolute null space by the characterization theorem, Theorem 1.20. □

1.5. More on existence of absolute null spaces

We have seen earlier that Hausdorff's m-convergence, a sufficient condition for the existence of uncountable absolute null spaces, was essentially a part of a characterization theorem. The characterization used a well ordering of $\aleph_1$-many disjoint subsets of an uncountable absolute measurable space. The well ordering idea was used to advantage by Recław [**130**] for subspaces of $\mathbb{R}$ and by Plewik [**127**, Lemma] for subspaces of $\{0, 1\}^{\mathbb{N}}$ to prove another sufficient condition for the existence of absolute null spaces. It was pointed out by Recław that, with the aid of $\mathfrak{B}$-homeomorphisms, $\mathbb{R}^n$ can replace the ambient space $\mathbb{R}$. Of course the use of $\mathfrak{B}$-homeomorphisms defeats the emphasis on topological homeomorphism if one can avoid the use of $\mathfrak{B}$-homeomorphisms. We shall present the results of Recław and Plewik in a setting more in line with the notions of absolute measurable space and absolute null space. This will permit us to state a more general characterization of absolute null spaces.

For a set X, recall that R is a *relation on Y* if $R \subset Y \times Y$ – that is, for s and t in Y, we say that *s is related to t*, written $s\,R\,t$, if $(s, t) \in R$. We say that *a subset R of $Y \times Y$ well orders a subset X of Y* if the relation $R \cap (X \times X)$ is a well ordering of X. So, if the ordinal number η is the order type of such a well ordered set X, then X has an indexing $\{ x_\alpha : \alpha < \eta \}$ such that $x_\alpha\, R\, x_\beta$ if and only if $\alpha < \beta < \eta$. Of course, the ordinal number η need not be a cardinal number.

Before we turn to Recław's theorem let us make an elementary observation about absolute null spaces X contained in a separable metrizable space Y. Note that $X \times X$ is an absolute null space contained in $Y \times Y$. Let $\{ x_\alpha : \alpha < \eta \}$ be a well ordering of X by an ordinal number η. This well ordering corresponds to a unique subset R of $X \times X$. As every subset of $X \times X$ is an absolute measurable space we have that R is an absolute measurable space that is contained in $Y \times Y$ such that X is well ordered by R.

Recław's theorem is the following.

THEOREM 1.52 (Recław). *Let R be an absolute measurable space contained in* $[0, 1] \times [0, 1]$. *Then, any subset X of* $[0, 1]$ *that is well ordered by the relation R is an absolute null space contained in* $[0, 1]$.

We shall prove the following more general form.

THEOREM 1.53. *Let Y be a separable metrizable space and let R be an absolute measurable space contained in $Y \times Y$. Then, any subset X of Y that is well ordered by the relation R is an absolute null space contained in Y.*

PROOF. There is no loss in assuming that Y is a subspace of the Hilbert cube $[0, 1]^{\mathbb{N}}$. Let X be a subset of Y that is well ordered by the relation R and denote its indexing by $\{ x_\alpha : \alpha < \eta \}$. We shall assume that it is not an absolute null space contained in Y and derive a contradiction. There is a continuous, complete Borel measure μ on $[0, 1]^{\mathbb{N}}$ such that $1 = \mu([0, 1]^{\mathbb{N}}) \geq \mu^*(X) > 0$. As X is well ordered, there is an η_0 such that $\mu(\{ x_\alpha : \alpha < \eta_0 \}) > 0$, and such that $\mu(\{ x_\alpha : \alpha < \beta \}) = 0$ whenever $\beta < \eta_0$. Denote $\{ x_\alpha : \alpha < \eta_0 \}$ by X_0. Let A be a Borel set in $[0, 1]^{\mathbb{N}}$ such that $A \supset X_0$ and such that the inner measure $\mu_*(A \setminus X_0)$ is 0. Let W denote the relation

$\{(x,y) \in A \times A : yRx\}$. As $W = (A \times A) \cap R^{-1}$ we have that W is an absolute measurable space. Hence there is a Borel set V such that $V \subset W$ and $(\mu \times \mu)(W \setminus V) = 0$. As usual, we denote $\{y \in [0,1]^{\mathbb{N}} : (x,y) \in V\}$ by V_x, a Borel set in $[0,1]^{\mathbb{N}}$. As the function $x \mapsto \mu(V_x)$ is Borel measurable, the set $\{x : \mu(V_x) > 0\}$ is a Borel subset of $[0,1]^{\mathbb{N}}$. We assert that this set is contained in $A \setminus X_0$. Indeed, for x in X_0, the set $D = \{y \in X : yRx\}$ is a μ-null set, whence there is a Borel set C in $[0,1]^{\mathbb{N}}$ that contains it with $\mu(C) = 0$. Since $V_x \cap (X \setminus D)$ is a subset of X and $R \cap (X \times X)$ is a well ordering of X we have $V_x \cap (X \setminus D) = \emptyset$. Consequently $V_x \setminus C \subset V_x \setminus X_0 \subset A \setminus X_0$. Now $V_x \setminus C$ is a Borel set and $\mu_*(A \setminus X) = 0$, whence $\mu(V_x \setminus C) = 0$. Thereby we have shown $\mu(V_x) = 0$ whenever $x \in X_0$. The assertion that the Borel set $\{x : \mu(V_x) > 0\}$ is a subset of $A \setminus X_0$ has been verified. We have $(\mu \times \mu)(V) \le \int_A \mu(V_x)\,d\mu \le \mu_*(A \setminus X_0) = 0$. Consequently $(\mu \times \mu)(W) = (\mu \times \mu)(W \setminus V) + (\mu \times \mu)(V) = 0$. We are ready to establish a contradiction. Observe that, for each y in A, the set $W^y = \{x \in X_0 : (x,y) \in W\}$ satisfies $W^y \supset \{x \in X_0 : yRx\} = X_0 \setminus \{x \in X_0 : xRy, x \ne y\}$. Hence $\mu(W^y) \ge \mu^*(X_0)$ whenever $y \in X_0$. As $\mu_*(A \setminus X_0) = 0$, by the Fubini theorem, we have the contradiction $0 = (\mu \times \mu)(W) = \int_A \mu(W^y)\,d\mu \ge (\mu^*(X_0))^2 > 0$. The theorem is proved. □

There is the following straightforward characterization of absolute null spaces that are contained in Y. The proof is left to the reader.

THEOREM 1.54. *Let Y be a separable metrizable space. In order that a subspace X of Y be an absolute null space it is necessary and sufficient that there exists an absolute measurable space R contained in $Y \times Y$ with the property that X is well ordered by the relation R.*

Theorem 1.52 generalizes a result of Plewik [**127**, Lemma] who assumes that the relation R is formed by a Borel subset of $[0,1] \times [0,1]$ – clearly R is an absolute measurable space. Actually, Plewik works in the collection $P(\omega) = \{X : X \subset \omega\}$, which is homeomorphic to the space $\{0,1\}^{\mathbb{N}}$. He devises a schema for constructing relations on $P(\omega)$ by beginning with relations $<_n$ on the subsets of $\{k : 0 \le k < n\} = [0,n) = n$, that is $<_n \subset P([0,n)) \times P([0,n))$, for each n in ω. He comments that, for each n in ω, the relation

$$<^n = \{(X,Y) : X \cap n <_n Y \cap n\}$$

on $P(\omega)$ is a simultaneously closed and open subset of $P(\omega) \times P(\omega)$. Define $X \prec Y$ for X and Y in $P(\omega)$ to mean that $X \cap n <_n Y \cap n$ holds for almost all n, that is, there is an m such that $X \cap n <_n Y \cap n$ holds whenever $n \ge m$. This, of course, defines a relation $\prec$ on $P(\omega)$. We now have the following proposition by Plewik.

PROPOSITION 1.55 (Plewik). *The relation $\prec$ is an F_σ subset of $P(\omega) \times P(\omega)$.*

The proof of the proposition is left to the reader as an exercise. Recall that F_σ sets of $\{0,1\}^{\mathbb{N}} \times \{0,1\}^{\mathbb{N}}$ are absolute Borel spaces, hence the above characterization theorem applies.

Plewik gives several applications of his proposition by defining appropriate relations $\prec$, that is, by devising ways to invoke his schema for the relation $\prec$. Indeed, he is able to include the Hausdorff Ω-Ω^* gap example of an absolute null space contained in $\{0, 1\}^{\mathbb{N}}$. He also shows that examples like that of Grzegorek's among others can be included in his setting. Recław also exhibits an example of a relation R on $[0, 1]$ that is an absolute measurable space and not an absolute Borel space, thereby sharpening Plewik's theorem (see Exercise 1.11 on page 29). The details of and more comments on these applications will be given in Chapter 6.

1.6. Comments

We close the chapter with a few comments.

1.6.1. Metric spaces. The results in Sections 1.1 and 1.2 are found in **[152]**. These early results concerning absolute measurable spaces were couched in the context of separable *metric* spaces, which differs from that of separable metrizable spaces used in our book. The definitions in the book are independent of the metric that can be assigned to a given metrizable space. Hence the notion of absolute measurable space as has been developed in the book is based solely on the Borel sets of a given separable, metrizable space, which are – of course – independent of the metrics that are associated with a given metrizable space. This observation was made by R. Shortt in **[138, 139]**.[12] The motivation for this change from metric to metrizable is the following definition, due to E. Szpilrajn-Marczewski **[152]**, of property M and its related property $\mathsf{M}(\text{rel } Y)$ for separable metric spaces Y.

DEFINITION 1.56 (Szpilrajn-Marczewski). *A subset X of a separable metric space Y is said to have* property $\mathsf{M}(\text{rel } Y)$ *if, for each finite Borel measure μ on X, there are Borel sets A and B of Y such that $A \subset X \subset B$ and $\mu(A) = \mu(B)$.*

DEFINITION 1.57 (Szpilrajn-Marczewski). *A separable metric space X is said to have* property M *if, for every separable metric space Y, each topological copy of X contained in Y has property* $\mathsf{M}(\text{rel } Y)$.

On inspection of the definition of property $\mathsf{M}(\text{rel } Y)$, one can easily see that only the topology that results from the metric on the space Y is used in the definition since the collection $\mathfrak{B}(Y)$ of all Borel sets of Y is the same for any metrization of a metrizable space Y. In the second definition, the embedding is not required to be an isometric embedding. Hence the definition of property M clearly is independent of the metric for metrizable spaces X. The appropriate modification of property M has resulted in our definition of absolute measurable spaces, the modifier *absolute* is used to emphasize the topological embedding feature of the definition. Since every metric space is metrizable, this change in definition will cause no loss in the analysis of any specific metric space.

The next chapter will be concerned with the notion of universally measurable sets X in a metrizable space Y. This notion is the appropriate modification of the property

[12] An extended discussion of Shortt's observation can be found in Appendix B.

$\mathsf{M}(\text{rel } Y)$. The modifier *universally* is used to indicate that the metrizable space Y is fixed, not the metric.

1.6.2. Absolutely measurable functions. It is natural to want to consider functions $f: X \to \mathbb{R}$ in the context of absolute measurable spaces, where X is a separable, metrizable space. Given such a space X, we define the σ-ring

$$\mathsf{ab}\,\mathfrak{M}(X) = \{M : M \subset X, M \in \mathsf{abMEAS}\}.$$

This σ-ring is a σ-algebra if and only if $X \in \mathsf{abMEAS}$. It is natural to say that a function $f: X \to \mathbb{R}$ is *absolutely measurable* if $M \cap f^{-1}[F]$ is in $\mathsf{ab}\,\mathfrak{M}(X)$ for every closed set F of $\mathbb{R}$ and every M in $\mathsf{ab}\,\mathfrak{M}(X)$. Observe, for an absolutely measurable $f: X \to \mathbb{R}$, that $f^{-1}[F] \in \mathsf{ab}\,\mathfrak{M}(X)$ for every closed set F if and only if $X \in \mathsf{ab}\,\mathfrak{M}(X)$. In the next chapter we shall investigate the completely different notion of universally measurable function, which will agree with that of absolutely measurable function whenever $X \in \mathsf{abMEAS}$. A general investigation of absolutely measurable functions will not be carried out, since the more interesting investigation will be in the context of universally measurable sets in a space X.

1.6.3. Historical references. Here we give credit to the various authors who have influenced the theorems proved in this chapter. Other references will also be made.

The paper [**152**] by Szpilrajn-Marczewski[13] is the source of Theorem 1.7, Proposition 1.12, and Theorems 1.16 and 1.17. Statement (δ) of Theorem 1.20 was observed by E. Grzegorek and C. Ryll-Nardzewski in [**71**]. We have already attributed Theorem 1.21 and Theorem 1.34 to Sierpiński and Szpilrajn [**142**]. The characterization[14] provided by Theorem 1.35 of the existence of absolute null subspaces of a space is motivated by Hausdorff [**73**]. Our development of the Ω-Ω^* gap example of Hausdorff follows the one given by Laver [**88**] with added details from the original paper by Hausdorff [**73**].[15] The development of the second example, which uses constituents of co-analytic spaces, is due to Sierpiński and Szpilrajn [**142**].

The partition theorem, Theorem 1.37, has other applications – in particular, the Sierpiński–Erdös theorem and the related Duality Principle. The reader is directed to the book [**120**] by Oxtoby for a nice discussion of these topics. The Duality Principle yields an intimate connection between sets of measure 0 and sets of the first category of Baire. See also C. G. Mendez [**107, 108**]. Theorem 1.39, which has been attributed to N. Lusin [**92**] and also proved by P. Mahlo [**95**], provides the existence of what is now called Lusin sets. A *Lusin set* in a space X is an uncountable set M with the property that every nowhere dense subset of X intersects M in an at most countable set. For more on Lusin and other singular sets see Brown and Cox [**18**] and A. W. Miller [**110, 111**].

[13] This paper is written in Polish. An English translation of it has been made by John C. Morgan II.

[14] This characterization anticipates the results of Recław [**130**] and Plewik [**127**] which are also discussed in Section 1.5 of this chapter.

[15] A nice survey article by M. Scheepers [**134**] contains a historical discussion of Hausdorff's work on gaps. More will be said about this article in Chapter 6.

Section 1.2.2 concerns what are called the Ulam numbers (see [**55**, page 58]) and what are called real-valued measurable cardinal numbers (see [**58**, definition 4.12, page 972]). A cardinal number κ is an *Ulam number* if every nonnegative, real-valued, continuous measure μ on a set X with $\mathrm{card}(X) = \kappa$ is the 0 measure whenever the collection $\mathfrak{M}(X,\mu)$ of μ-measurable sets is $P(X)$, that is the measure space $\big(X,\mu,P(X)\big)$ is trivial. Hence $\aleph_1$ is an Ulam number. A cardinal number κ is *real-valued measurable* if there exists a continuous probability measure μ on a set X with $\mathrm{card}(X) = \kappa$ and $\mathfrak{M}(X,\mu) = P(X)$; that is, $\mu(X) = 1$ and every subset of X is μ-measurable. Hence $\aleph_1$ is not real-valued measurable.

The cardinal number κ_{G} was used by Grzegorek [**68**] for the purpose of solving a problem proposed by Banach [**7**] concerning the existence of certain kinds of measures (see the footnote on page 19; more will be said about Banach's problem in Chapter 6). In that paper, Grzegorek used the nondescript symbol $\mathfrak{m}_1$ to denote this cardinal number, we have used κ_{G} to honor him. With the aid of this cardinal number he gave an affirmative answer to the problem of Banach. Grzegorek uses the notion of characteristic functions as developed by Marczewski (= Szpilrajn) in [**7, 150, 151**].[16] The use of Marczewski's development allows the Banach problem to be translated into a problem involving Borel measures on subsets of $\{0,1\}^{\mathbb{N}}$. In [**68**] Grzegorek used the closed interval $[0,1]$ and Lebesgue measure λ to define his cardinal number $\mathfrak{m}_0$ (see page 45). The present day literature uses $\textsf{non-}\mathbb{L}$ to denote $\mathfrak{m}_0$. The presentation given in our book uses the space $\{0,1\}^{\mathbb{N}}$ in place of $[0,1]$ to define the cardinal number κ_0 in Section 1.4.2. We shall show in the next chapter that Grzegorek's cardinal number $\mathfrak{m}_0$ is the same as κ_0. In a subsequent paper [**71**] Grzegorek and Ryll-Nardzewski used this cardinal number to show that the use of the continuum hypothesis by Darst in [**39**] was not necessary (we shall turn to Darst's theorem in the next chapter). We mention that κ_{G} is an Ulam number and is not real-valued measurable.

Darst [**37**] proved Theorem 1.50 for real-valued $\mathfrak{B}$-maps defined on Borel subsets X of $\mathbb{R}$. As Grzegorek's example shows, the requirement that the domain of $\mathfrak{B}$-maps be absolute Borel spaces cannot be avoided. Corollary 1.51 can be strengthened as follows. First define a map $f\colon X \to Y$ to be an $\textsf{ab}\,\mathfrak{M}$*-map* if $f[M]$ is an absolute measurable space whenever M is an absolute measurable space contained in X. Then the proof of Corollary 1.51 will result in the rather trivial

PROPOSITION 1.58. *If $f\colon X \to Y$ is an $\textsf{ab}\,\mathfrak{M}$-map, then $f[N] \in \textsf{abNULL}$ whenever $N \subset X$ and $N \in \textsf{abNULL}$.*

Note that f need not be Borel measurable and that X is arbitrary. Hence the definition of $\textsf{ab}\,\mathfrak{M}$-map appears to be somewhat contrived. Observe that a continuous bijection of an absolute measurable space X need not be such a map as witnessed by Grzegorek's example.

Exercises

1.1. Prove Proposition 1.12 on page 4.
1.2. Prove Theorem 1.19 on page 8.

[16] Marczewski's development is reproduced in many articles and books cited in our book. In particular, it appears in some form or other in [**12, 32, 145, 94**].

1.3. Let $f: X \to Y$ be a Borel measurable map, where X and Y are separable metrizable spaces. Prove that $f^{-1}[\{y\}]$ is an absolute null space for every y in Y if and only if $f_{\#}\mu$ is continuous for every continuous, complete, finite Borel measure μ on X.

1.4. Prove Lemma 1.24 on page 10.

1.5. Prove Proposition 1.26 on page 10, the σ-ring property of the class abNULL of absolute null spaces.

1.6. Prove Lemma 1.40 on page 18. Hint: Show that $\mu(X) < \infty$ may be assumed.

1.7. Show that Grzegorek's example (Theorem 1.44, page 21) leads to a solution of Banach's problem stated in the footnote on page 19. Hint: The example $\mathfrak{A}$ has the property that there is a countably generated subalgebra $\mathfrak{A}'$ that is measurable (obviously $\mathfrak{A} = \mathfrak{A} \cup \mathfrak{A}'$); use disjoint copies of the set S.

1.8. Let Y and Z be separable metrizable spaces. Let R be a subset of $Y \times Y$ and let $X = \{x_\alpha : \alpha < \eta\}$ be a well ordered subset of Y. Each homeomorphism h of Y into Z gives a natural homeomorphism $h \times h$ of $Y \times Y$ into $Z \times Z$, where $(h \times h)(s,t) = (h(s), h(t))$ for $(s,t) \in Y \times Y$. Denote $(h \times h)[R]$ by S, a subset of $Z \times Z$. Prove that $h[X] = \{h(x_\alpha) : \alpha < \eta\}$ is well ordered by the relation S if and only if $X = \{x_\alpha : \alpha < \eta\}$ is well ordered by the relation R. Clearly S is an absolute measurable space contained in $Z \times Z$ if and only if R is an absolute measurable space contained in $Y \times Y$.

1.9. Prove Theorem 1.54 on page 25.

1.10. Prove Proposition 1.55 on page 25.

1.11. Let F be a Borel set in $[0,1]^2$ such that $F_u = \{v : (u,v) \in F\}$ is not empty for each u in $[0,1]$. Define the Borel sets A and B in $[0,1]^3$ as follows: $A = \{(x,y,z) : (x,z) \in F\}$, $B = \{(x,y,z) : (y,z) \in F\}$.

(a) Show $\pi[A \setminus B] = \{(x,y) : F_x \setminus F_y \neq \emptyset\}$, where $\pi(x,y,z) = (x,y)$.

(b) Show $R = \{(x,y) : F_x \subset F_y\}$ is a co-analytic set that is a linear ordering of $[0,1]$ (though not necessarily the usual linear ordering).

(c) Show that there is an absolute measurable space R with $\operatorname{card}(R) = \mathfrak{c}$ such that it is contained in $[0,1]^2$, is not analytic or co-analytic or absolute null, and is a linear ordering of $\{x : (x,y) \in R\}$. Hint: Add Grzegorek's example.

2

The universally measurable property

The property of this chapter historically precedes that of absolute measurable spaces. The works of Sierpiński and Szpilrajn [**142**] and Szpilrajn-Marczewski [**152**] make more natural the introduction of absolute measurable spaces before the development of universally measurable sets in a space. The universally measurable property concerns sets in a fixed separable metrizable space rather than the property of topological embedding of a space into other spaces. This change of emphasis will be highlighted by switching the modifier "absolute" to "universally." Interesting situations arise when the fixed space is absolute measurable.

The notion of a universally measurable set in a space is more complicated than that of absolute measurable spaces. Emphasis will be placed on the interplay between universally measurable sets in a space and absolute measurable subspaces. Of particular importance is the coinciding of universally null sets in a space X and the absolute null subspaces of X. Included is a presentation of a sharpening, due to Darst and Grzegorek, of the Purves theorem.

A closure-like operation, called the universally positive closure, is introduced to facilitate the study of the topological support of measures on X. This closure operation is used to define positive measures, those whose topological supports are as large as possible. It is shown that the notion of universally measurable sets in X can be achieved by using only those measures that are positive.

The Grzegorek and Ryll-Nardzewski solution to the natural question of symmetric differences of Borel sets and universally null sets is given. Their solution has connections to the question in the Preface, due to Mauldin, concerning absolute measurable spaces contained in $\mathbb{R}$. (This connection will be discussed in Chapter 6, the chapter whose emphasis includes set theoretic considerations.)

The historically early results for $X = [0, 1]$ will be used to motivate the use of the group of homeomorphisms in the study of universally measurable sets. These results for $[0, 1]$ together with $\mathfrak{B}$-homeomorphism also lead to results about universally measurable sets in other separable metrizable spaces. Of course, topological properties are not preserved by $\mathfrak{B}$-homeomorphisms, properties that are of interest in this book. The use of $\mathfrak{B}$-homeomorphism will not be excluded if topological considerations are not involved. (In Appendix B, a brief discussion is given of universally measurable sets from the point of view of measure and probability theories which emphasizes $\mathfrak{B}$-homeomorphism.)

2.1. Universally measurable sets

Recall that MEAS is the collection of all Borel measure spaces $\mathrm{M}(X,\mu) = (X,\mu,\mathfrak{M}(X,\mu))$ that are σ-finite and complete, that $\mathfrak{M}(X,\mu)$ is the σ-algebra associated with a measure space, and that $\mathrm{MEAS}^{\mathrm{cont}}$ is the collection of those measure spaces in MEAS that are continuous. Let us establish notation that emphasizes a fixed space.

NOTATION 2.1. *Fix a separable metrizable space X and define the following collection of measures on X.*

$$\mathrm{MEAS}(X) = \{\, \mu\colon\ \mathrm{M}(X,\mu) \in \mathrm{MEAS}\,\}$$
$$\mathrm{MEAS}^{\mathrm{cont}}(X) = \{\, \mu\colon\ \mathrm{M}(X,\mu) \in \mathrm{MEAS}^{\mathrm{cont}}\,\}.$$

We shall now modify the definition of the property $\mathsf{M}(\mathrm{rel}\, X)$ defined by Szpilrajn-Marczewski (see Definition 1.56).

DEFINITION 2.2. *Let X be a fixed separable metrizable space. A subset M of X is said to be a* universally measurable set *in X if M is in $\mathfrak{M}(X,\mu)$ whenever $\mu \in \mathrm{MEAS}(X)$. The collection of all universally measurable sets in X will be denoted by $\mathsf{univ}\,\mathfrak{M}(X)$.*

DEFINITION 2.3. *Let X be a fixed separable metrizable space. A subset M of X is said to be a* universally null set *in X if M is in $\mathfrak{N}(X,\mu)$ whenever $\mu \in \mathrm{MEAS}^{\mathrm{cont}}(X)$. The collection of all universally null sets in X will be denoted by $\mathsf{univ}\,\mathfrak{N}(X)$.*

Obviously, the collections $\mathsf{univ}\,\mathfrak{M}(X)$ and $\mathsf{univ}\,\mathfrak{N}(X)$ are the intersections

$$\mathsf{univ}\,\mathfrak{M}(X) = \bigcap\{\, \mathfrak{M}(X,\mu)\colon\ \mu \in \mathrm{MEAS}(X)\,\}$$

and

$$\mathsf{univ}\,\mathfrak{N}(X) = \bigcap\{\, \mathfrak{N}(X,\mu)\colon\ \mu \in \mathrm{MEAS}^{\mathrm{cont}}(X)\,\}.$$

Analogous to Theorems 1.5 and 1.19, the σ-finite requirements in the above definitions may be replaced with finite with no changes in the collections. The proof is easy and is left to the reader.

2.1.1. Elementary relationships.

We have the obvious

PROPOSITION 2.4. *For separable metrizable spaces X, the collection $\mathsf{univ}\,\mathfrak{M}(X)$ is a σ-algebra of subsets of X such that $\mathfrak{B}(X) \subset \mathsf{univ}\,\mathfrak{M}(X)$, and the collection $\mathsf{univ}\,\mathfrak{N}(X)$ is a σ-ideal.*

For the next proposition recall from page 27 that $\mathsf{ab}\,\mathfrak{M}(X)$ is the collection consisting of all absolute measurable subspaces M contained in X. The proof of the proposition is left to the reader (see Theorem 1.17).

PROPOSITION 2.5. *For separable metrizable spaces X, $\mathsf{ab}\,\mathfrak{M}(X)$ is a σ-ring contained in $\mathsf{univ}\,\mathfrak{M}(X)$. Moreover, $\mathsf{univ}\,\mathfrak{M}(X) = \mathsf{ab}\,\mathfrak{M}(X)$ if and only if X is an absolute measurable space.*

We also have the following

Proposition 2.6. *Let Y be a separable metrizable space. If X is a universally measurable set in Y, then*

$$\mathsf{univ}\,\mathfrak{M}(X) = \{\, E \in \mathsf{univ}\,\mathfrak{M}(Y) : E \subset X \,\}.$$

Proof. Denote the inclusion map of X into Y by f. Let μ be a complete, finite Borel measure on X. Then $f_\#\mu$ is a complete, finite Borel measure on Y. Hence X is $(f_\#\mu)$-measurable. It follows that the restriction measure $(f_\#\mu)|X$ is the measure μ on the subspace X of Y. Consequently, if E is a universally measurable set in Y that is also a subset of X, then E is a universally measurable set in X. Next let ν be a complete, finite Borel measure on Y. Then $\nu|X$ is a complete finite Borel measure on X. It follows that $f_\#(\nu|X)$ is the limited measure $\nu \llcorner X$. So, if E is a universally measurable set in X, then E is a $(\nu \llcorner X)$-measurable set in Y, whence a ν-measurable set in Y. □

The following theorems are essentially due to Sierpiński and Szpilrajn [**142**].

Theorem 2.7. *Let M be a subset of a separable metrizable space X. Then $M \in \mathsf{univ}\,\mathfrak{N}(X)$ if and only if M is an absolute null space (that is, $M \in \mathsf{abNULL}$).*

Proof. It is clear that $M \in \mathsf{univ}\,\mathfrak{N}(X)$ whenever $M \in \mathsf{abNULL}$ and $M \subset X$. So let $M \in \mathsf{univ}\,\mathfrak{N}(X)$. By the definition of $\mathsf{univ}\,\mathfrak{N}(X)$ we infer from the statement (α) of Theorem 1.20 that $M \in \mathsf{abNULL}$. □

Theorem 2.8. *For separable metrizable spaces X and Y, let f be a $\mathfrak{B}$-homeomorphism of X onto Y. Then, for subsets M of X,*

(1) $f^{-1}[M] \in \mathfrak{B}(X)$ *if and only if* $M \in \mathfrak{B}(Y)$,
(2) $f^{-1}[M] \in \mathsf{univ}\,\mathfrak{M}(X)$ *if and only if* $M \in \mathsf{univ}\,\mathfrak{M}(Y)$,
(3) $f^{-1}[M] \in \mathsf{univ}\,\mathfrak{N}(X)$ *if and only if* $M \in \mathsf{univ}\,\mathfrak{N}(Y)$.

Proof. The first equivalence assertion follows easily from the definition of $\mathfrak{B}$-homeomorphism. The other two follow since $f_\#$ establishes a natural bijection between Borel measures on X and on Y. □

Remark 2.9. It is well-known that every separable metrizable space can be topologically embedded into the Hilbert cube $[0, 1]^{\mathbb{N}}$. Also, there is a $\mathfrak{B}$-homeomorphism φ of the Hilbert cube onto $\{0, 1\}^{\mathbb{N}}$, which is homeomorphic to the classical Cantor ternary set. Consequently, the study of $\mathsf{univ}\,\mathfrak{M}(X)$ only from the point of view of universally measurable sets in X can be carried out on subspaces of the Cantor set. The difficulty is that the $\mathfrak{B}$-homeomorphism φ does not preserve the topological structure of the space X. Nor does it preserve many other structures of X, for example, the order structure of the space $\mathbb{R}$. Consequently, the study of universally measurable sets in a separable metrizable space X does not end with the above theorem if one is interested in, for example, the topological structure of such sets in the ambient space X, or the geometric structure of such sets in the event that X is a metric space.

2.1.2. The theorem of Darst and Grzegorek. Darst and Grzegorek investigated the above Theorem 2.8 under the sole condition that $f: X \to Y$ be Borel measurable, that is, without the bijection requirement. We do this now.

It will be convenient to use the usual convention: If $f: X \to Y$ is any function and $\mathcal{M}_X$ and $\mathcal{M}_Y$ are collections of subsets of X and Y respectively, then $f[\mathcal{M}_X]$ and $f^{-1}[\mathcal{M}_Y]$ are, respectively, the collections $\{f[M]: M \in \mathcal{M}_X\}$ and $\{f^{-1}[M]: M \in \mathcal{M}_Y\}$.

With this convention we have that f is a Borel measurable map if and only if $f^{-1}[\mathfrak{B}(Y)] \subset \mathfrak{B}(X)$. It is well-known that there exists an absolute Borel space X such that $f[\mathfrak{B}(X)] \not\subset \mathfrak{B}(Y)$ for some Borel measurable map f.

By Definition 1.15, a map $f: X \to Y$, where X and Y are separable metrizable spaces, is a $\mathfrak{B}$-map if and only if $f^{-1}[\mathfrak{B}(Y)] \subset \mathfrak{B}(X)$ and $f[\mathfrak{B}(X)] \subset \mathfrak{B}(Y)$. Consequently, Purves's Theorem A.43 can be stated as *$f^{-1}[\mathfrak{B}(Y)] \subset \mathfrak{B}(X)$ and $f[\mathfrak{B}(X)] \subset \mathfrak{B}(Y)$ if and only if* $\operatorname{card}(U(f)) \leq \aleph_0$, *whenever X is an absolute Borel space*, where $U(f)$ is the set of uncountable order of f. The next theorem by Darst and Grzegorek is a sharpening of Purves's theorem.

THEOREM 2.10 (Purves–Darst–Grzegorek). *Let $f: X \to Y$ be a Borel measurable map from an absolute Borel space X into a separable metrizable space Y. Then the following conditions are equivalent.*

(1) *f is a $\mathfrak{B}$-map.*
(2) $\operatorname{card}(U(f)) \leq \aleph_0$, *where $U(f)$ is the set of uncountable order of f.*
(3) $f[\mathfrak{B}(X)] \subset \mathfrak{B}(Y)$.
(4) $f[\operatorname{univ}\mathfrak{M}(X)] \subset \operatorname{univ}\mathfrak{M}(Y)$.
(5) $f[\operatorname{univ}\mathfrak{N}(X)] \subset \operatorname{univ}\mathfrak{N}(Y)$.
(6) $f[\operatorname{univ}\mathfrak{N}(X)] \subset \operatorname{univ}\mathfrak{M}(Y)$.

PROOF. Purves's Theorem A.43 yields the equivalence of the first three conditions since f is a Borel measurable map. That condition (1) implies condition (4) follows from Theorem 1.50. Proposition 1.58 gives condition (4) implies condition (5). Obviously, condition (5) implies condition (6). It remains to prove that condition (6) implies condition (2). Suppose that condition (2) fails. Then, by Grzegorek's Theorem 1.49, condition (6) fails. This completes the proof. □

2.1.3. Universally positive closure. Let M be a subset of a separable metrizable space X. Denote by $\mathcal{V}$ the collection of all open sets V such that $V \cap M$ is a universally null set in X. As X is a Lindeloff space, there is a countable subcollection $V_0, V_1, \ldots,$ of $\mathcal{V}$ such that $V = \bigcup \mathcal{V} = \bigcup_{i=0}^{\infty} V_i$. Since $\operatorname{univ}\mathfrak{N}(X)$ is a σ-ideal, we have $V \cap M$ is a universally null set in X. We call the closed set $\mathrm{F}_X(M) = X \setminus V$ the *universally positive closure of M in X* (or, *positive closure* for short). F_X is not the topological closure operator Cl_X, but it does have the following properties. (Often the reference to the ambient space X will be dropped from the operators F_X and Cl_X whenever the context of a discussion permits it.)

Proposition 2.11. *Let X be a separable metrizable space. Then the following statements hold.*

(1) *If $M_1 \subset M_2 \subset X$, then $\mathrm{F}_X(M_1) \subset \mathrm{F}_X(M_2)$.*
(2) *If M_1 and M_2 are subsets of X, then*

$$\mathrm{F}_X(M_1 \cup M_2) = \mathrm{F}_X(M_1) \cup \mathrm{F}_X(M_2).$$

(3) *If $M \subset X$, then*

$$\mathrm{F}_X(M) = \mathrm{F}_X\big(M \cap \mathrm{F}_X(M)\big) = \mathrm{Cl}_X\big(M \cap \mathrm{F}_X(M)\big).$$

(4) *If $M \subset X$, then $\mathrm{F}_X\big(\mathrm{F}_X(M)\big) = \mathrm{F}_X(M)$.*

Proof. Statement (1) is obvious.

Let us prove statement (2). As

$$M_i \cap \big(X \setminus (\mathrm{F}(M_1) \cup \mathrm{F}(M_2))\big) \subset M_i \cap (X \setminus \mathrm{F}(M_i)) \in \mathsf{univ}\,\mathfrak{N}(X),$$

for $i = 1, 2$, we have

$$(M_1 \cup M_2) \cap \big(X \setminus (\mathrm{F}(M_1) \cup \mathrm{F}(M_2))\big) \in \mathsf{univ}\,\mathfrak{N}(X).$$

Hence $\mathrm{F}(M_1 \cup M_2) \subset \mathrm{F}(M_1) \cup \mathrm{F}(M_2)$. We infer $\mathrm{F}(M_1) \cup \mathrm{F}(M_2) \subset \mathrm{F}(M_1 \cup M_2)$ from statement (1), thereby statement (2) is proved.

Clearly, $\mathrm{F}(N) = \emptyset$ whenever N is a universally null set in X. Hence $\mathrm{F}(M) = \mathrm{F}\big(M \cap \mathrm{F}(M)\big)$ follows from statement (2). The open set $U = X \setminus \mathrm{Cl}\big(M \cap \mathrm{F}(M)\big)$ satisfies $U \cap M \subset X \setminus \mathrm{F}(M)$. As

$$U \cap M \subset (X \setminus \mathrm{F}(M)) \cap M \in \mathsf{univ}\,\mathfrak{N}(X)$$

we have $X \setminus \mathrm{Cl}\big(M \cap \mathrm{F}(M)\big) = U \subset X \setminus \mathrm{F}(M)$. Thereby $\mathrm{F}(M) = \mathrm{Cl}\big(M \cap \mathrm{F}(M)\big)$ is established.

Let us turn to statement (4). As $\mathrm{F}(M) \supset M \cap \mathrm{F}(M)$, from statements (1) and (3), we have $\mathrm{F}(\mathrm{F}(M)) \supset \mathrm{F}(M \cap \mathrm{F}(M)) = \mathrm{F}(M)$. Applying (3) again, we have $\mathrm{F}(\mathrm{F}(M)) = \mathrm{Cl}(\mathrm{F}(M) \cap \mathrm{F}(\mathrm{F}(M))) = \mathrm{Cl}(\mathrm{F}(M)) = \mathrm{F}(M)$. □

Proposition 2.12. *Let Y be a separable metrizable space. If $M \subset X \subset Y$, then $\mathrm{F}_X(M) = X \cap \mathrm{F}_Y(M)$.*

Proof. Observe that $(Y \setminus \mathrm{F}_Y(M)) \cap M \in \mathsf{abNULL}$. As $M \subset X$ we have $(Y \setminus \mathrm{F}_Y(M)) \cap M = (X \setminus \mathrm{F}_Y(M)) \cap M$. So, $\mathrm{F}_X(M) \subset X \cap \mathrm{F}_Y(M)$. Next let V be an open subset of Y such that $V \cap X = X \setminus \mathrm{F}_X(M)$. As $M \subset X$ we have $V \cap M = M \setminus \mathrm{F}_X(M) \in \mathsf{abNULL}$. Hence $Y \setminus V \supset \mathrm{F}_Y(M)$. The proposition follows because $\mathrm{F}_X(M) = X \cap (Y \setminus V) \supset X \cap \mathrm{F}_Y(M)$. □

Another property is the topological invariance of the positive closure operator.

Proposition 2.13. *For homeomorphisms $h\colon X \to Y$ of separable metrizable spaces X and Y, if $M \subset X$, then* $\mathrm{F}_Y(h[M]) = h[\mathrm{F}_X(M)]$.

Proof. Denote the open subset $X \setminus \mathrm{F}_X(M)]$ of X by U. Then $U \cap M$ is a universally null set in X. It follows that $h[U \cap M]$ is a universally null set in Y. As $h[U \cap M] = h[U] \cap h[M]$ and $h[U]$ is an open set in Y, we have $h[U \cap M] \subset Y \setminus \mathrm{F}_Y(h[M])$, whence $\mathrm{F}_Y(h[M]) \subset h[M \cap \mathrm{F}_X(M)]$. Also $\mathrm{Cl}_Y(h[M \cap \mathrm{F}_X(M)]) = h[\mathrm{Cl}_X(M \cap \mathrm{F}_X(M))] = h[\mathrm{F}_X(M)]$. Hence $\mathrm{F}_Y(h[M]) \subset h[\mathrm{F}_X(M)]$. From the last inclusion we have $h[\mathrm{F}_X(M)] = h[\mathrm{F}_X(h^{-1}h[M])] \subset h[h^{-1}[\mathrm{F}_Y(h[M])]] = \mathrm{F}_Y(h[M])$. □

2.2. Positive measures

Let X be a separable metrizable space. Since $X \setminus \mathrm{F}_X(X)$ is an absolute null space, it is immediate that $\mathrm{support}(\mu) \subset \mathrm{F}_X(X)$ for every continuous, complete, σ-finite Borel measure μ on X. A useful class of continuous, complete, σ-finite Borel measures on X consists of those that are designated as positive. A μ is defined to be a *positive measure* if $\mathrm{support}(\mu) = \mathrm{F}_X(X) \neq \emptyset$. We denote this class of measures by $\mathrm{MEAS}^{\mathrm{pos}}(X)$, that is,

$$\mathrm{MEAS}^{\mathrm{pos}}(X) = \{\, \mu \in \mathrm{MEAS}^{\mathrm{cont}}(X) \colon \emptyset \neq \mathrm{support}(\mu) = \mathrm{F}_X(X) \,\},$$

where $\mathrm{MEAS}^{\mathrm{cont}}(X)$ is the collection of all continuous, complete, σ-finite Borel measures μ on X. (Note: $\mathrm{MEAS}^{\mathrm{pos}}(X) \subset \mathrm{MEAS}^{\mathrm{cont}}(X)$ is assumed.) It is obvious that $\mathrm{MEAS}^{\mathrm{pos}}(X) \neq \emptyset$ does not imply that X is an absolute measurable space. Indeed, consider any subspace of $\mathbb{R}$ that is non Lebesgue measurable. Equally obvious is that $\mathrm{MEAS}^{\mathrm{pos}}(X) = \emptyset$ if X is an absolute null space. We turn to the converse next.

2.2.1. Existence of positive measures. It is not immediate that positive measures exist if $\mathrm{F}_X(X) \neq \emptyset$. To this end, we have

Theorem 2.14. *For separable metrizable spaces X,* $\mathrm{MEAS}^{\mathrm{pos}}(X)$ *is not empty if and only if* $\mathrm{F}_X(X)$ *is not empty.*

Proof. We shall use F for F_X in the proof. Suppose $\mathrm{MEAS}^{\mathrm{pos}}(X)$ is not empty. Then there exists a measure μ such that $\mathrm{support}(\mu) = \mathrm{F}(X) \neq \emptyset$.

Suppose that $\mathrm{F}(X) \neq \emptyset$. Note that $\mu\big(X \setminus \mathrm{F}(X)\big) = 0$ for every continuous, complete, finite Borel measure μ on X. Let $U_0, U_1, \ldots,$ be a countable base for the open sets of X. From the definition of $\mathrm{F}(X)$, we have $\mu_n\big(U_n \cap \mathrm{F}(X)\big) > 0$ for some continuous, complete, finite Borel measure μ_n on X whenever $U_n \cap \mathrm{F}(X) \neq \emptyset$. We may assume $\mu_n\big(U_n \cap \mathrm{F}(X)\big) < 2^{-n}$. Let $\nu_n = \mu_n \llcorner \big(U_n \cap \mathrm{F}(X)\big)$ for each n. Then, for each Borel set B, we have $\nu(B) = \sum_{n=0}^{\infty} \nu_n(B) < 2$. Also, $\nu(\{x\}) = 0$ for every point x of X. Hence ν determines a continuous, complete, finite Borel measure on X. We already know $\mathrm{support}(\nu) \subset \mathrm{F}(X)$. Let U be an open set such that $U \cap \mathrm{F}(X) \neq \emptyset$. There exists an n such that $U \supset U_n \cap \mathrm{F}(X) \neq \emptyset$, whence $\nu(U) > 0$. Hence $\mathrm{F}(X) \subset \mathrm{support}(\nu)$. □

Corollary 2.15. *Let X be a separable metrizable space. If M is a subset of X with $\mathrm{F}_X(M) \neq \emptyset$, then* $\mathrm{support}(\mu) = \mathrm{F}_X(M)$ *for some continuous, complete, finite Borel measure μ on X.*

Proof. We have $\mathrm{F}_M(M) = M \cap \mathrm{F}_X(M) \neq \emptyset$. Hence there is a measure ν in $\mathrm{MEAS}^{\mathrm{pos}}(M)$. Let μ be an extension of ν such that $\mathrm{support}(\mu) = \mathrm{F}_X(M)$. Such an extension will exist with the aid of the inclusion map of $\mathrm{F}_M(M)$ into X. □

Of course, there are spaces for which the existence of positive measures is obvious – for example, the unit n-cube $[0,1]^n$ has the Lebesgue measure.

Lemma 2.16. *Let X be a separable metrizable space and let μ be a continuous, complete, finite Borel measure on X. If $\mathrm{F}_X(X) \neq \emptyset$, then there is a positive, continuous, complete, finite Borel measure ν on X such that* $\nu \llcorner \mathrm{support}(\mu) = \mu \llcorner \mathrm{support}(\mu)$.

Proof. Let $U = X \setminus \mathrm{support}(\mu)$ and let $\sigma \in \mathrm{MEAS}^{\mathrm{pos}}(X)$. Then $\nu = \mu \llcorner \big(\mathrm{support}(\mu)\big) + \sigma \llcorner U$ is a positive measure that fulfills the requirements. Indeed, to show that ν is positive, observe that $\mathrm{F}(X) = \mathrm{support}(\mu) \cup \big(U \cap \mathrm{F}(X)\big)$ since $\mathrm{support}(\mu) \subset \mathrm{F}(X)$. Hence, if V is an open set such that $V \cap \mathrm{F}(X) \neq \emptyset$, then either $V \cap \mathrm{support}(\mu) \neq \emptyset$ or $V \cap \big(U \cap \mathrm{F}(X)\big) \neq \emptyset$, whence $\nu(V) > 0$ and thereby ν is positive. The remaining part of the proof is trivial. □

2.2.2. A characterization of $\mathsf{univ}\,\mathfrak{M}(X)$.

Let X be a separable metrizable space and recall that the collection of all positive measures on X is denoted by $\mathrm{MEAS}^{\mathrm{pos}}(X)$. Define the two collections

$$\mathsf{univ}\,\mathfrak{M}^{\mathrm{pos}}(X) = \bigcap\{\, \mathfrak{M}(X,\mu) \colon \mu \in \mathrm{MEAS}^{\mathrm{pos}}(X) \,\},$$
$$\mathsf{univ}\,\mathfrak{N}^{\mathrm{pos}}(X) = \bigcap\{\, \mathfrak{N}(X,\mu) \colon \mu \in \mathrm{MEAS}^{\mathrm{pos}}(X) \,\}.$$

Clearly, $\mathsf{univ}\,\mathfrak{M}(X) \subset \mathsf{univ}\,\mathfrak{M}^{\mathrm{pos}}(X)$ and $\mathsf{univ}\,\mathfrak{N}(X) \subset \mathsf{univ}\,\mathfrak{N}^{\mathrm{pos}}(X)$. We have the following characterization.

Theorem 2.17. *Let X be a separable metrizable space. Then*

$$\mathsf{univ}\,\mathfrak{M}(X) = \mathsf{univ}\,\mathfrak{M}^{\mathrm{pos}}(X) \text{ and } \mathsf{univ}\,\mathfrak{N}(X) = \mathsf{univ}\,\mathfrak{N}^{\mathrm{pos}}(X).$$

This characterization will turn out to be quite useful in the investigation of the unit n-cube $[0,1]^n$. We shall see later in this chapter that the homeomorphism group of the space $[0,1]$ will play a nice role in a characterization of $\mathsf{univ}\,\mathfrak{M}\big([0,1]\big)$.

Proof of Theorem. If $\mathrm{F}(X) = \emptyset$, then $\mathrm{MEAS}^{\mathrm{pos}}(X) = \emptyset$ and $X \in \mathsf{abNULL}$. Hence $\mathsf{univ}\,\mathfrak{M}^{\mathrm{pos}}(X) = \bigcap\emptyset$ which, by the usual convention, is equal to $P(X) = \{\,E : E \subset X\,\}$. Moreover, $\mathsf{univ}\,\mathfrak{N}(X) = \mathsf{univ}\,\mathfrak{M}(X) = P(X)$.

Assume $\mathrm{F}(X) \neq \emptyset$. Let us prove that $\mathsf{univ}\,\mathfrak{M}(X) \supset \mathsf{univ}\,\mathfrak{M}^{\mathrm{pos}}(X)$. To this end, let $M \in \mathsf{univ}\,\mathfrak{M}^{\mathrm{pos}}(X)$ and let μ be a continuous, complete, finite Borel measure on X.

By Lemma 2.16 there is a positive, continuous, complete, finite Borel measure ν on X such that

$$\mu \llcorner \big(\mathrm{support}(\mu)\big) = \nu \llcorner \big(\mathrm{support}(\mu)\big).$$

Since $M \in \mathfrak{M}(X, \nu)$, we have $M \cap \mathrm{support}(\mu) \in \mathfrak{M}(X, \nu)$; and from $M \cap \mathrm{support}(\mu) \subset \mathrm{support}(\mu)$ we conclude $M \cap \mathrm{support}(\mu) \in \mathfrak{M}(X, \mu)$. Also, from the completeness of μ, we have $M \setminus \mathrm{support}(\mu) \in \mathfrak{N}(X, \mu)$. We have shown $M \in \mathfrak{M}(X, \mu)$ for every μ. Therefore $M \in \mathsf{univ}\,\mathfrak{M}(X)$, and the first equation of the theorem is proved.

The proof of the second equation is left to the reader.

2.3. Universally measurable maps

Let X and Y be separable metrizable spaces and consider a map $f : X \to Y$. Recall, for a complete, σ-finite Borel measure space $\mathrm{M}(X, \mu)$, that the map f is said to be μ-measurable if $f^{-1}[U]$ is μ-measurable for each open set U of Y. The collection of all such μ-measurable maps will be denoted by $\mathsf{MAP}(X, \mu\,; Y)$.

At the end of Chapter 1 we defined the notion of absolute measurable functions on the σ-ring $\mathsf{ab}\,\mathfrak{M}(X)$, which may be properly contained in the σ-algebra $\mathsf{univ}\,\mathfrak{M}(X)$ (see Section 1.6.2 on page 27). Here we turn our attention to universally measurable maps $f : X \to Y$.

2.3.1. Definitions. There are two possible ways to define universally measurable maps. The following is preferred.

DEFINITION 2.18. *Let X and Y be separable metrizable spaces. A map $f : X \to Y$ is said to be* universally measurable *if it is a member of the collection*

$$\mathsf{univ}\,\mathsf{MAP}(X; Y) = \bigcap \big\{ \mathsf{MAP}(X, \mu\,; Y) : \mu \in \mathrm{MEAS}(X) \big\}.$$

We have the following obviously equivalent condition.

PROPOSITION 2.19. *Let X and Y be separable metrizable spaces. A map $f : X \to Y$ is universally measurable if and only if $f^{-1}[B]$ is in $\mathsf{univ}\,\mathfrak{M}(X)$ whenever B is in $\mathfrak{B}(Y)$.*

Clearly every Borel measurable map is a universally measurable map. We have the obvious proposition on composition of maps.

PROPOSITION 2.20. *For separable metrizable spaces X, Y and Z let $g : Y \to Z$ be a Borel measurable map. Then gf is in $\mathsf{univ}\,\mathsf{MAP}(X; Z)$ whenever f is in $\mathsf{univ}\,\mathsf{MAP}(X; Y)$.*

Improvements of the last two propositions can be found in Section 4.1 of Chapter 4.

A sequence of maps f_n, $n = 1, 2, \ldots$, defined on a set X into a metrizable space Y is said to be *pointwise convergent* if the sequence $f_n(x)$, $n = 1, 2, \ldots$, is convergent for every x in X; the resulting map is denoted as $\lim_{n\to\infty} f_n$. The following is easily proved.

THEOREM 2.21. *For separable metrizable spaces X and Y, if a sequence f_n, $n = 1, 2, \ldots$, in* $\mathsf{univ\,MAP}(X;Y)$ *is pointwise convergent, then* $\lim_{n\to\infty} f_n$ *is in* $\mathsf{univ\,MAP}(X;Y)$.

2.3.2. Graph of universally measurable map. Suppose that X and Y are separable metrizable spaces and let $f: X \to Y$ be a universally measurable map. Then for each complete, finite Borel measure ν on X there is a Borel class 2 map $g: X \to Y$ such that $f = g$ ν-almost everywhere (see Appendix A). As graph(g) is a Borel set in $X \times Y$, the graph of g is a universally measurable set in $X \times Y$. The same is true for the graph of f, which will be shown next.

THEOREM 2.22. *If $f: X \to Y$ is a universally measurable map, where X and Y are separable metrizable spaces, then* graph(f) *is a universally measurable set in $X \times Y$. Additionally, if X is an absolute measurable space, then so is* graph(f).

PROOF. Let μ be a complete, finite Borel measure on $X \times Y$ and denote the natural projection of $X \times Y$ onto X by π. As $\nu = \pi_{\#}\mu$ is a complete, finite Borel measure on X, there is a Borel measurable map $g: X \to Y$ such that $E = \{x: f(x) \neq g(x)\}$ has ν measure equal to 0. Let A be a Borel set in X such that $E \subset A$ and $\nu(A) = 0$. We have that $\mathrm{graph}(f) = \big(\pi^{-1}[A] \cap \mathrm{graph}(f)\big) \cup \big(\pi^{-1}[X \setminus A] \cap \mathrm{graph}(g)\big)$. As the first summand has μ measure equal to 0 and the second summand is a Borel set in $X \times Y$, we have that graph(f) is μ-measurable. Therefore graph(f) is a universally measurable set in $X \times Y$.

Suppose further that X is an absolute measurable space. Let Y' be a completely metrizable space that contains Y, π' be the natural projection of $X \times Y'$ onto X, and let $\mu' = \varphi_{\#}\mu$, where φ is the inclusion map of $X \times Y$ into $X \times Y'$. With $\nu = \pi'_{\#}\mu'$ there is an absolute Borel space B contained in X such that $\nu(X \setminus B) = 0$. As $B \times Y'$ is an absolute Borel space, we have that $\pi^{-1}[B] \cap \mathrm{graph}(g)$ is an absolute Borel space. Moreover, $\mu(\pi^{-1}[X \setminus B]) = 0$. Thereby graph$(f)$ is an absolute measurable space. □

It is known that if $f: X \to Y$ is a map such that graph(f) is an analytic space, then X is an analytic space and f is a Borel measurable map. Also, if graph(f) is an absolute Borel space, then X is an absolute Borel space and f is a Borel measurable map. (See Section A.2 of Appendix A.) It is tempting to conjecture that the same can be said if graph(f) is an absolute measurable space. But this cannot be as witnessed by Grzegorek's example of an absolute null space Y with $\mathrm{card}(Y) = \mathsf{non}\text{-}\mathbb{L}$. Let X be a non-Lebesgue measurable subset of $\mathbb{R}$ with $\mathrm{card}(X) = \mathsf{non}\text{-}\mathbb{L}$. Then each bijection $f: X \to Y$ has the property that graph(f) is an absolute null space whence an absolute measurable space. Suppose that f is universally measurable and let ν be a complete, finite Borel measure on X, say the naturally induced measure on X by the Lebesgue measure on $\mathbb{R}$. Then the map φ given by $x \mapsto \big(x, f(x)\big)$ will be universally measurable and $\varphi_{\#}\nu$ is the zero measure because graph(f) is an absolute null space. So we will have $0 = \varphi_{\#}\nu\big(\mathrm{graph}(f)\big) = \nu(X) = \lambda^*(X) > 0$ and a contradiction will occur.

In [**85**, Theorem 2, page 489] it is shown that if X is an analytic space and $f: X \to Y$ is a map whose graph is also an analytic space then f is necessarily a Borel measurable map. This leads to the question

QUESTION. Suppose that X is an absolute measurable space and Y is a separable metrizable space. Let $f: X \to Y$ be a map whose graph is an absolute measurable space. Is f necessarily universally measurable?

2.3.3. Real-valued functions. We have been using the notation fg for the composition of maps f and g. Now we want to introduce real-valued functions f and g and the pointwise products of them. This presents a notational difficulty since products of real numbers r and s are usually indicated by rs. To avoid confusion, we shall use $f \cdot g$ to denote the product of real-valued functions f and g. We shall use rf for the scalar multiple of a real number r and a real-valued function f. The same conventions will be used for complex-valued functions. (The set of complex numbers will be denoted by $\mathbb{C}$.)

In addition to the usual operations of addition, subtraction, multiplication and division of real-valued functions we will deal with the lattice operations of pointwise maximum and minimum of finite sets of real-valued functions and also the supremum and infimum of countable collections of real-valued functions (of course, extended real-valued functions may result in the last two operations). We shall use the usual symbols for these operations. A collection of real-valued functions is said to be pointwise bounded above (or simply bounded above) by a function f if every member of the collection is pointwise less than or equal to f.

THEOREM 2.23. *Let X be a separable metrizable space.*

(1) *If $r \in \mathbb{R}$ and $f \in \mathsf{univ\,MAP}(X;\mathbb{R})$, then $rf \in \mathsf{univ\,MAP}(X;\mathbb{R})$.*
(2) *If $f \in \mathsf{univ\,MAP}(X;\mathbb{R})$, then $|f| \in \mathsf{univ\,MAP}(X;\mathbb{R})$.*
(3) *If f and g are in $\mathsf{univ\,MAP}(X;\mathbb{R})$, then so are $f+g$, $f \cdot g$, $f \vee g$ and $f \wedge g$ in $\mathsf{univ\,MAP}(X;\mathbb{R})$.*
(4) *If f_n, $n = 1, 2, \ldots$, is a pointwise bounded sequence of functions in $\mathsf{univ\,MAP}(X;\mathbb{R})$, then $\bigvee_{n=1}^{\infty} f_n$ and $\bigwedge_{n=1}^{\infty} f_n$ are in $\mathsf{univ\,MAP}(X;\mathbb{R})$.*

Of course, one may replace $\mathbb{R}$ in the above theorem with a complete normed linear space of functions and the statements (1), (2), and the appropriate parts of (3) will remain valid; and, if the linear space has a suitable lattice structure as well, with appropriate changes, the remaining statements will remain valid.

2.4. Symmetric difference of Borel and null sets

For complete, σ-finite Borel measure spaces $\mathrm{M}(X,\mu)$, it is well-known that each μ-measurable set M is the symmetric difference of a Borel set B and a μ-null set N. (The symmetric difference of A and B is the set $A \,\Delta\, B = (A \setminus B) \cup (B \setminus A)$.) It was shown by E. Grzegorek and C. Ryll-Nardzewski [**70**] that no such representation exists for universally measurable sets and universally null sets in an uncountable, separable completely metrizable space X. That is, for such a space X and a universally measurable set M in X, it can be false that M is the symmetric difference of a Borel set B and a universally null set N in X. This section is devoted to their proof of this result.

2.4.1. ***Properties of the symmetric difference.*** Let X be a separable metrizable space. Denote by $\mathsf{univ}\,\mathfrak{M}_\Delta(X)$ the collection of all sets in X that have a symmetric difference representation $X = B \,\Delta\, N$ for some B in $\mathfrak{B}(X)$ and for some N in $\mathsf{univ}\,\mathfrak{N}(X)$. Clearly, $\mathsf{univ}\,\mathfrak{M}_\Delta(X) \subset \mathsf{univ}\,\mathfrak{M}(X)$.

Suppose that M is in $\mathsf{univ}\,\mathfrak{M}_\Delta(X)$. Then $M = B \,\Delta\, N$, where B is some Borel set and N is some absolute null space. Hence there are absolute null spaces N_1 and N_2 such that

(1) $N_1 \subset B$,
(2) $B \cap N_2 = \emptyset$,
(3) $M = (B \setminus N_1) \cup N_2$.

So, $N_2 = M \cap (X \setminus B)$ is a Borel subset of the subspace M, and both N_1 and N_2 are totally imperfect spaces.

Recall from Chapter 1 that $\mathsf{ANALYTIC}$ is the collection of all analytic spaces.

PROPOSITION 2.24. *If* $A \in \mathsf{ANALYTIC} \cap \mathsf{univ}\,\mathfrak{M}_\Delta(X)$*, then there is a Borel set* B_0 *such that* $A \subset B_0$ *and* $N = B_0 \setminus A$ *is a totally imperfect space.*

PROOF. With M replaced by A, from condition (2) above, N_2 is a Borel subset of an analytic set. As N_2 is also totally imperfect, we have that N_2 is a countable set. Let $B_0 = B \cup N_2$, a Borel set, and let $N = N_1$. Then $B_0 \setminus A = N_1$, and the proposition is proved. □

2.4.2. ***Main theorem.*** The main theorem concerns the equality $\mathsf{univ}\,\mathfrak{M}(X) = \mathsf{univ}\,\mathfrak{M}_\Delta(X)$.

Employing the equation (A.1) on page 181, we have

LEMMA 2.25. *Let* X *be a separable metrizable space and let* Y *be a subset of* X *such that* Y *is an uncountable absolute* G_δ *space. If* A *is a subset of* Y *such that* $A \in \mathsf{ANALYTIC} \cap \mathsf{univ}\,\mathfrak{M}_\Delta(X)$*, then the constituent decomposition*

$$Y \setminus A = \bigcup_{\alpha<\omega_1} A_\alpha,$$

of the co-analytic space $Y \setminus A$ *has the property that there exists a* β *with* $\beta < \omega_1$ *such that* $\operatorname{card}(A_\alpha) \leq \aleph_0$ *whenever* $\alpha > \beta$.

PROOF. By the above proposition, there is a Borel set B_0 such that $A \subset B_0$ and $N = B_0 \setminus A$ is totally imperfect. There is no loss in assuming that $B_0 \subset Y$. By Theorem A.6, there is a β with $\beta < \omega_1$ such that the Borel subset $Y \setminus B_0$ is contained in $\bigcup_{\alpha \leq \beta} A_\alpha$. So, $\bigcup_{\alpha > \beta} A_\alpha \subset N$. As the constituents are mutually disjoint, we have $A_\alpha = A_\alpha \cap N$ for $\alpha > \beta$. So the absolute Borel space A_α is totally imperfect, whence countable, whenever $\alpha > \beta$. □

THEOREM 2.26 (Grzegorek–Ryll-Nardzewski). *If X is an uncountable, separable, completely metrizable space, then* $\operatorname{univ}\mathfrak{M}_\Delta(X)$ *is not equal to* $\operatorname{univ}\mathfrak{M}(X)$.

PROOF. In the space X there is an analytic space A such that the co-analytic space $X \setminus A$ has a constituent decomposition $X \setminus A = \bigcup_{\alpha<\omega_1} A_\alpha$ with the property that the Borel orders of the constituents A_α are unbounded (see the sentence that immediately follows Theorem A.5). By the lemma, such an analytic set is not in $\operatorname{univ}\mathfrak{M}_\Delta(X)$. □

As a corollary we have

COROLLARY 2.27. *If X is a separable metrizable space that is not totally imperfect, then* $\operatorname{univ}\mathfrak{M}_\Delta(X) \neq \operatorname{univ}\mathfrak{M}(X)$.

PROOF. If X is not totally imperfect, then it contains a nonempty, compact, perfect subset Y. Hence $\operatorname{univ}\mathfrak{M}(Y) = \operatorname{ab}\mathfrak{M}(Y) \subset \operatorname{univ}\mathfrak{M}(X)$. Suppose that $\operatorname{univ}\mathfrak{M}_\Delta(X) = \operatorname{univ}\mathfrak{M}(X)$. With the aid of the theorem there exists an M in $\operatorname{univ}\mathfrak{M}(Y) \setminus \operatorname{univ}\mathfrak{M}_\Delta(Y)$. Obviously $M \subset Y$ and $M \in \operatorname{univ}\mathfrak{M}(X)$. Clearly, if $M \subset Y$ and $M \in \operatorname{univ}\mathfrak{M}_\Delta(X)$, then $M \in \operatorname{univ}\mathfrak{M}_\Delta(Y)$. So a contradiction will occur if $M \in \operatorname{univ}\mathfrak{M}_\Delta(X)$. □

The completely metrizable space requirement in the theorem may be replaced by the requirement that X be an absolute measurable space as the next corollary shows.

COROLLARY 2.28. *Let X be an absolute measurable space. Then* $\operatorname{univ}\mathfrak{M}_\Delta(X) \neq \operatorname{univ}\mathfrak{M}(X)$ *if and only if* $\mathrm{F}_X(X) \neq \emptyset$.

PROOF. Let $\mathrm{F}_X(X) \neq \emptyset$ for an absolute measurable space X. By the definition of the positive closure operator, X is not an absolute null space, whence X is not totally imperfect.

Let $\mathrm{F}_X(X) = \emptyset$ for an absolute measurable space X. Then X is an absolute null space. So, $\operatorname{univ}\mathfrak{M}(X) = \operatorname{univ}\mathfrak{N}(X) = \{M : M \subset X\} = \operatorname{univ}\mathfrak{M}_\Delta(X)$. □

Observe that the disjoint topological union of a space X_1 that is not absolute measurable and a space X_2 that is absolute measurable but not absolute null yields the following proposition.

PROPOSITION 2.29. *There is a separable metrizable space X that is a non absolute measurable space with* $\operatorname{univ}\mathfrak{M}_\Delta(X) \neq \operatorname{univ}\mathfrak{M}(X)$.

2.4.3. More continuum hypothesis. Let us turn to the equality question for spaces X that are not absolute measurable spaces. The following example uses a space constructed by Sierpiński. The existence of this example is assured by the continuum hypothesis. Sierpiński showed that there is a subset X of $\mathbb{R}$ with $\operatorname{card}(X) = \mathfrak{c} = \aleph_1$ such that every subset M of X with $\operatorname{card}(M) = \mathfrak{c}$ has positive outer Lebesgue measure. Of course the continuum hypothesis then implies that a subset N of X has Lebesgue measure 0 if and only if $\operatorname{card}(N) \leq \aleph_0$.

We shall give a proof of the existence of a Sierpiński set with the aid of the partition theorem, Theorem 1.37. To this end, consider the collection $\mathcal{K}$ of all subsets N

of $\mathbb{R}$ with $\lambda(N) = 0$. It is easily seen that this collection satisfies the hypothesis of Theorem 1.37. Hence there is a partition X_α, $\alpha < \omega_1$, of $\mathbb{R}$ such that $\operatorname{card}(X_\alpha) = \aleph_1$ and $X_\alpha \in \mathcal{K}$ for each α. That is, $\mathbb{R} = \bigcup_{\alpha<\omega_1} X_\alpha$ and $X_\alpha \cap X_\beta = \emptyset$ whenever $\alpha \neq \beta$. From each X_α select a point x_α and let $X = \{x_\alpha : \alpha < \omega_1\}$.

Proposition 2.30. *Assume that the continuum hypothesis holds. The subset X of $\mathbb{R}$ defined above is a Sierpiński set, that is,* $\operatorname{card}(X) = \aleph_1$ *and X has the property that a subset M of X has $\lambda(M) = 0$ if and only if* $\operatorname{card}(M) < \aleph_1$.

Proof. Suppose that M is a subset of X with $\lambda(M) = 0$. As $M \in \mathcal{K}$, there is a β with $\beta < \omega_1$ such that $M \subset \bigcup_{\alpha<\beta} X_\alpha$. Then $M \subset \{x_\alpha : \alpha < \beta\}$, whence $\operatorname{card}(M) < \aleph_1$. □

Proposition 2.31. *Assume that the continuum hypothesis holds and let X be a Sierpiński subset of $\mathbb{R}$. Then X is not an absolute measurable space and* $\operatorname{univ}\mathfrak{M}_\Delta(X) = \operatorname{univ}\mathfrak{M}(X)$.

Proof. Suppose $M \in \operatorname{univ}\mathfrak{M}(X)$. Then M is $(\lambda|X)$-measurable. Hence there are Borel subsets of A and B of X such that $A \supset M \supset B$ and $\lambda^*(A \setminus B) = 0$, whence $\operatorname{card}(A \setminus B) \leq \aleph_0$. From $M = A \setminus (A \setminus M)$ we infer $M \in \operatorname{univ}\mathfrak{M}_\Delta(X)$. Hence $\operatorname{univ}\mathfrak{M}_\Delta(X) = \operatorname{univ}\mathfrak{M}(X)$.

Suppose that $X \in \mathsf{abMEAS}$. As $\lambda(X) > 0$, we have $\mathrm{F}_X(X) \neq \emptyset$. Consequently, $\operatorname{univ}\mathfrak{M}_\Delta(X) \neq \operatorname{univ}\mathfrak{M}(X)$ and a contradiction has been reached. Thereby we have $X \notin \mathsf{abMEAS}$. □

Sierpiński sets will be discussed further in Chapter 6.

2.5. Early results

In a summary of the early works [**15**] (written by S. Braun and E. Szpilrajn in collaboration with K. Kuratowski in 1937), the following result concerning subsets M of the interval $[0, 1]$ is presented. As we shall see, this theorem provides the motivation for a large part of the book.

Theorem 2.32. *Let M be a subset of $[0, 1]$. Then the following statements concerning the Lebesgue measure λ are equivalent.*

(1) *If $f : \mathbb{R} \to \mathbb{R}$ is a nondecreasing function, then $\lambda(f[M]) = 0$.*
(2) *If f is a homeomorphism from M onto a subset of $\mathbb{R}$, then $\lambda(f[M]) = 0$.*
(3) *If f is a $\mathfrak{B}$-homeomorphism of M into $\mathbb{R}$, then $\lambda(f[M]) = 0$.*
(4) *If f is a bijection of M into $\mathbb{R}$ for which f^{-1} is Borel measurable, then $\lambda(f[M]) = 0$.*
(5) *M is an absolute null space.*
(6) *If $f : \mathbb{R} \to \mathbb{R}$ is an orientation preserving homeomorphism, then $\lambda(f[M]) = 0$.*

Proof. Let us prove that statement (5) implies statement (4). Let $M \in \mathsf{abNULL}$ and let μ be the restriction measure $\mu = \lambda|\big(f[M]\big)$. As f^{-1} is an injective Borel measurable map, the induced measure $\nu = f^{-1}{}_{\#}\mu$ is a continuous, complete, finite Borel measure.

Hence $\nu(M) = 0$. Also, $\nu(M) = \big(f^{-1}{}_{\#}\mu\big)(M) = \mu\big(f[M]\big)$. Therefore the Lebesgue outer measure of $f[M]$ is equal to 0, thereby statement (4) is verified.

That statement (4) implies statement (3), and that statement (3) implies statement (2) are quite trivial.

It is obvious that statement (1) implies statement (6) and that statement (2) also implies statement (6).

Let us prove that statement (6) implies statement (5). Let ν be a positive, continuous, complete, finite Borel measure. There is no loss in assuming that $\nu\big([0,1]\big) = 1$. Let $f\colon [0,1] \to [0,1]$ be the increasing function defined by

$$f(x) = \begin{cases} \nu\big([0,x]\big), & \text{if } 0 < x \leq 1; \\ 0, & \text{if } x = 0. \end{cases}$$

Clearly f is a homeomorphism and can be extended to an increasing homeomorphism of $\mathbb{R}$. Observe that $\lambda\big([0,y]\big) = \nu\big(f^{-1}\big[[0,y]\big]\big)$ whenever $0 \leq y \leq 1$. Hence $\lambda(B) = \nu\big(f^{-1}[B]\big)$ for every Borel set B contained in $[0,1]$. By statement (6), $\lambda\big(f[M]\big) = 0$. Let B be a Borel subset of $\mathbb{R}$ such that $f[M] \subset B$ and $\lambda(B) = 0$. Then $0 = \lambda(B) = \nu\big(f^{-1}[B]\big) \geq \nu^*\big(f^{-1}\big[f[M]\big]\big) = \nu^*(M)$.

To complete the proof, let us prove that statement (5) implies statement (1). Let $f\colon \mathbb{R} \to \mathbb{R}$ be a nondecreasing function. There is a countable subset D of $\mathbb{R}$ such that f restricted to $\mathbb{R} \setminus f^{-1}[D]$ is a homeomorphism. It follows that $\lambda\big(f\big[M \setminus f^{-1}[D]\big]\big) = 0$ and $\lambda\big(f\big[M \cap f^{-1}[D]\big]\big) = 0$, whence $\lambda(f[M]) = 0$. □

Actually the statement (6) was not part of the summary mentioned above. We have included it here to illustrate how the homeomorphism group of [0, 1] plays a role in this theorem. Indeed, in the proof of (6) implies (5) we have actually shown that, for each positive, continuous, complete Borel measure ν with $\nu\big([0,1]\big) = 1$, there is an increasing homeomorphism f of $[0,1]$ such that $\lambda|[0,1] = f_{\#}\nu$. Clearly the last requirement on ν is not a serious one since the formula can be corrected by the insertion of a suitable coefficient before the measure $\lambda|[0,1]$.

2.6. The homeomorphism group of [0, 1]

For a topological space X, the group of homeomorphisms of X onto X will be denoted by $\mathsf{HOMEO}(X)$.

2.6.1. Elementary general properties. Observe that, for any Borel measure μ on a separable metrizable space X and for any positive number c, we have the σ-algebra equality $\mathfrak{M}(X,\mu) = \mathfrak{M}(X, c\,\mu)$. Hence the measures μ used in the definitions of $\mathsf{univ}\,\mathfrak{M}(X)$ and $\mathsf{univ}\,\mathfrak{M}^{\mathrm{pos}}(X)$ may be required to have the added condition that $\mu(X) = 1$ whenever $\mu(X) \neq 0$ without any change in the resulting collections. As usual, measure spaces $\mathrm{M}(X,\mu)$ with $\mu(X) = 1$ are called *probability spaces.*

LEMMA 2.33. *For a separable metrizable space X, if μ is a complete, σ-finite Borel measure on X and if $h \in \mathsf{HOMEO}(X)$, then $h_{\#}\mu$ is a complete, σ-finite Borel measure on X. Moreover, $h^{-1}[M] \in \mathfrak{M}(X,\mu)$ if and only if $M \in \mathfrak{M}(X, h_{\#}\mu)$.*

Proof. The first statement follows from Proposition A.36. Let us prove the second statement. Observe that A and B are Borel sets such that $A \subset M \subset B$ if and only if $h^{-1}[A]$ and $h^{-1}[B]$ are Borel sets such that $h^{-1}[A] \subset h^{-1}[M] \subset h^{-1}[B]$. We infer the second statement from $h_{\#}\mu(B \setminus A) = \mu\big(h^{-1}[B \setminus A]\big)$. □

Observe that $h_{\#}\big[\mathrm{MEAS}^{\mathrm{cont}}(X)\big] = \mathrm{MEAS}^{\mathrm{cont}}(X)$ whenever h is in $\mathsf{HOMEO}(X)$. Hence we have invariance under action of $\mathsf{HOMEO}(X)$.

Theorem 2.34. *Let X be a separable metrizable space and let h be in $\mathsf{HOMEO}(X)$. For subsets M of X, M is an absolute measurable subspace if and only if $h^{-1}[M]$ is absolute measurable subspace. Also $M \in \mathsf{univ}\,\mathfrak{M}(X)$ if and only if $h^{-1}[M] \in \mathsf{univ}\,\mathfrak{M}(X)$. Hence, for continuous, complete, finite Borel measures μ on X,*

$$h^{-1}[\mathsf{univ}\,\mathfrak{M}(X)] = \mathsf{univ}\,\mathfrak{M}(X) \subset \bigcap\{\,\mathfrak{M}(X, h_{\#}\mu) \colon h \in \mathsf{HOMEO}(X)\,\}.$$

Proof. The first statement is the result of the topological embedding property of absolute measurable spaces. The second statement follows from the above lemma. The final statement follows easily. □

We remind the reader of the definition of $\mathsf{ab}\,\mathfrak{M}(X)$, it is the collection of all subsets of X that are absolute measurable spaces (see page 27). The first statement of the theorem is the invariance of the collection $\mathsf{ab}\,\mathfrak{M}(X)$ under the action of $\mathsf{HOMEO}(X)$.

The universally positive closure operator F_X has the following nice property. The proof is immediate from Proposition 2.13.

Proposition 2.35. *Let X be a separable metrizable space. If M is a subset of X, then $h[\mathrm{F}_X(M)] = \mathrm{F}_X(h[M])$ whenever $h \in \mathsf{HOMEO}(X)$. Hence $h[\mathrm{F}_X(X)] = \mathrm{F}_X(X)$ for every h in $\mathsf{HOMEO}(X)$.*

2.6.2. The space $[0,1]$.

The observation made in Section 2.5 about probability measures will now be stated as a lemma.

Lemma 2.36. *If μ is a positive, continuous, complete, Borel probability measure on $[0,1]$, then there exists an h in $\mathsf{HOMEO}([0,1])$ such that $\mu = h_{\#}\big(\lambda|[0,1]\big)$. Additionally, h may be assumed to be increasing, that is, orientation preserving.*

Thus we have

Theorem 2.37. *A necessary and sufficient condition for M to be in $\mathsf{univ}\,\mathfrak{M}\big([0,1]\big)$ is that M be in $\mathfrak{M}\big([0,1], h_{\#}(\lambda|[0,1])\big)$ for every h in $\mathsf{HOMEO}([0,1])$.*

Proof. The necessary condition is clear since $[0,1]$ is an absolute measurable space.

For the sufficiency, let μ be a positive, continuous, complete Borel probability measure on $[0,1]$. Let h be such that $\mu = h_{\#}(\lambda|[0,1])$. As M is in $\mathfrak{M}\big([0,1], h_{\#}(\lambda|[0,1])\big)$ we have that $h^{-1}[M]$ is in $\mathfrak{M}\big([0,1],\mu\big)$. Consequently, $h^{-1}[M] \in \mathsf{univ}\,\mathfrak{M}^{\mathrm{pos}}([0,1]) = \mathsf{univ}\,\mathfrak{M}\big([0,1]\big)$. □

The above lemma also yields the next theorem.

THEOREM 2.38. *Let μ and ν be any pair of positive, continuous, complete, finite Borel measures on* $[0,1]$. *Then there exists an h in* $\mathsf{HOMEO}([0,1])$ *such that* $\mu = h_{\#}\nu$ *whenever* $\mu([0,1]) = \nu([0,1])$. *Moreover, h may be assumed to be orientation preserving.*

PROOF. There exist h_1 and h_2 in $\mathsf{HOMEO}([0,1])$ such that $\mu = h_{1\#}\big(c\,\lambda|[0,1]\big)$ and $\nu = h_{2\#}\big(c\,\lambda|[0,1]\big)$. We have $\mu = h_{1\#}\big(c\,\lambda|[0,1]\big) = h_{1\#}\big(h_2{}^{-1}\big)_{\#}\nu = \big(h_1 h_2{}^{-1}\big)_{\#}\nu$. As $h = h_1 h_2{}^{-1} \in \mathsf{HOMEO}([0,1])$, the theorem is proved. □

The following theorem is the case $n = 1$ of the Oxtoby–Ulam theorem, the subject of Chapter 3. We leave its verification to the reader.

THEOREM 2.39. *In order that a complete Borel measure μ on* $[0,1]$ *be such that* $\lambda = h_{\#}\mu$ *for some h in* $\mathsf{HOMEO}([0,1])$, *where λ is the Lebesgue measure, it is necessary and sufficient that*

(1) $\mu(U) > 0$ *for every nonempty, open set U of the space* $[0,1]$,
(2) $\mu\big(\{x\}\big) = 0$ *for every x in* $[0,1]$,
(3) $\mu([0,1]) = 1$.

It also may be required that $h|\partial[0,1]$ *is the identity map.*

2.6.3. non-$\mathbb{L}$ and κ_{G} revisited. We promised in the comment section of the previous chapter a proof of the equality of the two cardinal numbers κ_{G} and non-$\mathbb{L}$ where

$$\text{non-}\mathbb{L} = \min\{\,\mathrm{card}(E)\colon E \subset [0,1] \text{ with } \lambda^*(E) > 0\,\}.$$

Here λ^* is the Lebesgue outer measure on $[0,1]$. In Section 1.4 we defined another cardinal number κ_0. The idea for the definition of κ_0 (which has already been shown to be equal to κ_{G}) is the same as that used in the definition of non-$\mathbb{L}$. We have the equality of the three cardinal numbers.

PROPOSITION 2.40. non-$\mathbb{L} = \kappa_{\mathrm{G}} = \kappa_0$.

PROOF. Let E be a subset of $[0,1]$ with $\lambda^*(E) > 0$. We may assume $E \cap \mathbb{Q} = \emptyset$, where $\mathbb{Q}$ is the set of rational numbers, whence $E \subset \mathcal{N}$. Let f be a topological embedding of $\mathcal{N}$ into $\{0,1\}^{\mathbb{N}}$. Then $\mu = f_{\#}(\lambda|\mathcal{N})$ is a continuous, complete, finite Borel measure on $\{0,1\}^{\mathbb{N}}$. It is easily seen that $f[E]$ is a set with $\mu^*\big(f[E]\big) > 0$. Hence non-$\mathbb{L} \geq \kappa_0 = \kappa_{\mathrm{G}}$.

Next suppose that μ is a continuous, complete, finite Borel measure on $\{0,1\}^{\mathbb{N}}$ and E is a set with $\mu^*(E) > 0$. Let f be a homeomorphism from $\{0,1\}^{\mathbb{N}}$ onto the Cantor ternary set $\mathcal{C}$, whence f is a continuous map into $[0,1]$. Hence $f_{\#}\mu$ is a continuous, complete, finite Borel measure defined on $[0,1]$. Let ν be the positive, continuous, complete Borel measure $f_{\#}\mu + \lambda|\big([0,1] \setminus \mathrm{support}(f_{\#}\mu)\big)$. There is a homeomorphism h of $[0,1]$ such that $h_{\#}\nu = c\lambda$ where c is a positive number. It is easily shown that $E' = h\big[f[E]\big]$ is a subset of $[0,1]$ such that $\lambda^*(E') > 0$. Indeed, let B be a Borel set that contains E'. As $(hf)^{-1}[B] \supset E$, we have $c\lambda(B) = h_{\#}\nu(B) =$

$\nu(h^{-1}[B]) \geq f_\# \mu(h^{-1}[B]) = \mu((hf)^{-1}[B]) \geq \mu^*(E) > 0$. Thereby we have shown that $\kappa_G = \kappa_0 \geq \text{non-}\mathbb{L}$. □

We now have the important example of Grzegorek that shows the existence of an absolute null subspace X of $\mathbb{R}$ with $\text{card}(X) = \text{non-}\mathbb{L}$.

THEOREM 2.41. *There exists a subset X of $\mathbb{R}$ such that* $\text{card}(X) = \text{non-}\mathbb{L}$ *and X is an absolute null space.*

PROOF. By Corollary 1.45 there is a subset Y of $\{0,1\}^{\mathbb{N}}$ such that $Y \in \text{abNULL}$ and $\text{card}(Y) = \text{non-}\mathbb{L}$. Let $h\colon \{0,1\}^{\mathbb{N}} \to \mathbb{R}$ be a continuous injection. The set $X = h[Y]$ fulfills the requirement. □

The reader will find several exercises on non-$\mathbb{L}$ in the Exercise section at the end of the chapter. Clearly, $\text{card}(\bigcup_{\alpha<\text{non-}\mathbb{L}} E_\alpha) \leq \text{non-}\mathbb{L}$ if and only if $\text{card}(E_\alpha) \leq \text{non-}\mathbb{L}$ whenever $\alpha < \text{non-}\mathbb{L}$.

It can happen that a set E can be small in the Lebesgue measure sense even though $\lambda^*(E) > 0$ and $\text{card}(E) = \text{non-}\mathbb{L}$. It can also happen that such a subset E of $[0,1]$ is such that $[0,1] \setminus E$ contains no subset of positive Lebesgue measure. As usual, the *Lebesgue inner measure* of a set E is

$$\lambda_*(E) = \sup\{\, \lambda(M) \colon M \subset E, M \text{ is a Borel set} \,\}.$$

A subset E of $[0,1]$ is said to have *full Lebesgue measure* in $[0,1]$ if and only if $\lambda_*([0,1] \setminus E) = 0$ (note that E need not be Lebesgue measurable). Let us state the assertion as a proposition.

PROPOSITION 2.42. *There are subsets E of* $[0,1]$ *with* $\text{card}(E) = \text{non-}\mathbb{L}$ *that have full Lebesgue measure in* $[0,1]$.

The next theorem will follow easily and is left as an exercise.

THEOREM 2.43. *Let X be an absolute measurable space. If μ is a continuous, complete, finite Borel measure on X with $\mu(X) > 0$, then there is a subset Y of X such that* $\text{card}(Y) = \text{non-}\mathbb{L}$ *and such that the μ inner measure $\mu_*(X \setminus Y)$ is equal to* 0.

2.7. The group of $\mathfrak{B}$-homeomorphisms

Let X be a separable metrizable space. The collection of all $\mathfrak{B}$-homeomorphisms $f\colon X \to X$ forms a group which will be denoted by $\mathfrak{B}\text{-HOMEO}(X)$. The collection $\text{MEAS}^{\text{pos}}([0,1])$ of all positive, continuous, complete, finite Borel measures on $[0,1]$ is not invariant under the action of $\mathfrak{B}\text{-HOMEO}([0,1])$. That is, there exists an f in $\mathfrak{B}\text{-HOMEO}([0,1])$ such that $f_\#(\lambda|[0,1])$ is not a positive measure. Indeed, let $\mathcal{C}$ be the Cantor ternary set. There is a $\mathfrak{B}$-homeomorphism f such that $f[\mathcal{C}] = (\frac{1}{2}, 1]$ and $f[[0,1] \setminus \mathcal{C}] = [0, \frac{1}{2}]$. For this f we have that $(\frac{1}{2}, 1]$ is a set whose $f_\#(\lambda|[0,1])$ measure is 0. Despite this fact, there is an analogue of the homeomorphism group property for the group $\mathfrak{B}\text{-HOMEO}(X)$ for absolute measurable spaces X which will be proved in Section 2.7.2.

2.7.1. Positive measures and $\mathfrak{B}$-homeomorphisms. Let Y be an absolute Borel space with $\mathrm{F}_Y(Y) \neq \emptyset$. Then each open set V with $V \cap \mathrm{F}_Y(Y) \neq \emptyset$ is an uncountable absolute Borel space. Clearly, there is a countable collection V_i, $i = 1, 2, \ldots$, of mutually disjoint, nonempty, open sets of Y such that $V_i \cap \mathrm{F}_Y(Y) \neq \emptyset$. Let $U_0 = Y \setminus \bigcup_{i=2}^{\infty} V_i$ and $U_i = V_{i+1}$, $i > 1$. Each U_i is an uncountable absolute Borel space. Next let F_i, $i = 1, 2, \ldots$, be a sequence of mutually disjoint topological copies of the Cantor ternary set such that each nonempty open set of $[0, 1]$ contains an F_i for some i. Define $g \colon Y \to [0, 1]$ to be a $\mathfrak{B}$-homeomorphism such that $g[U_0] = [0, 1] \setminus \bigcup_{i=1}^{\infty} F_i$ and $g[U_i] = F_i$ for $i = 1, 2, \ldots$. With the aid of this $\mathfrak{B}$-homeomorphism we have

LEMMA 2.44. *Let Y be an absolute Borel space and μ be a positive, complete, finite Borel measure on Y. Then there is a $\mathfrak{B}$-homeomorphism $g \colon Y \to [0, 1]$ such that $g_\# \mu$ is a positive, complete, finite Borel measure on $[0, 1]$. If μ is also continuous, then $g_\# \mu$ is also continuous.*

PROOF. Let $\mu \in \mathrm{MEAS}^{\mathrm{pos}}(Y)$. Then $\mathrm{F}_Y(Y) = \mathrm{support}(\mu) \neq \emptyset$. Hence $\mu(V) > 0$ whenever V is an open set with $V \cap \mathrm{F}_Y(Y) \neq \emptyset$. Let $g \colon Y \to [0, 1]$ be as defined above. Then $g_\# \mu(F_i) > 0$ for every i and therefore $g_\# \mu$ is in $\mathrm{MEAS}^{\mathrm{pos}}([0, 1])$. □

2.7.2. $\mathfrak{B}$-homeomorphism group. Here is a $\mathfrak{B}$-homeomorphism analogue of Theorem 2.38 for absolute measurable spaces X.

THEOREM 2.45. *Let X be an absolute measurable space. If μ and ν are positive, continuous, complete, finite Borel measures on X such that $\mu(X) = \nu(X)$, then there is an f in $\mathfrak{B}$-HOMEO(X) such that $f_\# \nu = \mu$.*

PROOF. Since X is an absolute measurable space, there are absolute Borel spaces Y_μ and Y_ν contained in $\mathrm{F}_X(X)$ such that $\mu\big(Y_\mu\big) = \mu(X)$ and $\nu\big(Y_\nu\big) = \nu(X)$. Let $Y = Y_\mu \cup Y_\nu$. Then $Y \subset \mathrm{F}_X(X)$, $\mu(X \setminus Y) = 0$ and $\nu(X \setminus Y) = 0$. Let $g \colon Y \to [0, 1]$ be a $\mathfrak{B}$-homeomorphism provided by the above lemma. As $g_\#(\nu|Y)$ and $g_\#(\mu|Y)$ are positive, continuous, complete Borel measures on $[0, 1]$ with $g_\#(\nu|Y)\big([0, 1]\big) = g_\#(\mu|Y)\big([0, 1]\big)$, there is an h in HOMEO$([0, 1])$ such that $h_\# g_\#(\nu|Y) = g_\#(\mu|Y)$. Consequently, $g^{-1}{}_\# h_\# g_\#(\nu|Y) = \mu|Y$. Let $f \colon X \to X$ be the $\mathfrak{B}$-homeomorphism that is equal to $g^{-1} h g$ on Y and to the identity map on $X \setminus Y$. Then we have $f_\# \nu = \mu$. □

Let us now relax the positive measure requirement of Theorem 2.45.

LEMMA 2.46. *Let X be an absolute measurable space and μ be a continuous, complete, finite Borel measure on X with $0 < \mu(X)$. Then there is a φ in $\mathfrak{B}$-HOMEO(X) such that $\varphi_\# \mu$ is a positive, continuous, complete, finite Borel measure on X.*

PROOF. Since $\mu(X) > 0$ we have that $\mathrm{F}_X(X)$ is not empty. Hence, for every open set U with $U \cap \mathrm{F}_X(X) \neq \emptyset$, there is an uncountable absolute Borel set contained in $U \cap \mathrm{F}_X(X)$. From this we infer that there is a countable collection F_i, $i = 0, 1, 2, \ldots$, of mutually disjoint topological copies of $\{0, 1\}^{\mathbb{N}}$ such that each F_i is a nowhere dense subset of $\mathrm{F}_X(X)$ and such that $Y_1 = \bigcup_{i=0}^{\infty} F_i$ is dense in $\mathrm{F}_X(X)$. There is no loss in assuming that $\mu(F_i) = 0$ for each i. As X is an absolute measurable space, there is an absolute Borel space Y_0 contained in $\mathrm{F}_X(X)$ such that $\mu(X) = \mu(Y_0)$. Then $Y = Y_0 \cup Y_1$ is an absolute Borel space. Let $Z_1 = Y \setminus \mathrm{support}(\mu)$ and denote by U_0 the uncountable absolute Borel space $F_0 \cup Z_1$. Observe that $\mu(U_0) = 0$ holds. There exists a collection U_i, $i = 1, 2, \dots$, of mutually disjoint, uncountable absolute Borel spaces such that $\mu(U_i) > 0$ for each i and such that $Y \setminus U_0 = \bigcup_{i=1}^{\infty} U_i$. Let φ be a $\mathfrak{B}$-homeomorphism of X such that $\varphi|(X \setminus Y)$ is the identity map on $X \setminus Y$, and $\varphi[U_i] = F_i$ for $i = 0, 1, 2, \dots$. Then $\mathrm{support}(\varphi_{\#}\mu) = \mathrm{F}_X(X)$ and the lemma is proved. □

THEOREM 2.47. *Let X be an absolute measurable space. If μ and ν are continuous, complete, finite Borel measures on X with $\mu(X) = \nu(X)$, then there is a φ in $\mathfrak{B}$-HOMEO(X) such that $\varphi_{\#}\nu = \mu$.*

PROOF. If $\mu(X) = 0$, then let φ be the identity map. Next suppose $\mu(X) > 0$. Then $\nu(X) > 0$. There are φ_1 and φ_2 in $\mathfrak{B}$-HOMEO(X) such that $\varphi_{1\#}\mu$ and $\varphi_{2\#}\nu$ are positive measures with $\varphi_{1\#}\mu(X) = \varphi_{2\#}\nu(X)$. Hence there is a φ_0 in $\mathfrak{B}$-HOMEO(X) such that $\varphi_{1\#}\mu = \varphi_{0\#}\varphi_{2\#}\nu$. So $\mu = {\varphi_1{}^{-1}}_{\#}\varphi_{0\#}\varphi_{2\#}\nu$. Let $\varphi = \varphi_1{}^{-1}\varphi_0\varphi_2$ to complete the proof. □

THEOREM 2.48. *Suppose X is an absolute measurable space that contains a topological copy K of $\{0,1\}^{\mathbb{N}}$. Let $h\colon \{0,1\}^{\mathbb{N}} \to K$ be a homeomorphism and let μ be a positive, continuous, complete, finite Borel measure on $\{0,1\}^{\mathbb{N}}$. If ν is a continuous, complete, finite Borel measure on X with $\mu\big(\{0,1\}^{\mathbb{N}}\big) = \nu(X)$, then there is a φ in $\mathfrak{B}$-HOMEO(X) such that $\varphi_{\#}\nu = h_{\#}\mu$.*

PROOF. As $h_{\#}\mu$ is a continuous, complete, finite Borel measure on X, the proof follows immediately from the previous theorem. □

2.7.3. An example. For each absolute measurable space X, the $\mathfrak{B}$-HOMEO(X) equivalence classes of continuous, complete, finite Borel measures on X are precisely those determined by the values $\mu(X)$. The following example shows that the values $\mu(X)$ does not characterize the HOMEO(X) equivalence classes of continuous, complete, finite Borel measures on X. Consider the absolute Borel space $X = [0,1] \times [0,1]$. Denote the algebraic boundary of X by ∂X. If $h \in$ HOMEO(X), then $h[\partial X] = \partial X$. For an f in $\mathfrak{B}$-HOMEO(X), it may happen that $f[\partial X]$ is not the same as ∂X. The homeomorphism property of $[0,1] \times [0,1]$ will be investigated in Chapter 3.

2.7.4. An application. A straightforward application of Theorem 2.45 will result in the following theorem.

THEOREM 2.49. *Let X be an absolute measurable space and μ be a positive, continuous, complete, finite Borel measure on X. If f is a μ-measurable, extended real-valued function on X such that f is real-valued μ-almost everywhere on X, then there is a φ in $\mathfrak{B}$-HOMEO(X) such that $f\varphi$ is μ-measurable and $\int_X f\varphi\, d\mu$ exists.*

We will consider the measure ν on X given by

$$\nu(E) = \int_E \frac{k}{1+|f|}\, d\mu, \qquad E \in \mathfrak{M}(X, \mu),$$

where k is such that $\nu(X) = \mu(X)$. Clearly, $\nu(E) \leq k\, \mu(E)$ whenever $E \in \mathfrak{M}(X, \mu)$. Moreover, $\nu(E) = 0$ if and only if $\mu(E) = 0$ because f is real-valued μ-almost everywhere on X. So ν is a positive, continuous, complete, finite Borel measure on X. By Theorem 2.45 there is a φ in $\mathfrak{B}$-HOMEO(X) such that $\varphi_{\#}\mu = \nu$. For E in $\mathfrak{B}(X)$ we have $\nu(E) = \varphi_{\#}\mu(E) = \mu\big(\varphi^{-1}[E]\big)$. So, $\nu(E) = 0$ if and only if $\mu\big(\varphi^{-1}[E]\big) = 0$. Let us show that a set is μ-measurable if and only if it is ν-measurable. If E is μ-measurable, then it is the union of a Borel set M and a set Z with $\mu(Z) = 0$, whence $\nu(Z) = 0$. Thereby E is ν-measurable. Conversely, suppose that E is ν-measurable. Then $E = M \cup Z$ where M is a Borel set and $\nu(Z) = 0$. Hence $\mu(Z) = 0$, that is, E is μ-measurable.

Lemma 2.50. *Let X be an absolute Borel space and let f, μ and ν be as in the discussion above. If g is a Borel measurable, real-valued function on X with $\int_X |g|\, d\nu < \infty$, then $\int_X g\, d\nu = \int_X g\varphi\, d\mu$ where φ is $\mathfrak{B}$-homeomorphism with $\varphi_{\#}\mu = \nu$.*

Proof. Clearly $\nu\big(g^{-1}[U]\big) = \mu\big((g\varphi)^{-1}[U]\big)$ whenever $U \in \mathfrak{B}(\mathbb{R})$. Hence $\int_X g\, d\nu = \int_X g\varphi\, d\mu$ since $\int_X |g|\, d\nu < \infty$. □

Proof of Theorem 2.49. As there is an absolute Borel space X' such that $\mu(X \setminus X') = 0$ we may assume that X is an absolute Borel space. Since f is μ-measurable and is real-valued μ-almost everywhere on X we have that f is ν-measurable and real-valued ν-almost everywhere. There is a Borel measurable, real-valued function g such that $f = g$ ν-almost everywhere. Now $\nu(Z) = 0$ and $\mu(\varphi^{-1}[Z]) = 0$, where $Z = \{x : f(x) \neq g(x)\}$. As $\{x : f\varphi(x) \neq g\varphi(x)\} = \varphi^{-1}[Z]$ we have $f\varphi = g\varphi$ μ-almost everywhere. Since $\int_X \frac{k\,|g|}{1+|f|}\, d\mu = \int_X \frac{k\,|f|}{1+|f|}\, d\mu < \infty$, we have $\int_X f\, d\nu = \int_X g\, d\nu = \int_X g\varphi\, d\mu = \int_X f\varphi\, d\mu$ and the theorem is proved.

2.8. Comments

Comments on universally measurable sets in a space X cannot be isolated from comments on absolute measurable spaces. So we shall comment on both of them here.

Similar to the notion of absolute Borel space, the notion of absolute measurable space is based on invariance under topological embedding – in this case, invariance of μ-measurability of topologically embedded copies of the space into any complete, finite Borel measure space $\big(Y, \mu, \mathfrak{M}(Y, \mu)\big)$. This notion is an extension of that of absolute Borel space in the sense that every absolute Borel space is an absolute measurable space. The topological nature of the definition of both absolute Borel space and absolute measurable space will naturally lead to topological questions about such spaces. As every separable metrizable space is topologically embeddable into the Hilbert cube $[0, 1]^{\mathbb{N}}$, one might wish to study only subspaces of this space. But the richer structure of Borel measurability of mappings is also available in the investigation of these spaces. The role of Borel measurable injections of absolute Borel spaces and absolute measurable spaces has been studied in the previous chapter. It has been shown that absolute measurable spaces are not invariant under Borel measurable injections. But they are preserved under $\mathfrak{B}$-homeomorphic embeddings. Employing such embeddings, one finds that absolute measurable spaces may be investigated by considering only subspaces of the space $\{0, 1\}^{\mathbb{N}}$ or the space $[0, 1]$ if the topological structure of the absolute measurable spaces is not of special interest. An example of this situation is found in Section 1.5

2.8.1. R. M. Shortt's observation. We have mentioned earlier in Chapter 1 that Shortt observed that universally measurable sets in a space X is independent of the metric that corresponds to the topology of the separable metrizable space (see [**139**]). Actually he made a stronger claim, namely that the universally measurable sets in X depend only on the fact that the σ-algebra $\mathfrak{A}$ is generated by a countable subcollection $\mathfrak{E}$ of $\mathfrak{A}$. The setting for this claim is measure theory and probability theory in contrast to the setting for the present book which concerns topological embedding and the role of homeomorphism in the notion of absolute measurable spaces. In general, measure theory and probability theory deal with a set X together with a given σ-algebra $\mathfrak{A}$ of subsets of the set X. The isomorphisms are required to preserve σ-algebra structures. Usually no topological structures are assumed. Hence questions that are of interest to us often do not appear. That part of measure theory and probability theory that concerned Shortt assumed a condition that induced a metric structure, namely countably generated σ-algebras. The resulting topological structure need not be unique since the σ-algebra may be countably generated in many ways. Of course, under this condition the isomorphisms generally will not be homeomorphisms or bi-Lipschitzian homeomorphisms. See Appendix B for a development of the probability theoretic approach to universally measurable spaces.

2.8.2. Historical references. In the Darst and Grzegorek Theorem 2.10, the equivalence of conditions (1), (2) and (3) is Purves's theorem (see Theorem A.43) which was proved in 1966 [**129**]. Hence their theorem sharpens Purves's result. For $X = \mathbb{R}$, Darst proved in 1970 [**37**] that condition (1) is equivalent to condition (4) and in 1971 [**39**] proved the equivalence of conditions (1) through (5) by assuming the continuum hypothesis in both papers. In 1981 Grzegorek and Ryll-Nardzewski [**71**] eliminated the continuum hypothesis from Darst's proof. The inclusion of condition (6) into the theorem was made by Grzegorek [**69**] in 1981.

The investigation of the symmetric difference property of universally measurable sets in a space was carried out by Grzegorek and Ryll-Nardzewski in [**70**]. The main theorem (Theorem 2.26) was first proved by Marczewski [**97**] with the aid of the continuum hypothesis. In Corollary 2.28, absolute G_δ spaces are replaced by absolute measurable spaces. The continuum hypothesis reappears in the investigation of the symmetric difference property by way of the Sierpiński set in $\mathbb{R}$. Sierpiński proved in 1924 [**141**, pages 80, 82] the existence of his set under the assumption of the continuum hypothesis. The Lusin sets and the Sierpiński sets are intimately connected by the Sierpiński–Erdös duality principle. For a nice discussion of this duality, see the book by Oxtoby [**120**]. We have given a proof of the existence of Sierpiński sets by means of the partition theorem, Theorem 1.37, without the aid of the duality.

The early results about subsets of $[0, 1]$ were summarized in [**15**]. This article concerned not only universally null sets of $[0, 1]$ but also many other singular sets such as Lusin sets, Sierpiński sets, concentrated sets and others. A good survey about singular sets can be found in the article by Brown and Cox [**18**] and in two articles by Miller [**110, 111**]. The 1937 article [**152**] by Marczewski deals with spaces more general than $[0, 1]$, that is, separable metric spaces. The notions of absolute measurable spaces and universally measurable sets in separable metrizable spaces lead to the singular sets called absolute null spaces and universally null sets, which are the same as was shown in Theorem 2.7. Though the book is about absolute measurable spaces and absolute null spaces, a few other singular sets will be included in later chapters as they are absolute null spaces with additional useful properties that will be exploited.

The homeomorphism group $\mathsf{HOMEO}([0, 1])$ was seen to be important very early in the study of universally measurable sets in $[0, 1]$. Indeed, the group reduced the investigation of all positive, complete, continuous, finite Borel measures on $[0, 1]$ to only the Lebesgue measure $\lambda|[0, 1]$. Analogues of this phenomenon are presented in Chapter 3.

The cardinal number κ_{G} was first shown to equal $\mathsf{non}\text{-}\mathbb{L}$ in [**68**]. The homeomorphism group property of universally measurable sets in $[0, 1]$ is used to establish the equivalence of the Grzegorek's approach to the cardinal number $\mathsf{non}\text{-}\mathbb{L}$ and the approach of Section 1.4. Observe that the proof of the equivalence of the two approaches also implicitly uses a $\mathfrak{B}$-homeomorphism into $\{0, 1\}^{\mathbb{N}}$ to prove $\kappa_0 = \kappa_{\mathrm{G}}$.

2.8.3. Positive measures and groups of maps. For separable metrizable spaces X, the groups $\mathsf{HOMEO}(X)$ and $\mathfrak{B}\text{-}\mathsf{HOMEO}(X)$ have natural roles in the book. The collections $\mathsf{univ}\,\mathfrak{M}(X)$ and $\mathsf{univ}\,\mathfrak{N}(X)$ are invariant under the action of the group $\mathsf{HOMEO}(X)$. It was shown that the collection $\mathsf{univ}\,\mathfrak{M}(X)$ is determined by the collection of positive measures on X, that is, $\mathsf{univ}\,\mathfrak{M}(X) = \mathsf{univ}\,\mathfrak{M}^{\mathrm{pos}}(X)$ and $\mathsf{univ}\,\mathfrak{N}(X) = \mathsf{univ}\,\mathfrak{N}^{\mathrm{pos}}(X)$. This fact is facilitated by the closure-like operation F_X which connects the absolute null subspaces of X and the topology of X. The group $\mathsf{HOMEO}(X)$ preserves the collection $\mathrm{MEAS}^{\mathrm{pos}}(X)$ of all positive, continuous, complete, σ-finite Borel measures on X. This is due to the identity $\mathrm{F}_X(h[M]) = h[\mathrm{F}_X(M)]$ for every h in $\mathsf{HOMEO}(X)$ and every subset M of X. Of course the identity does not hold for every h in $\mathfrak{B}\text{-}\mathsf{HOMEO}(X)$.

The group $\mathfrak{B}\text{-}\mathsf{HOMEO}(X)$ does not preserve topological properties; in particular, we have seen that $\mathrm{MEAS}^{\mathrm{pos}}(X)$ is not invariant under this group. Indeed, if X is an absolute measurable space that is not an absolute null space and if μ is a positive, continuous, complete, finite Borel measure on X, then the collection $\{\varphi_{\#}\mu : \varphi \in \mathfrak{B}\text{-}\mathsf{HOMEO}(X)\}$ is precisely the collection of all continuous, complete Borel measures ν on X with $\nu(X) = \mu(X)$.

Exercises

2.1. Let X be the union of a totally imperfect, non Lebesgue measurable subset of $[0,1]$ and the interval $[2,3]$. X is not an absolute measurable space. Describe the collection $\mathsf{ab}\,\mathfrak{M}(X)$. Describe the collection $\mathfrak{B}(X) \setminus \mathsf{ab}\,\mathfrak{M}(X)$.

2.2. Prove: $\mathsf{non}\text{-}\mathbb{L} = \min\{\operatorname{card}(E) : E \subset [0,1] \text{ with } \mu^*(E) > 0\}$ whenever μ is a positive, continuous, complete, finite Borel measure on $[0,1]$.

2.3. Prove that if M is a Lebesgue measurable subset of $[0,1]$ with $\lambda(M) > 0$, then there is a subset E of M such that $\lambda^*(E) = \lambda(M)$ and $\operatorname{card}(E) = \mathsf{non}\text{-}\mathbb{L}$. *Hint*: There is a topological copy G of $\mathcal{N}$ contained in M such that $\lambda|G$ is a positive measure on G and $\lambda(G) > \frac{1}{2}\lambda(M)$.

2.4. Prove (see page 46 for λ_*): If a sequence E_n, $n \in \omega$, is such that $E_n \subset E_{n+1} \subset [0,1]$ for every n, then

$$\lambda_*([0,1] \setminus \textstyle\bigcup_{n\in\omega} E_n) = \inf\{\lambda_*([0,1] \setminus E_n) : n \in \omega\}.$$

2.5. Prove Theorem 2.43. Hint: There exists an absolute Borel space B contained in X such that $\mu|B$ is a positive measure and $\mu(X \setminus B) = 0$.

3

The homeomorphism group of X

The collection $\mathsf{univ}\,\mathfrak{M}(X)$ of universally measurable sets in a space X has been shown to be those subsets of X that are μ-measurable for every μ in the collection $\mathrm{MEAS}^{\mathrm{cont}}(X)$ of all continuous, complete, σ-finite Borel measures on X. We have seen that the collection $\mathrm{MEAS}^{\mathrm{cont}}(X)$ can be replaced by the smaller collection $\mathrm{MEAS}^{\mathrm{pos}}(X)$ of those measures μ in $\mathrm{MEAS}^{\mathrm{cont}}(X)$ that are also positive – that is, $\mathrm{support}(\mu) = \mathrm{F}_X(X) \neq \emptyset$.

Very early in the history of universally measurable sets in $[0,1]$ it was seen that the Lebesgue measure λ on $\mathbb{R}$ determined the σ-algebra $\mathsf{univ}\,\mathfrak{M}([0,1])$. That is, the measure $\lambda|[0,1]$ and the group $\mathsf{HOMEO}([0,1])$ generated all of $\mathsf{univ}\,\mathfrak{M}([0,1])$ in the sense that

$$\mathsf{univ}\,\mathfrak{M}([0,1]) = \bigcap\{M \in \mathfrak{M}\big([0,1], h_\#(\lambda|[0,1])\big) : h \in \mathsf{HOMEO}([0,1])\}.$$

This was made possible because of the elementary fact

$$\begin{aligned}&\{\mu \in \mathrm{MEAS}^{\mathrm{pos}}([0,1]) : \mu([0,1]) < \infty\}\\ &\quad = \textstyle\bigcup_{c>0}\bigcup\{c\,h_\#(\lambda|[0,1]) : h \in \mathsf{HOMEO}([0,1])\}.\end{aligned}$$

The aim of the chapter is to investigate these phenomena for spaces X other than $[0,1]$.

For a separable metrizable space X it will be convenient to denote the collection of all finite measures in $\mathrm{MEAS}^{\mathrm{pos}}(X)$ by $\mathrm{MEAS}^{\mathrm{pos,fin}}(X)$, that is,

$$\mathrm{MEAS}^{\mathrm{pos,fin}}(X) = \{\mu \in \mathrm{MEAS}^{\mathrm{pos}}(X) : \mu(X) < \infty\}. \tag{3.1}$$

Associated with this collection are two actions of the group $\mathsf{HOMEO}(X)$ on measures. In particular, for a fixed μ in $\mathrm{MEAS}^{\mathrm{pos,fin}}(X)$,

$$\begin{gathered}\textstyle\bigcup_{c>0}\bigcup\{c\,h_\#\mu : h \in \mathsf{HOMEO}(X)\} \subset \mathrm{MEAS}^{\mathrm{pos,fin}}(X),\\ \mathsf{univ}\,\mathfrak{M}(X) \subset \bigcap\{\mathfrak{M}(X, h_\#\mu) : h \in \mathsf{HOMEO}(X)\}.\end{gathered}$$

NOTATION 3.1. *For a separable metrizable space let $\mathcal{G}(X)$ be a nonempty subset of* $\mathsf{HOMEO}(X)$ *and let μ be a continuous, complete, σ-finite Borel measure on X. The collection $\{h_\#\mu : h \in \mathcal{G}(X)\}$ will be denoted by $\mathcal{G}(X)_\#\mu$.*

The above two inclusions lead to

DEFINITION 3.2. *For a separable metrizable space X let μ be a measure in* $\mathrm{MEAS}^{\mathrm{pos,fin}}(X)$ *and let $\mathcal{G}(X)$ be a subgroup of* $\mathsf{HOMEO}(X)$. *μ and $\mathcal{G}(X)$ are said to* generate $\mathrm{MEAS}^{\mathrm{pos,fin}}(X)$ *if*

$$\mathrm{MEAS}^{\mathrm{pos,fin}}(X) = \bigcup_{c>0} \bigcup \{ c\,\nu : \nu \in \mathcal{G}(X)_{\#}\mu \}. \tag{3.2}$$

μ and $\mathcal{G}(X)$ are said to generate $\mathsf{univ}\,\mathfrak{M}(X)$ *if*

$$\mathsf{univ}\,\mathfrak{M}(X) = \bigcap \{ \mathfrak{M}(X, h_{\#}\mu) : h \in \mathcal{G}(X) \}. \tag{3.3}$$

Of special interest is the action of homeomorphisms on the Lebesgue measure on the space $[0,1]^n$. Suppose that φ is a homeomorphism of $[0,1]^n$ onto X and λ_0 is the Lebesgue measure on $[0,1]^n$. Then $\varphi_{\#}\lambda_0$ is a positive, continuous, complete, finite Borel measure on X with $\varphi_{\#}\lambda_0\big(\varphi\big[\partial[0,1]^n\big]\big) = 0$. This leads to the following definition.

DEFINITION 3.3. *A measure μ on an n-cell*[1] *X is said to be* Lebesgue-like *if*

(1) $\mu \in \mathrm{MEAS}^{\mathrm{pos,fin}}(X)$,
(2) $\mu(\partial X) = 0$, *where ∂X is the algebraic boundary of X.*

Note that not every measure μ in $\mathrm{MEAS}^{\mathrm{pos,fin}}([0,1]^n)$ is Lebesgue-like if $n > 1$.

In the context of $\mathsf{HOMEO}(X)$ it will be convenient to define the notion of homeomorphic measures on X, which is related to Definition 1.4.

DEFINITION 3.4. *Borel measures μ and ν on X are said to be* homeomorphic *if there is an h in* $\mathsf{HOMEO}(X)$ *such that $\nu = h_{\#}\mu$.*

Let us begin by introducing a metric ρ on $\mathsf{HOMEO}(X)$.

3.1. A metric for HOMEO(X)

On choosing a bounded metric d for a separable metrizable space X, one will realize a useful metric ρ on the collection $\mathsf{HOMEO}(X)$. Although most of the results on relationships between the group $\mathsf{HOMEO}(X)$ and the σ-algebra $\mathsf{univ}\,\mathfrak{M}(X)$ do not refer to a metric on $\mathsf{HOMEO}(X)$, we will often use such a metric in many constructions that appear in this chapter. Fortunately, the constructions are made on compact spaces X. In this setting the metric ρ on $\mathsf{HOMEO}(X)$ is complete. This completeness will avail us with the Baire category theorem.

DEFINITION 3.5. *Let* d *be a bounded metric on X. For each h in* $\mathsf{HOMEO}(X)$, *its* norm, *denoted by $\|h\|$, is defined to be*

$$\|h\| = \sup\{ \mathrm{d}(h(x), x) : x \in X \}.$$

[1] An *n-cell* X is a topological copy of $[0,1]^n$. Its algebraic boundary ∂X is the topological copy of $\partial[0,1]^n$. Of course, the topological boundary of X is always empty. Clearly, $\mathrm{F}_X(X) = X \neq \emptyset$.

For f and g in HOMEO(X)*, define the* distance $\rho(f,g)$ *by the formula*

$$\rho(f,g) = \|fg^{-1}\| + \|f^{-1}g\|.$$

Obviously $\rho(f,g) \geq \mathrm{d}(f(x),g(x)) + \mathrm{d}(f^{-1}(x),g^{-1}(x))$ for each x in X. So, if f_n, $n = 1,2,\ldots$, is a Cauchy sequence in HOMEO(X), then $f_n(x)$ and $f_n{}^{-1}(x)$, $n = 1,2,\ldots$, are Cauchy sequences in X with respect to the metric d. From this we infer that ρ is a complete metric on HOMEO(X) whenever d is a bounded, complete metric for X.

PROPOSITION 3.6. *Suppose that X is a compact metrizable space and let* d *be a metric for X. The following are group properties of the norm $\|\cdot\|$ and the metric ρ on* HOMEO(X).

(1) $\|f\| = \|f^{-1}\|$,
(2) $\rho(f,g) = \rho(f^{-1},g^{-1})$, *whence* $\rho(f,\mathrm{id}) = \rho(f^{-1},\mathrm{id})$,
(3) $\rho(gfg^{-1},ghg^{-1}) \leq 2\,\omega\big(g\colon \rho(f,h)\big)$, *where $\omega(g\colon \eta)$ is the usual modulus of uniform continuity*[2] *of g,*
(4) $\rho(gf,g) \leq \|f\| + \omega\big(g\colon \|f\|\big)$,
(5) $\rho(fg,g) \leq \|f\| + \omega\big(g^{-1}\colon \|f\|\big)$.

The verifications of these properties are simple exercises left to the reader. The next proposition, which concerns the hyperspace[3] of a compact metric space X, is also left as an exercise for the reader.

PROPOSITION 3.7. *Let X be a compact metric space and let the hyperspace 2^X of nonempty closed subsets of X be endowed with the Hausdorff metric. The map $(F,h) \mapsto h^{-1}[F]$ is a continuous map of $2^X \times$* HOMEO(X) *into 2^X.*

Let F be a subset of a metric space X. A homeomorphism h is said to keep F *fixed* if $h(x) = x$ whenever $x \in F$, and is said to keep F *invariant* if $h[F] = F$. Consider the subgroups

$$\mathrm{HOMEO}(X;F\ \text{fixed}) = \{h \in \mathrm{HOMEO}(X)\colon h^{-1}(x) = x,\ x \in F\},$$

$$\mathrm{HOMEO}(X;F\ \text{inv}) = \{h \in \mathrm{HOMEO}(X)\colon h^{-1}[F] = F\}$$

of HOMEO(X). As the reader can easily verify, the first one is always closed, and the second one is closed whenever F is a compact subset of X.

Let us introduce a continuous, finite, Borel measure μ into the discussion. We assume that F is a compact subset of a separable metric space X. There is no loss in assuming that X is a subspace of the Hilbert cube and that the metric on X is induced by a metric on the Hilbert cube. The map $f \mapsto f_{\#}\mu(F)$ is a real-valued function on the metric space HOMEO(X). We claim that this map is upper semi-continuous. Indeed, let α be a real number and f be such that $f_{\#}\mu(F) < \alpha$. Let U be an open

[2] For a function $f\colon X \to Y$, where X and Y are metric spaces, the *modulus of uniform continuity* of f is $\omega(f\colon \eta) = \sup\{\mathrm{d}_Y(f(x),f(x'))\colon \mathrm{d}_X(x,x') \leq \eta\}$, where $\eta > 0$.

[3] See Appendix A, page 196, for the definition of the hyperspace 2^X and its Hausdorff metric.

neighborhood of F in the space X such that $f_{\#}\mu(U) < \alpha$. Let U' be an open set in the Hilbert cube such that $U = U' \cap X$. As F is compact there is a positive number δ such that $h^{-1}[F] \subset U'$ whenever $\rho(h, \mathrm{id}) < \delta$. Hence there is neighborhood V of f in $\mathsf{HOMEO}(X)$ such that $g_{\#}\mu(F) < \alpha$ whenever $g \in V$, thereby the upper semi-continuity at f follows. Let us summarize this discussion as a lemma.

LEMMA 3.8. *Let X be a metric space with a totally bounded metric. If F is a compact subset of X and μ is a continuous, finite Borel measure on X, then the function $f \mapsto f_{\#}\mu(F)$ is an upper semi-continuous real-valued function on the metric space* $\mathsf{HOMEO}(X)$.

Here is a simple proposition that will be used often. Its proof is left to the reader. Note that no metric on X is assumed.

PROPOSITION 3.9. *Suppose that X is a separable metrizable space. Let U and V be disjoint open sets and let F be $X \setminus (U \cup V)$. If h_U is in $\mathsf{HOMEO}(X \setminus V; F \text{ fixed})$ and h_V is in $\mathsf{HOMEO}(X \setminus U; F \text{ fixed})$, then there is an h in $\mathsf{HOMEO}(X; F \text{ fixed})$ such that $h|(X \setminus V) = h_U$ and $h|(X \setminus U) = h_V$.*

3.2. General properties

There are several assertions that hold for spaces more general than those investigated in this chapter. The first theorem follows easily from definition.

THEOREM 3.10. *For a separable metrizable space X, if μ is a measure in $\mathrm{MEAS}^{\mathrm{pos,fin}}(X)$ and $\mathcal{G}(X)$ is a subgroup of $\mathsf{HOMEO}(X)$ that generate $\mathrm{MEAS}^{\mathrm{pos,fin}}(X)$, then ν and $\mathcal{G}(X)$ generate $\mathrm{MEAS}^{\mathrm{pos,fin}}(X)$ for every ν in $\mathrm{MEAS}^{\mathrm{pos,fin}}(X)$.*

We have the following lemma.

LEMMA 3.11. *For a separable metrizable space X, let μ be a measure in $\mathrm{MEAS}^{\mathrm{pos,fin}}(X)$ and $\mathcal{G}(X)$ be a subgroup of $\mathsf{HOMEO}(X)$. If $\mathrm{MEAS}^{\mathrm{pos,fin}}(X)$ is generated by μ and $\mathcal{G}(X)$, then*

$$\mathsf{univ}\,\mathfrak{M}(X) = \bigcap\{\mathfrak{M}(X, h_{\#}\mu) : h \in \mathcal{G}(X)\},$$

that is, μ and $\mathcal{G}(X)$ generate $\mathsf{univ}\,\mathfrak{M}(X)$.

PROOF. Suppose μ satisfies equation (3.2). Let ν be any measure in $\mathrm{MEAS}^{\mathrm{pos,fin}}(X)$. Then there is a h in $\mathcal{G}(X)$ and a positive c such that $c\, h_{\#}\mu = \nu$. Note $\mathfrak{M}(X, c\, h_{\#}\mu) = \mathfrak{M}(X, h_{\#}\mu)$. The proof is easily completed by an application of Theorem 2.17. □

The next countable union theorem will prove quite useful.

THEOREM 3.12. *Suppose that X is an absolute measurable space and that X_i, $i = 1, 2, \ldots$, is a sequence in $\mathsf{univ}\,\mathfrak{M}(X)$ such that $X = \bigcup_{i=1}^{\infty} X_i$. Let μ be a positive, continuous, complete, finite Borel measure on X and $\mathcal{G}(X)$ be a subgroup of*

$\mathsf{HOMEO}(X)$. *If, for each i,* $\mu|X_i$ *and*

$$\mathcal{G}_i(X) = \{h|X_i : h \in \mathcal{G}(X) \text{ and } h[X_i] = X_i\}$$

generate $\mathsf{univ}\,\mathfrak{M}(X_i)$*, then* μ *and* $\mathcal{G}(X)$ *generate* $\mathsf{univ}\,\mathfrak{M}(X)$.

PROOF. Only $\bigcap\{\mathfrak{M}(X, h_{\#}\mu) : h \in \mathcal{G}(X)\} \subset \mathsf{univ}\,\mathfrak{M}(X)$ must be shown. Suppose that M is a member of the left-hand side and fix an i. As X_i is a universally measurable set in X, we have, by Proposition 2.5, that X_i is an absolute measurable space. Hence the set $E_i = M \cap X_i$ satisfies $E_i \in \mathfrak{M}\big(X_i, (h|X_i)_{\#}(\mu|X_i)\big)$ whenever $h \in \mathcal{G}(X)$ and $h[X_i] = X_i$. As $\mu|X_i$ and $\mathcal{G}_i(X)$ generate $\mathsf{univ}\,\mathfrak{M}(X_i)$, we have that E_i is in $\mathsf{univ}\,\mathfrak{M}(X_i)$. Hence E_i is an absolute measurable space. As $M = \bigcup_{i=1}^{\infty} E_i$ is an absolute measurable space, it now follows that M is in $\mathsf{univ}\,\mathfrak{M}(X)$. □

The proof of the last theorem leads to the observation: *If a subset* $\mathcal{G}_0(X)$ *of* $\mathsf{HOMEO}(X)$ *is such that* $\bigcup_{c>0}\{c\,\nu : \nu \in \mathcal{G}_0(X)_{\#}\mu\} = \mathrm{MEAS}^{\mathrm{pos,fin}}(X)$*, then* μ *and any subgroup* $\mathcal{G}(X)$ *of* $\mathsf{HOMEO}(X)$ *that contains* $\mathcal{G}_0(X)$ *will generate* $\mathrm{MEAS}^{\mathrm{pos,fin}}(X)$. A similar observation can be made for generating $\mathsf{univ}\,\mathfrak{M}(X)$.

Here are simple observations whose proofs are left to the reader. Recall the definition of $\mathsf{ab}\,\mathfrak{M}(X)$; it is the collection of all subsets of X that are absolute measurable spaces (see page 27).

PROPOSITION 3.13. *Let* X *and* Y *be absolute measurable spaces with* $X \subset Y$*. If* $M \in \mathsf{ab}\,\mathfrak{M}(X)$*, then* $M \in \mathsf{ab}\,\mathfrak{M}(Y)$*. And, if* $M \in \mathsf{ab}\,\mathfrak{M}(Y)$*, then* $M \cap X \in \mathsf{ab}\,\mathfrak{M}(X)$*. Consequently, a subset* M *of* X *is in* $\mathsf{univ}\,\mathfrak{M}(X)$ *if and only if* M *is in* $\mathsf{univ}\,\mathfrak{M}(Y)$.

PROPOSITION 3.14. *Let* X *be a separable metrizable space. A subgroup* $\mathcal{G}(X)$ *of* $\mathsf{HOMEO}(X)$ *and a continuous, complete, finite Borel measure* μ *on* X *will generate* $\mathsf{univ}\,\mathfrak{M}(X)$ *if and only if there is a subcollection* $\mathcal{G}_0(X)$ *of* $\mathcal{G}(X)$ *such that*

$$\mathsf{univ}\,\mathfrak{M}(X) \supset \bigcap\{\mathfrak{M}(X, h_{\#}\mu) : h \in \mathcal{G}_0(X)\}.$$

3.3. One-dimensional spaces

Let us begin with the simplest of one-dimensional spaces, namely the connected one-dimensional manifolds M_1. Topologically, there are four of them: $[0, 1]$, $[0, 1)$, $(0, 1)$, and $\partial([0, 1] \times [0, 1])$, the algebraic boundary of the two-cell $[0, 1] \times [0, 1]$. We shall derive the desired results from the theorem for the Lebesgue measure $\lambda|[0, 1]$ and the group $\mathsf{HOMEO}([0, 1])$.

3.3.1. Universally measurable sets in M_1. Since one-dimensional manifolds M_1 are absolute Borel spaces we have $\mathsf{univ}\,\mathfrak{M}(M_1) = \mathsf{ab}\,\mathfrak{M}(M_1)$.

Consequently we have

PROPOSITION 3.15. *Let* X *be* $(0, 1)$ *or* $[0, 1)$*. For subsets* E *of* X*,* E *is in* $\mathsf{univ}\,\mathfrak{M}(X)$ *if and only if* E *is in* $\mathsf{univ}\,\mathfrak{M}\big([0, 1]\big)$.

Proposition 3.16. *Let X be $(0,1)$ or $[0,1)$ and let $f: X \to [0,1]$ be the inclusion map. If μ is a positive, continuous, complete, finite Borel measure on X, then $(f_\# \mu)|X = \mu$. Also, if ν is a positive, continuous, complete, finite Borel measure on $[0,1]$, then $f_\#(\nu|X) = \nu$.*

Proof. The first implication is obvious. The second implication follows easily since the measure ν is continuous. □

For completeness we state the theorem for the manifold $[0,1]$.

Theorem 3.17. *The group* $\mathsf{HOMEO}([0,1]; \partial[0,1] \text{ fixed})$ *and the restricted Lebesgue measure $\lambda|[0,1]$ generate* $\mathrm{MEAS}^{\mathrm{pos,fin}}([0,1])$ (*the collection of all positive, continuous, complete, finite Borel measures on* $[0,1]$) *and thereby generate* $\mathsf{univ}\,\mathfrak{M}([0,1])$.

Of course, the issue is whether some μ in $\mathrm{MEAS}^{\mathrm{pos,fin}}(M_1)$ and the group $\mathsf{HOMEO}(M_1)$ will generate $\mathrm{MEAS}^{\mathrm{pos,fin}}(M_1)$. Let us begin with M_1 being either $(0,1)$ or $[0,1)$.

Lemma 3.18. *If X is either $(0,1)$ or $[0,1)$, then $\mathsf{HOMEO}(X)$ and $\lambda_0 = \lambda|X$ generate $\mathrm{MEAS}^{\mathrm{pos,fin}}(X)$. Hence*

$$\mathsf{univ}\,\mathfrak{M}(X) = \bigcap\{\mathfrak{M}(X, h_\#\lambda_0) : h \in \mathsf{HOMEO}(X)\},$$

that is, λ_0 and $\mathsf{HOMEO}(X)$ generate $\mathsf{univ}\,\mathfrak{M}(X)$.

Proof. Consider the commutative diagram

$$\begin{array}{ccc} X & \xrightarrow[\subset]{f} & [0,1] \\ {\scriptstyle h|X}\uparrow & & \uparrow{\scriptstyle h} \\ X & \xrightarrow[\subset]{f} & [0,1] \end{array}$$

where f is the inclusion map and h is an orientation preserving homeomorphism. We have $f_\#\lambda_0 = \lambda_1$, where $\lambda_1 = \lambda|[0,1]$. Let ν be in $\mathrm{MEAS}^{\mathrm{pos,fin}}(X)$ with $\nu(X) = 1$. As $f_\#\nu$ is in $\mathrm{MEAS}^{\mathrm{pos,fin}}([0,1])$, there is an orientation preserving homeomorphism h in $\mathsf{HOMEO}\big([0,1]\big)$ such that $f_\#\nu = h_\#\lambda_1 = h_\# f_\#\lambda_0$. From the commutative diagram we have $f_\#\nu = h_\# f_\#\lambda_0 = f_\#(h|X)_\#\lambda_0$. Note that $f^{-1}[M] = M$ whenever $M \subset X$. Hence $\nu(M) = f_\#\nu(M) = f_\#(h|X)_\#\lambda_0(M) = (h|X)_\#\lambda_0(M)$ whenever $M \in \mathfrak{B}(X)$. The lemma now follows. □

Let us turn to the one-dimensional manifold $\mathbb{S}_1$, the algebraic boundary of the planar set $\{(x,y) : \|(x,y)\|_2 \le 1\}$. We denote the one-dimensional Hausdorff measure by H_1. Let I_1 and I_2 be two topological copies of $[0,1]$ in $\mathbb{S}_1$ such that $\mathbb{S}_1 = (I_1 \setminus \partial I_1) \cup (I_2 \setminus \partial I_2)$. Then $\mathsf{H}_1\,|I_i$ and $\mathsf{HOMEO}(I_i)$ generate $\mathrm{MEAS}^{\mathrm{pos,fin}}(I_i)$ for each i. Also, for each i, $\mu|I_i \in \mathrm{MEAS}^{\mathrm{pos,fin}}(I_i)$ whenever $\mu \in \mathrm{MEAS}^{\mathrm{pos,fin}}(\mathbb{S}_1)$. We are now ready to prove

LEMMA 3.19. *The collection* $\mathrm{MEAS}^{\mathrm{pos,fin}}(\mathbb{S}_1)$ *is generated by the measure* $\mathsf{H}_1|\mathbb{S}_1$ *and the group* $\mathsf{HOMEO}(\mathbb{S}_1)$. *Hence*

$$\mathsf{univ}\,\mathfrak{M}(\mathbb{S}_1) = \bigcap\{\mathfrak{M}(\mathbb{S}_1, h_\#(\mathsf{H}_1|\mathbb{S}_1)) : h \in \mathsf{HOMEO}(\mathbb{S}_1)\},$$

that is, $\mathsf{H}_1|\mathbb{S}_1$ *and* $\mathsf{HOMEO}(\mathbb{S}_1)$ *generate* $\mathsf{univ}\,\mathfrak{M}(\mathbb{S}_1)$.

PROOF. For convenience let $\mu = \mathsf{H}_1|\mathbb{S}_1$. Let ν be a positive measure such that $\nu(\mathbb{S}_1) = \mu(\mathbb{S}_1)$. With the notation that precedes the statement of the lemma, there is no loss in assuming $\nu(I_1) \geq \mu(I_1)$. Hence

$$\nu_0 = \mu \llcorner (I_1 \setminus I_2) + \tfrac{\nu(I_1)-\mu(I_1\setminus I_2)}{\nu(I_1\cap I_2)}\, \nu \llcorner (I_1 \cap I_2) + \nu \llcorner (I_2 \setminus I_1)$$

is a positive Borel measure. Observe $\nu_0(I_1) = \nu(I_1)$ and $\nu_0(\mathbb{S}_1) = \nu(\mathbb{S}_1) = \mu(\mathbb{S}_1)$. There is an h_1 in $\mathsf{HOMEO}(I_1; \partial I_1 \text{ fixed})$ such that $\nu_0|I_1 = h_{1\#}(\nu|I_1)$. Let H_1 be the map h_1 on I_1 and the identity map on $\mathbb{S}_1 \setminus I_1$. Clearly, $H_1 \in \mathsf{HOMEO}(\mathbb{S}_1)$. Let $f_1 : I_1 \to \mathbb{S}_1$ be the inclusion map. Then $H_{1\#}\nu = f_{1\#}h_{1\#}(\nu_0|I_1) + \nu \llcorner (\mathbb{S}_1 \setminus I_1) = \nu_0$.

Let us work on I_2. As $\nu_0(I_1 \setminus I_2) = \mu(I_1 \setminus I_2)$ we have $\nu_0(I_2) = \mu(I_2)$. Hence there is an h_2 in $\mathsf{HOMEO}(I_2; \partial I_2 \text{ fixed})$ such that $h_{2\#}(\nu_0|I_2) = \mu|I_2$. Let H_2 be the map h_2 on I_2 and the identity map on $\mathbb{S}_1 \setminus I_2$. Clearly, H_2 is in $\mathsf{HOMEO}(\mathbb{S}_1)$. Let $f_2 : I_2 \to \mathbb{S}_1$ be the inclusion map. Then $H_{2\#}\nu_0 = \mu \llcorner (\mathbb{S}_1 \setminus I_2) + f_{2\#}h_{2\#}(\nu_0 \llcorner I_2) = \mu$. Finally, $h = H_2H_1$ is in $\mathsf{HOMEO}(\mathbb{S}_1)$ and $h_\#\nu = \mu$. The remainder of the proof is easily completed. □

A second proof can be produced by selecting a point $*$ in $\mathbb{S}_1$ and considering the subgroup $\mathsf{HOMEO}(\mathbb{S}_1; \{*\} \text{ fixed})$ of $\mathsf{HOMEO}(\mathbb{S}_1)$. Let $\varphi : [0,1] \to \mathbb{S}_1$ be a continuous surjection such that $\varphi|(0,1)$ is a homeomorphism and $\varphi[\partial I] = \{*\}$. Observe that each μ in $\mathrm{MEAS}^{\mathrm{pos,fin}}(\mathbb{S}_1)$ corresponds to a unique measure μ_0 in $\mathrm{MEAS}^{\mathrm{pos,fin}}([0,1])$ with $\varphi_\#\mu_0 = \mu$ and that φh_0 is in $\mathsf{HOMEO}(\mathbb{S}_1; \{*\} \text{ fixed})$ whenever $h_0 \in \mathsf{HOMEO}([0,1])$. The reader is asked to show that if μ and ν are in $\mathrm{MEAS}^{\mathrm{pos,fin}}(\mathbb{S}_1)$ then there is an h in $\mathsf{HOMEO}(\mathbb{S}_1; \{*\} \text{ fixed})$ such that $\nu = h_\#\mu$ whenever $\mu(\mathbb{S}_1) = \nu(\mathbb{S}_1)$. As a consequence we have that $\mathrm{MEAS}^{\mathrm{pos,fin}}(\mathbb{S}_1)$ is generated by μ and $\mathsf{HOMEO}(\mathbb{S}_1; \{*\} \text{ fixed})$. The reader will see this approach can be used to advantage in the case of the n-dimensional sphere $\mathbb{S}_n$ (see page 73).

Let us summarize the theorem for $[0,1]$ and the last two lemmas into the following.

THEOREM 3.20. *If* M_1 *is a connected one-dimensional manifold, then some* μ *in* $\mathrm{MEAS}^{\mathrm{pos,fin}}(M_1)$ *and the group* $\mathsf{HOMEO}(M_1)$, *indeed the subgroup* $\mathsf{HOMEO}(M_1; \partial M_1 \text{ fixed})$, *will generate* $\mathrm{MEAS}^{\mathrm{pos,fin}}(M_1)$. *Hence*

$$\mathsf{univ}\,\mathfrak{M}(M_1) = \bigcap\{\mathfrak{M}(M_1, h_\#\nu) : h \in \mathsf{HOMEO}(M_1)\},$$

whenever $\nu \in \mathrm{MEAS}^{\mathrm{pos,fin}}(M_1)$.

3.3.2. The simple triod. This example will show the utility of Definition 3.2. A simple triod is a one-dimensional space that is homeomorphic to the letter T. That is, a simple triod is the union of three planar line segments I_1, I_2 and I_3 with exactly one common point which is one of the two end-points of each of the line segments. Denote the common point by p and the other end-point of I_i by a_i, $i = 1,2,3$. For each φ in HOMEO(T), observe that $\varphi\big[\{a_1,a_2,a_3\}\big] = \{a_1,a_2,a_3\}$ and $\varphi(p) = p$. Denote the set $\bigcup_{i=1}^{3} \partial I_i$ by V.

We shall assume that the line segments I_i, $i = 1,2,3$, each have length equal to 1. Consider the restricted Hausdorff measure $\mu = \mathsf{H}_1 \,|\mathsf{T}$. We have the following proposition.

PROPOSITION 3.21. *Let* $\mu = \mathsf{H}_1 \,|\mathsf{T}$ *and* $\mathcal{G}(\mathsf{T}) = \mathsf{HOMEO}(\mathsf{T}; V \text{ fixed})$. *Then* μ *and* $\mathcal{G}(\mathsf{T})$ *generate* univ $\mathfrak{M}(\mathsf{T})$, *and hence* μ *and* HOMEO(T) *generate* univ $\mathfrak{M}(\mathsf{T})$. *The measure* μ *and the group* HOMEO(T) *do not generate* $\mathrm{MEAS}^{\mathrm{pos,fin}}(\mathsf{T})$.

PROOF. The above notation gives $\mathsf{T} = I_1 \cup I_2 \cup I_3$. Let $\mathcal{G}_i(\mathsf{T}) = \{h|I_i \colon h \in \mathsf{HOMEO}(\mathsf{T}; V \text{ fixed}), h[I_i] = I_i\}$ for $i = 1,2,3$. As $\mathcal{G}_i(\mathsf{T})$ is $\mathsf{HOMEO}(X_i; \partial I_i \text{ fixed})$, the measure $\mu|I_i$ and the group $\mathcal{G}_i(\mathsf{T})$ generate univ $\mathfrak{M}(X_i)$; Theorem 3.12 establishes the first statement of the proposition. To establish the second statement consider the measure $\nu = 1\,\mu\llcorner I_1 + 2\,\mu\llcorner I_2 + 3\,\mu\llcorner I_3$. The reader is asked to verify the second statement with the aid of μ and ν. □

It is clear that the argument in the above proof will apply to the more general one-dimensional finitely triangulable space $|K_1|$ (that is, roughly speaking, $|K_1|$ is formed from a finite number of vertices and a finite number of arcs that join pairs of distinct vertices with at most one arc joining such pairs; the higher dimensional case will be defined later). The space $|K_1|$ need not be connected. Let $|K_1^0|$ denote the collection of all vertices of $|K_1|$.

THEOREM 3.22. *Let* $|K_1|$ *be a one-dimensional finitely triangulable space. Then there is a* μ *in* $\mathrm{MEAS}^{\mathrm{pos,fin}}(|K_1|)$ *such that* μ *and the group* $\mathsf{HOMEO}(|K_1|; |K_1^0| \text{ fixed})$ *generate* univ $\mathfrak{M}(|K_1|)$. *Hence the same* μ *and* $\mathsf{HOMEO}(|K_1|)$ *also generate* univ $\mathfrak{M}(|K_1|)$.

A one-dimensional finitely triangulable space can be realized by vertices and straight line segments in $\mathbb{R}^k$ for a sufficiently large k. The proof of the theorem is left to the reader.

For a space X, the equivalence classes $\{h_{\#}\mu \colon h \in \mathsf{HOMEO}(X)\}$ of the collection $\mathrm{MEAS}^{\mathrm{pos,fin}}(X)$, where μ is in $\mathrm{MEAS}^{\mathrm{pos,fin}}(X)$, can be quite complicated. Let us investigate these equivalence classes for some simple examples. We have already determined them for the connected one-dimensional manifolds M_1 and the simple triod T. That is, for M_1, the equivalence classes are characterized by the pair (μ, r) where μ is a measure in $\mathrm{MEAS}^{\mathrm{pos,fin}}(M_1)$ and $\mu(M_1) = r$. We leave the description of the equivalence classes for the simple triod T as an exercise for the reader.

3.3.3. *More examples.* Up to now all of the examples in this chapter have been locally connected. The next two examples are the usual connected but not locally connected spaces.

EXAMPLE (**Graph of** $\sin(1/x)$). In the space $\mathbb{R}^2$, let $X = X_0 \cup X_1$ where $X_0 = \{(x_1,x_2)\colon x_1 = 0, -1 \le x_2 \le 1\}$ and X_1 is the graph of $x_2 = g(x_1) = \sin(1/x_1)$, where $0 < x_1 \le \pi^{-1}$. As X is locally connected at x if and only if $x \in X_1$, we have $h[X_0] = X_0$ whenever h is in $\mathsf{HOMEO}(X)$. Hence a necessary condition on μ and ν to satisfy $h_\#\mu = \nu$ is that $\mu(X_0) = \nu(X_0)$ and $\mu(X_1) = \nu(X_1)$. But this condition is not sufficient as we shall show. Let $p_0 = \left(\frac{1}{\pi}, 0\right)$. The points of maximum of the graph X_1 will be listed as p_i, $i = 1, 2, \ldots$, where the first coordinates of the sequence form a decreasing sequence in $(0, \pi^{-1}]$. Denote by I_i the arc in X_1 that joins p_{i-1} to p_i, $i = 1, 2, \ldots$. Also, let J_k be the arc in X_1 that joins $\left(\frac{1}{k\pi}, 0\right)$ to $\left(\frac{1}{(k+1)\pi}, 0\right)$, $k = 1, 2, \ldots$. Select positive, continuous, complete Borel measures μ and ν on X such that $\mu(I_i) = 2^{-i}$ and $\nu(J_k) = 2^{-k}$ and $\mu(X_0) = \nu(X_0) = 1$. Then $A_n = \bigcup_{i=1}^n I_i$ and $B_n = \bigcup_{k=1}^n J_k$ are arcs such that p_0 is an end point of both A_n and B_n and such that $\mu(A_n) = \nu(B_n) = 1 - 2^{-n}$ for each n. Suppose that h is a homeomorphism in $\mathsf{HOMEO}(X)$ such that $\nu = h_\#\mu$. Note that $h^{-1}[B_n]$ is a unique arc in X_1 that contains p_0. As $\nu(B_n) = h_\#\mu(B_n) = \mu(h^{-1}[B_n])$ we have $A_n = h^{-1}[B_n]$, whence $B_n = h[A_n]$. Consequently, $h(p_i) = \left(\frac{1}{(i+1)\pi}, 0\right)$. Hence $h(p_i)$ converges to $(0,0)$ as $i \to \infty$. But p_i converges to $(0,1)$ and $h\big((0,1)\big)$ is either $(0,1)$ or $(0,-1)$. A contradiction has appeared. Consequently, there are no h in $\mathsf{HOMEO}(X)$ such that $\nu = h_\#\mu$.

EXAMPLE (**Warsaw circle**). The Warsaw circle W is a well-known example in topology. It is a one-dimensional subset of $\mathbb{R}^2$ formed from the space X of the above example by joining the points $(0,-1)$ and $(\pi^{-1}, 0)$ with a topological arc X_2 in $\mathbb{R}^2$ so that $X \cap X_2$ is precisely the set consisting of these two points. The space $W = X_0 \cup X_1 \cup X_2$ is the Warsaw circle. The above analysis of X can be adjusted to apply to the Warsaw circle. Clearly, $h[X_0] = X_0$ and $h[X_1 \cup X_2] = X_1 \cup X_2$ whenever $h \in \mathsf{HOMEO}(W)$. Observe that a continuous, complete, finite Borel measure μ on W is positive if and only if $\mu|(X_1 \cup X_2)$ is positive.

We leave the proof of the following theorem as an exercise.

THEOREM 3.23. *Let W be the Warsaw circle. For some measure μ in* $\mathrm{MEAS}^{\mathrm{pos,fin}}(W)$, *$\mu$ and* $\mathsf{HOMEO}(W)$ *generate* $\mathsf{univ}\,\mathfrak{M}(W)$.

An analogous theorem holds for the $\sin(1/x)$ example above.

3.4. The Oxtoby–Ulam theorem

We have seen that the collection $\mathrm{MEAS}^{\mathrm{pos,fin}}([0,1])$ of all positive, continuous, complete, finite Borel measures on $[0,1]$ can be characterized by the group $\mathsf{HOMEO}([0,1])$ and the Lebesgue measure λ on $[0,1]$. Indeed, it was shown that the equivalence classes of $\mathrm{MEAS}^{\mathrm{pos,fin}}([0,1])$ are determined by the nonnegative real numbers c (that is, $c\,\lambda$). The natural generalization of this fact to the unit n-cell $[0,1]^n$

was proved by J. C. Oxtoby and S. M. Ulam in [**122**]. Rather than just cite the result we include a proof for the benefit of the reader.

THEOREM 3.24 (Oxtoby–Ulam). *Let λ be the Lebesgue measure on the n-cell* $[0,1]^n$. *In order for a Borel measure μ on* $[0,1]^n$ *be such that there is an h in* $\mathsf{HOMEO}([0,1]^n)$ *with $\mu = h_{\#}\lambda$ it is necessary and sufficient that*

(1) *μ be a positive, continuous, complete Borel measure,*
(2) $\mu([0,1]^n) = 1$,
(3) $\mu(\partial[0,1]^n) = 0$.

Moreover, the homeomorphism h may have the property that it is the identity map on $\partial[0,1]^n$.

The proof will be divided into several parts. Before embarking on the proof let us state a consequence of the Oxtoby–Ulam theorem for n-dimensional finitely triangulable spaces.

THEOREM 3.25. *Let $|K_n|$ be an n-dimensional finitely triangulable space. Then there exists a measure μ in* $\mathrm{MEAS}^{\mathrm{pos,fin}}(|K_n|)$ *such that μ and* $\mathsf{HOMEO}(|K_n|)$ *generate* $\mathsf{univ}\,\mathfrak{M}(|K_n|)$.

We shall first prove the Oxtoby–Ulam theorem, delaying the proof of this consequence to the end of the section. The definition of a finite-dimensional finitely triangulable space is also delayed to the end of this section (see the footnote on page 72).

3.4.1. Proofs of the Oxtoby–Ulam theorem.

The literature contains two proofs of the Oxtoby–Ulam theorem. They are essentially the same since both rely on the same key lemma concerning the existence of a homeomorphism that possesses a special property. The fact is that the two proofs of the key lemma are very different. The original proof by Oxtoby and Ulam relies on a complete metric on $\mathsf{HOMEO}([0,1]^n)$ and the Baire category theorem, and the subsequent proof by C. Goffman and G. Pedrick [**63**] relies on a measure theoretic property of σ-finite Borel measures on $\mathbb{R}^n$. We shall give both proofs of the key lemma.

It is time to state the key Lemma 3.26. Observe that the universally positive closure $\mathrm{F}_M(M)$ of an n-dimensional manifold is M. We shall call a measure μ on a compact, connected manifold M (with or without boundary) *Lebesgue-like* if μ is a positive, continuous, complete, finite Borel measure on M with $\mu(\partial M) = 0$.

LEMMA 3.26. *For an m-cell J let μ be a Lebesgue-like Borel measure on* $I = J \times [-1,1]$ *and let α_1 and α_2 be positive numbers such that* $\alpha_1 + \alpha_2 = \mu(I)$. *Then there is a φ in* $\mathsf{HOMEO}(I;\ \partial I\ \text{fixed})$ *such that $\varphi_{\#}\mu$ is Lebesgue-like on both* $R_1 = J \times [-1,0]$ *and* $R_2 = J \times [0,1]$ *and such that* $\varphi_{\#}\mu(R_1) = \alpha_1$ *and* $\varphi_{\#}\mu(R_2) = \alpha_2$.

We begin with the proof by Goffman and Pedrick. The proof uses a "parallel slicing lemma" for continuous, σ-finite Borel measures μ on $\mathbb{R}^n$ (proved by Goffman and

Pedrick in [63, Lemma 1]) which assures that every slice parallel to some fixed hyperplane has μ measure equal to 0. (See also [62].)

A *hyperplane* E *of* $\mathbb{R}^n$ is the set of all solutions of the equation $\langle x - b, e\rangle = 0$, where e is in the unit sphere $\mathbb{S}_{n-1}$ and b is in $\mathbb{R}^n$, and $\langle\cdot,\cdot\rangle$ is the usual inner product of $\mathbb{R}^n$. The vector e is, of course, a unit normal vector of the hyperplane E. By a k-flat we mean a nonempty intersection of $n-k$ hyperplanes whose normals are linearly independent. Each k-flat determines a family consisting of all hyperplanes that contains the k-flat. This family yields a set of unit normals that forms a closed nowhere dense subset of $\mathbb{S}_{n-1}$ whose $(n-1)$-dimensional Hausdorff measure is 0.

For a finite Borel measure μ on $\mathbb{R}^n$ (not necessarily continuous) and for each integer k with $0 < k < n$ let $\mathcal{F}_k$ be the collection of all k-flats F such that $\mu(F) > 0$ and such that F contains no j-flats with positive μ measure with $j < k$. It is easily seen that $\mathcal{F}_1$ is a countable set. Indeed, let us assume the contrary. Then there is an uncountably infinite number of 1-flats F with the property that $\mu(F) > 0$ and $\mu(\{p\}) = 0$ for each p in F. Clearly we may assume that there is a positive number ε with $\mu(F) \geq \varepsilon$ for these uncountably many 1-flats F. Now select a sequence F_m, $m = 1, 2, \ldots$, of distinct members from this collection. Clearly, $\mu(F_m \cap \bigcup_{j<m} F_j) = 0$ for each m. So we have $m\,\varepsilon \leq \mu(\bigcup_{j\leq m} F_j) \leq \mu(\mathbb{R}^n) < \infty$ for each m, a contradiction. Analogously one can show that each $\mathcal{F}_k$ is a countable set. Let $\mathcal{F}_\mu = \bigcup_{j=1}^{n-1} \mathcal{F}_j$. Then, by the Baire category theorem, the set

$$\mathcal{E}_\mu = \textstyle\bigcup_{F\in\mathcal{F}_\mu}\{e \in \mathbb{S}_{n-1} : e \text{ is normal to } F\}$$

is of the first Baire category in $\mathbb{S}_{n-1}$. It is easily seen that $\mathsf{H}_{n-1}(\mathcal{E}_\mu) = 0$, also. We have the following lemma.

Lemma 3.27. *If μ is a continuous, σ-finite Borel measure on $\mathbb{R}^n$, then, except for points e in a subset $\mathcal{E}_\mu$ of the first Baire category in $\mathbb{S}_{n-1}$ with $\mathsf{H}_{n-1}(\mathcal{E}_\mu) = 0$,*

$$\mu(E) = 0 \textit{ for every hyperplane } E \textit{ for which } e \textit{ is a normal.}$$

Proof. First assume that μ is finite and let $e \in \mathbb{S}_{n-1} \setminus \mathcal{E}_\mu$. Suppose that $\mu(E) > 0$ for some hyperplane E with e as its normal. From $e \in \mathbb{S}_{n-1} \setminus \mathcal{E}_\mu$ we infer $E \notin \mathcal{F}_{n-1}$. Let E_{n-2} be an $(n-2)$-flat contained in E such that some j-flat contained in E_{n-2} has positive μ measure, where $j < n-1$. Clearly E_{n-2} is not in $\mathcal{F}_{n-2}$ since e is a normal to E_{n-2} and $e \notin \mathcal{E}_\mu$. After finitely many steps we will get a 1-flat E_1 contained in E such that some 0-flat of E_1 has positive measure. This shows that μ is not a continuous measure. Hence, if μ is a continuous, finite Borel measure and $e \in \mathbb{S}_{n-1} \setminus \mathcal{E}_\mu$, then $\mu(E) = 0$ for every hyperplane E whose normal is e. Moreover, $\mathcal{E}_\mu$ is a set of first Baire category in $\mathbb{S}_{n-1}$ with $\mathsf{H}_{n-1}(\mathcal{E}_\mu) = 0$.

For a σ-finite measure μ, write μ as a sum $\sum_{m=1}^{\infty} \mu_m$, where the summands are finite measures. Then $\mathcal{E} = \bigcup_{m=1}^{\infty} \mathcal{E}_{\mu_m}$ is a subset of the first Baire category of $\mathbb{S}_{n-1}$ with $\mathsf{H}_{n-1}(\mathcal{E}) = 0$. If e is in $\mathbb{S}_{n-1} \setminus \mathcal{E}$ and E is a hyperplane with normal e, then $\mu(E) = \sum_{m=1}^{\infty} \mu_m(E) = 0$. This completes the proof. □

The Goffman–Pedrick construction of a homeomorphism with certain special properties uses the following simple observations. Let $I = J \times [-1, 1]$, where J is a Cartesian product of m intervals. We begin with a continuous function $f : J \to [0, 1)$ that satisfies $f(u) = 0$ if and only if $u \in \partial J$. Then $u \mapsto \big(f(u)\big)^r$, where $u \in J$ and $0 < r$, is a one-parameter family of continuous functions on J such that

$$\operatorname{graph}(f^s) \cap \operatorname{graph}(f^t) = \partial J \text{ whenever } s \neq t.$$

Also, for each positive number ε and for each ν in $\mathrm{MEAS}^{\mathrm{pos,fin}}(I)$, there is a t_0 and an s_0 such that the sets

$$M_1 = \{(x, y) \in I : y \leq -f^{t_0}(x)\} \text{ and } M_2 = \{(x, y) \in I : f^{s_0}(x) \leq y\}$$

satisfy $\nu(M_1) < \varepsilon$ and $\nu(M_2) < \varepsilon$ and such that

$$\nu(\operatorname{graph}(-f^{t_0})) = \nu(\operatorname{graph}(f^{s_0})) = 0.$$

Let us go to the Goffman–Pedrick proof.

Goffman–Pedrick Proof of Lemma 3.26. Let $f_1 : J \to [0, 1)$ be a continuous function such that $M_1 = \{(x, y) \in I : y \leq -f_1(x)\}$ satisfies $\mu(M_1) < \alpha_1/2$. By means of continuous piecewise linear maps on the line segments $\{x\} \times [-1, 1]$ that map -1, $-f_1(x)$ and 1 respectively to -1, 0 and 1, one can construct a ψ_1 in $\mathsf{HOMEO}(I, \partial I \text{ fixed})$. We then have $\psi_{1\#}\mu(R_1) < \alpha_1/2$.

Let $f_2 : J \to [0, 1)$ be a continuous function whose graph satisfies $\psi_{1\#}\mu(\operatorname{graph}(f_2)) = 0$ and is such that $M_2 = \{(x, y) \in I : y \leq f_2(x)\}$ satisfies $\psi_{1\#}\mu(M_2) > \alpha_1$. Select next a positive number y_0 such that $\psi_{1\#}\mu(J \times [-1, y_0]) < \alpha_1$, and let $y_1 = \max\{f_2(x) : x \in J\}$. Obviously, $0 < y_0 < y_1 < 1$. In view of Lemma 3.27 there is a unit normal e in $\mathbb{R}^{m+1}$ such that some hyperplane H_0 with normal e separates $J \times [-1, 0]$ and $J \times [y_0, 1]$, and some hyperplane H_1 with normal e separates $J \times [-1, y_1]$ and $J \times \{1\}$, and every hyperplane with normal e intersects I with $\psi_{1\#}\mu$ measure equal to 0.

For the above unit normal e in $\mathbb{R}^{m+1}$ let $h_t : \mathbb{R}^m \to \mathbb{R}$ be a linear function such that e is normal to $\operatorname{graph}(h_t)$ and $h_t(0) = t$ for each t in $\mathbb{R}$. There is a t_0 such that $\operatorname{graph}(h_{t_0}) = H_0$, and there is a t_1 such that $\operatorname{graph}(h_{t_1}) = H_1$. The map

$$t \mapsto \psi_{1\#}\mu(\{(u, v) \in I : 0 \leq v \leq h_t \wedge f_2(u)\}),\ t_0 \leq t \leq t_1,$$

is a continuous function whose value at t_0 is less than α_1 and whose value at t_1 is greater than α_1. Consequently there is a continuous function $g : J \to [0, 1)$ such that $g(u) = 0$ if and only if $u \in \partial J$, and $\psi_{1\#}\mu(\{(u, v) \in I : 0 \leq v \leq g(u)\}) = \alpha_1$, and $\psi_{1\#}\mu(\operatorname{graph}(g)) = 0$. In a manner similar to the construction of ψ_1 we can construct a ψ_2 in $\mathsf{HOMEO}(I; \partial I \text{ fixed})$ such that

$$\psi_2{}^{-1}[R_1] = \{(u, v) \in I : 0 \leq v \leq g(u)\}$$

and

$$\psi_2{}^{-1}[J \times \{0\}] = \text{graph}(g).$$

Hence $(\psi_2\psi_1)_\#\mu(R_1) = \alpha_1$ and $(\psi_2\psi_1)_\#\mu(J \times \{0\}) = 0$. The composition $\varphi = \psi_2\psi_1$ clearly satisfies $\varphi \in \mathsf{HOMEO}(I; \partial I \text{ fixed})$, $\varphi_\#\mu(R_1) = \alpha_1$ and $\varphi_\#\mu(R_2) = \alpha_2$. □

We give next the Oxtoby–Ulam proof of the key Lemma 3.26. Their proof contains the germ of the proof of Lemma 3.46 due to J. C. Oxtoby and V. S. Prasad [**121**] which allows the extension of the Oxtoby–Ulam theorem to the Hilbert cube $[0, 1]^{\mathbb{N}}$, the subject of Section 3.6.

As stated earlier, the Oxtoby–Ulam proof relies on the Baire category theorem applied to a suitable nonempty complete metric space H that contains the collection

$$\mathsf{F} = \{g \in \mathsf{HOMEO}(I; \partial I \text{ fixed}) \colon g_\#\mu(R_1) = \alpha_1,\ g_\#\mu(R_2) = \alpha_2\}$$

as a dense G_δ subset. Recall that every nonempty closed subspace of a complete metric space is complete. We have shown earlier that the map $g \mapsto g_\#\mu(X)$ is an upper semicontinuous function on the complete metric space $\mathsf{HOMEO}(I; \partial I \text{ fixed})$ whenever X is a closed subset of I. Hence a suitable closed subspace of $\mathsf{HOMEO}(I; \partial I \text{ fixed})$ that contains F is

$$\begin{aligned} \mathsf{H} = \{g \in \mathsf{HOMEO}(I; \partial I \text{ fixed}) \colon g_\#\mu(R_1) \geq \alpha_1\} \\ \cap \{g \in \mathsf{HOMEO}(I; \partial I \text{ fixed}) \colon g_\#\mu(R_2) \geq \alpha_2\}. \end{aligned}$$

Since it is not immediate that H is not empty, we shall show in the next paragraph the existence of an element in H. It is interesting that this proof of existence does not use the Baire category theorem. Observe that

$$\mathsf{F} = \mathsf{H} \setminus \bigcup_{i=1}^{2} \bigcup_{n=1}^{\infty} \left\{g \in \mathsf{HOMEO}(I; \partial I \text{ fixed}) \colon g_\#\mu(R_i) \geq \alpha_i + \tfrac{1}{n}\right\}$$

and that every set

$$\mathsf{H}_{in} = \left\{g \in \mathsf{HOMEO}(I; \partial I \text{ fixed}) \colon g_\#\mu(R_i) \geq \alpha_i + \tfrac{1}{n}\right\}$$

is closed in $\mathsf{HOMEO}(I; \partial I \text{ fixed})$. Hence the Baire category theorem will apply after we show that $\mathsf{H} \cap \mathsf{H}_{in}$ is nowhere dense in the space H.

To show that H is nonempty, first note that if both $\mu(R_1) < \alpha_1$ and $\mu(R_2) < \alpha_2$ fail then the identity map is in H. So assume $\mu(R_1) < \alpha_1$. As in the Goffman–Pedrick proof, select a continuous function $f \colon J \to [0, 1)$ such that $f(x) > 0$ if and only if $x \notin \partial J$. For each positive number t define the two sets

$$\begin{aligned} A_t &= \{(x,y) \in I \colon -1 \leq y \leq f^t(x)\}, \\ B_t &= \{(x,y) \in I \colon f^t(x) \leq y \leq 1\}. \end{aligned}$$

There is a ψ_t in HOMEO($I; \partial I$ fixed) such that ${\psi_t}^{-1}[R_1] = A_t$ and ${\psi_t}^{-1}[R_2] = B_t$. As $A_t \supset R_1$ and $B_t \subset R_2$ we have

$$ {\psi_s}^{-1}[R_1] \supset {\psi_t}^{-1}[R_1] \supset R_1 \text{ whenever } 0 < s < t, \tag{3.4} $$

$$ {\psi_t}^{-1}[R_2] \subset {\psi_s}^{-1}[R_2] \subset R_2 \text{ whenever } 0 < s < t. \tag{3.5} $$

The strictly decreasing function $t \mapsto \psi_{t\#}\mu(R_1)$ converges to $\mu(R_1)$ as $t \to \infty$ and converges to $\mu(I)$ as $t \to 0$. From equation (3.4), it is continuous from the left, and also $\{s\colon \psi_{s\#}\mu(R_1) \geq \alpha_1\}$ is equal to $(0, s_0]$ for some real number s_0 because $\alpha_1 > \mu(R_1)$. Hence $\psi_{s_0\#}\mu(R_1) \geq \alpha_1$. Let $t > s_0$. Then $\alpha_1 > \psi_{t\#}\mu(R_1)$. From equation (3.5) we have $\psi_{t\#}\mu(R_2) \leq \psi_{s_0\#}\mu(R_2)$. Hence

$$ \begin{aligned} \alpha_1 + \alpha_2 = \psi_{t\#}\mu(I) &\leq \psi_{t\#}\mu(R_1) + \psi_{t\#}\mu(R_2) \\ &\leq \psi_{t\#}\mu(R_1) + \psi_{s_0\#}\mu(R_2). \end{aligned} $$

Consequently, $\alpha_2 \leq \psi_{s_0\#}\mu(R_2)$ as well as $\alpha_1 \leq \psi_{s_0\#}\mu(R_1)$. Thereby we have shown that $g = \psi_{s_0}$ is in H if $\mu(R_1) < \alpha_1$. A similar argument applies to the case $\mu(R_2) < \alpha_2$, hence $\mathsf{H} \neq \emptyset$.

First, a preliminary lemma is needed for the Oxtoby–Ulam proof of the key lemma.

Lemma 3.28. *Let μ be a continuous, finite Borel measure on a compact metrizable space X, let $0 \leq \alpha < \beta \leq \mu(X)$ be given, and let F be a closed set with $\mu(F) = 0$. Then there exists an open set G such that $G \cap F = \emptyset$ and $\alpha < \mu(G) < \beta$.*

Proof. As μ is continuous, each point x of $X \setminus F$ has an open neighborhood U_x with $\mu(U_x) < \beta - \alpha$ and $U_x \cap F = \emptyset$. Let K be a compact set such that $K \cap F = \emptyset$ and $\mu(K) > \alpha$. The above open cover of $X \setminus F$ contains a finite open cover $U_1, U_2, \ldots, U_m$ of K. Let $G_i = \bigcup_{j \leq i} U_j$. There is no loss in assuming $U_i \setminus \bigcup_{j<i} U_j \neq \emptyset$ for every i. Let k be such that $\mu(G_k) > \alpha$ and $\mu(G_{k-1}) \leq \alpha$. Then $\mu(G_k) \leq \mu(G_{k-1}) + \mu(U_k) < \alpha + (\beta - \alpha) = \beta$. The open set G_k fulfills the requirement of the lemma. □

Oxtoby–Ulam Proof of Lemma 3.26. It remains to be shown that $\mathsf{H}_{in} \cap \mathsf{H}$ is closed and nowhere dense in H. As we already know that H_{in} is a closed subset of HOMEO($I; \partial I$ fixed), we need to show that $\mathsf{H} \setminus \mathsf{H}_{in}$ is a dense in H. We consider the case $i = 1$. Let $g \in \mathsf{H}_{1n} \cap \mathsf{H}$ and $0 < \varepsilon < 1$. We seek a g' in $\mathsf{H} \setminus \mathsf{H}_{1n}$ such that $\rho(g', g) < \varepsilon$. To this end, observe that $g \in \mathsf{H}_{1n} \cap \mathsf{H}$ implies

$$ g_{\#}\mu(R_1) \geq \alpha_1 + \tfrac{1}{n} \quad \text{and} \quad g_{\#}\mu(R_1) - \alpha_1 \leq g_{\#}\mu(J \times \{0\}), $$

where the second inequality holds because of the identity

$$ \alpha_1 + \alpha_2 = g_{\#}\mu(R_1) + g_{\#}\mu(R_2) - g_{\#}\mu(J \times \{0\}) $$

and because $g \in \mathsf{H}$ yields $g_{\#}\mu(R_1) \geq \alpha_1$ and $g_{\#}\mu(R_2) \geq \alpha_2$. By Lemma 3.28 applied to $g_{\#}\mu|(J \times \{0\})$ with $F = (\partial J) \times \{0\}$, $\alpha = g_{\#}\mu(R_1) - \alpha_1 - \frac{1}{n}$ and $\beta = g_{\#}\mu(R_1) - \alpha_1$, there is a set G that is open relative to $J \times \{0\}$ such that $G \cap F = \emptyset$ and $\alpha < g_{\#}\mu(G) < \beta$. With $\eta = \min\{\varepsilon, \omega(g\colon \varepsilon)\}$, where ω is the modulus of uniform continuity, select a continuous function $f: J \times \{0\} \to [0, \eta/2]$ such that $f(x) = 0$ if and only if $x \in J \times \{0\} \setminus G$. For each x in $J \times \{0\}$ and for each δ with $0 < \delta \leq 1$, let $l_\delta(x, \cdot)$ be a continuous linear function that respectively maps the intervals $[-1, -f(x)]$, $[-f(x), -\delta f(x)]$, $[-\delta f(x), f(x)]$, and $[f(x), 1]$ onto the intervals $[-1, -f(x)]$, $[-f(x), 0]$, $[0, f(x)]$, and $[f(x), 1]$. These maps $l_\delta(x, \cdot)$ define a homeomorphism $h_\delta(x,t) = (x, l_\delta(x,t))$ of I onto I such that h_δ is fixed on $\partial I \cup \{(x,t) \in J \times [-1,1]: (x,0) \notin G\}$. Observe that $h_\delta{}^{-1}[R_2] \supset R_2$, and $\bigcup_{0<\delta\leq 1} h_\delta{}^{-1}[R_1] = R_1 \setminus G$. We now have $(h_\delta g)_\# \mu(R_2) \geq g_\# \mu(R_2) \geq \alpha_2$ and $\alpha_1 + \frac{1}{n} > \lim_{\delta\to 0}(h_\delta g)_\# \mu(R_1) = g_\# \mu(R_1) - g_\# \mu(G) > \alpha_1$. Select a δ such that $g' = h_\delta g$ satisfies $\alpha_1 < g'_\# \mu(R_1) < \alpha_1 + \frac{1}{n}$. We have $g' \in \mathsf{H}$, $\rho(g', g) < \varepsilon$, and g' is in the open subset $\{f \in \mathsf{HOMEO}(I; \partial I \text{ fixed}): f_\# \mu(R_1) < \alpha_1 + \frac{1}{n}\}$ of $\mathsf{HOMEO}(I; \partial I \text{ fixed})$. Hence $\mathsf{H}_{1n} \cap \mathsf{H}$ is nowhere dense in H. To prove the same for H_{2n}, use the isometry φ defined by the map $(x,t) \mapsto (x, -t)$ for (x,t) in $J \times [-1,1]$. □

This concludes the two very different proofs of the key Lemma 3.26. We turn to the remainder of the proof of the Oxtoby–Ulam theorem that results from the key lemma.

Sometimes it will be convenient to work with n-cells. By an *n-cell subdivision $\mathcal{P}$ of an n-cell X* we mean a finite collection of nonoverlapping compact subsets σ of X that are n-cells such that the union of the members of $\mathcal{P}$ is X. As usual, two subsets A and B of X are said to be *nonoverlapping* if $\mathrm{Int}_X(A) \cap \mathrm{Int}_X(B) = \emptyset$, where Int_X is the usual interior operator in the topological space X. If $\mathcal{P}$ is an n-cell subdivision of an n-cell X, then the *mesh of $\mathcal{P}$* is defined to be

$$\mathrm{mesh}(\mathcal{P}) = \max\{\mathrm{diam}(\sigma): \sigma \in \mathcal{P}\}.$$

An n-cell subdivision $\mathcal{P}'$ of X is said to *refine* an n-cell subdivision $\mathcal{P}$ of X if each member of $\mathcal{P}'$ is contained in some member of $\mathcal{P}$.

By a *rectangular subdivision* $\mathcal{P}$ of an an n-cell $I = \times_{i=1}^{n}[a_i, b_i]$ we mean an n-cell subdivision of I whose members are n-dimensional rectangles with edges that are parallel to the coordinate axes of I. The collection of all end-points of the i-th intervals that form the subdivision $\mathcal{P}$ can be used to construct another rectangular subdivision $\mathcal{P}'$ of I that refines $\mathcal{P}$. Let $\mathcal{P}'(\sigma)$ be the collection of those σ' in $\mathcal{P}'$ that are contained in σ whenever $\sigma \in \mathcal{P}$.

Here is an elementary construction. With $n > 1$ let I^n be the Cartesian product of $I = [0,1]$. Let $x_{i,j}$, $j = 0, 1, \ldots, k$, be partition points of the i-th coordinate interval I of I^n. The coordinate hyperplanes of $\mathbb{R}^n$ determined by these partition points will form a rectangular subdivision of I^n, which we will denote by $\mathcal{P}$. Let μ and ν be Lebesgue-like Borel measures on I^n such that $\mu(I^n) = \nu(I^n)$. We select the partition points of the coordinate intervals I in such a way that $\nu(\partial\sigma) = 0$ for every σ in $\mathcal{P}$. Repeated applications of the key lemma, one coordinate hyperplane at a time, will lead to a homeomorphism φ in $\mathsf{HOMEO}(I^n; \partial I^n \text{ fixed})$ such that $(\varphi_\# \mu)|\sigma$ is Lebesgue-like on σ and $\varphi_\# \mu(\sigma) = \nu(\sigma)$ whenever $\sigma \in \mathcal{P}$. A note of caution: the subdivision $\varphi^{-1}[\mathcal{P}] = \{\varphi^{-1}[\sigma]: \sigma \in \mathcal{P}\}$ need not be a rectangular subdivision of I^n. This is not of concern since it is the fact that $\mathcal{P}$ is a rectangular subdivision that matters. The next proposition follows easily from this elementary construction.

PROPOSITION 3.29. *For $\varepsilon > 0$ and Lebesgue-like Borel measures μ and ν on an n-dimensional rectangle X with $\mu(X) = \nu(X)$, there is a rectangular subdivision $\mathcal{P}$ of X and a φ in* $\mathsf{HOMEO}(X; \partial X \text{ fixed})$ *such that*

(1) $\operatorname{mesh}(\mathcal{P}) < \varepsilon$,
(2) $(\varphi_{\#}\mu)|\sigma$ *and* $\nu|\sigma$ *are Lebesgue-like on* σ *whenever* $\sigma \in \mathcal{P}$,
(3) $\varphi_{\#}\mu(\sigma) = \nu(\sigma)$ *whenever* $\sigma \in \mathcal{P}$.

PROOF. It will be convenient to work with the n-cell I^n. So let $\vartheta : X \to I^n$ be a linear homeomorphism. Consider the Lebesgue-like Borel measures $\vartheta_{\#}\mu$ and $\vartheta_{\#}\nu$ on I^n. If $\mathcal{P}'$ is a rectangular subdivision of I^n, then $\mathcal{P} = \vartheta^{-1}[\mathcal{P}']$ is a rectangular subdivision of X whose mesh will be small if $\operatorname{mesh}(\mathcal{P}')$ is sufficiently small. From the previous paragraph we can construct a rectangular subdivision $\mathcal{P}'$ with $\operatorname{mesh}(\mathcal{P}') < \varepsilon$ and a homeomorphism φ' of I^n such that, for each σ' in $\mathcal{P}'$, $(\varphi'_{\#}\vartheta_{\#}\mu)|\sigma'$ and $(\vartheta_{\#}\nu)|\sigma'$ are Lebesgue-like on σ' and $\varphi'_{\#}\vartheta_{\#}\mu(\sigma') = \vartheta_{\#}\nu(\sigma')$, and such that $\varphi'(x) = x$ whenever $x \in \partial I^n$. Then $\varphi = \vartheta^{-1}\varphi'\vartheta$ is in $\mathsf{HOMEO}(X)$ and $\mathcal{P} = \vartheta^{-1}[\mathcal{P}']$ is a rectangular subdivision of X such that $(\varphi_{\#}\mu)|\sigma$ and $\nu|\sigma$ are Lebesgue-like on σ and $\varphi_{\#}\mu(\sigma) = \nu(\sigma)$ for each σ in $\mathcal{P}$ and such that $\varphi(x) = x$ whenever $x \in \partial X$. □

Note that the n-cell subdivision $\varphi^{-1}[\mathcal{P}]$ need not have small mesh. Our final lemma for the proof of the Oxtoby–Ulam theorem overcomes this deficiency. The following property will facilitate the statement of the lemma. A pair μ and ν of Lebesgue-like Borel measures on an n-cell X and an n-cell subdivision $\mathcal{P}$ of X are said to satisfy the *property* $\mathsf{P}(\mu, \nu\,; \mathcal{P})$ if $\mu(\sigma) = \nu(\sigma)$ and $\mu(\partial\sigma) = \nu(\partial\sigma) = 0$ whenever $\sigma \in \mathcal{P}$. Observe the following: $\mathsf{P}(\mu, \nu\,; \mathcal{P})$ if and only if $\mathsf{P}(\nu, \mu\,; \mathcal{P})$, and $\mathsf{P}(\varphi_{\#}\mu, \nu\,; \mathcal{P})$ if and only if $\mathsf{P}(\mu, \varphi^{-1}{}_{\#}\nu\,; \varphi^{-1}[\mathcal{P}])$ whenever φ is in $\mathsf{HOMEO}(X)$.

LEMMA 3.30. *Suppose that μ and ν are Lebesgue-like Borel measures on an n-dimensional rectangle X and $\mathcal{P}$ is a rectangular subdivision of X. If μ, ν and $\mathcal{P}$ satisfy* $\mathsf{P}(\mu, \nu\,; \mathcal{P})$ *and if $\varepsilon > 0$, then there is a φ in* $\mathsf{HOMEO}(X)$ *and a rectangular subdivision $\mathcal{P}'$ of X such that $\varphi_{\#}\mu$, ν and $\mathcal{P}'$ satisfy* $\mathsf{P}(\varphi_{\#}\mu, \nu\,; \mathcal{P}')$ *and*

(1) $\varphi|\sigma \in \mathsf{HOMEO}(\sigma; \partial\sigma \text{ fixed})$ *whenever* $\sigma \in \mathcal{P}$,
(2) $\mathcal{P}'$ *refines* $\mathcal{P}$,
(3) $\operatorname{mesh}(\mathcal{P}') < \varepsilon$.

PROOF. After applying the last proposition to each σ in $\mathcal{P}$, we have a φ in $\mathsf{HOMEO}(X)$ and a rectangular subdivision $\mathcal{P}'$ of X such that $\mathsf{P}(\varphi_{\#}\mu, \nu\,; \mathcal{P}')$ is satisfied, $\varphi|\sigma \in \mathsf{HOMEO}(\sigma; \partial\sigma \text{ fixed})$ whenever $\sigma \in \mathcal{P}$, $\operatorname{mesh}(\mathcal{P}') < \varepsilon$, and $\mathcal{P}'$ refines $\mathcal{P}$. □

PROOF OF THE OXTOBY–ULAM THEOREM. Let μ satisfy the three conditions given in the Oxtoby–Ulam Theorem 3.24. We shall inductively construct two sequences of homeomorphisms φ_j, ψ_j, $j = 1, 2, \ldots$, and two sequences of rectangular subdivisions $\mathcal{P}_j, \mathcal{P}'_j, j = 1, 2, \ldots,$ of I^n. The following diagram indicates the steps of the construction.

$$\mathsf{P}(\Phi_{j\#}\mu, \Psi_{(j-1)\#}\lambda; \mathcal{P}_j) \Longrightarrow \mathsf{P}(\Phi_{j\#}\mu, \Psi_{j\#}\lambda; \mathcal{P}'_j) \Longrightarrow$$
$$\mathsf{P}(\Phi_{(j+1)\#}\mu, \Psi_{j\#}\lambda; \mathcal{P}_{j+1}) \Longrightarrow \mathsf{P}(\Phi_{(j+1)\#}\mu, \Psi_{(j+1)\#}\lambda; \mathcal{P}'_{j+1}),$$

for $j = 1, 2, \ldots$, where $\Psi_0 = \text{id}$, $\mathcal{P}'_0$ is undefined, and

(1) $\Phi_j = \varphi_j \varphi_{j-1} \cdots \varphi_1$ and $\text{mesh}(\mathcal{P}_j) < 2^{-j}$,
(2) $\Psi_j = \psi_j \psi_{j-1} \cdots \psi_1$ and $\text{mesh}(\mathcal{P}'_j) < 2^{-j}$,
(3) $\varphi_j | \sigma' \in \mathsf{HOMEO}(\sigma'; \partial\sigma' \text{ fixed})$ whenever $\sigma' \in \mathcal{P}'_{j-1}$,
(4) $\psi_j | \sigma \in \mathsf{HOMEO}(\sigma; \partial\sigma \text{ fixed})$ whenever $\sigma \in \mathcal{P}_j$,
(5) $\text{mesh}({\Phi_j}^{-1}[\mathcal{P}'_j]) < 2^{-j}$,
(6) $\text{mesh}({\Psi_{j-1}}^{-1}[\mathcal{P}_j]) < 2^{-j}$,
(7) $\mathcal{P}'_j$ refines $\mathcal{P}_j$,
(8) $\mathcal{P}_j$ refines $\mathcal{P}'_{j-1}$.

By the above lemma, with μ, $\nu = \lambda$, $\mathcal{P} = \{I^n\}$ and $\varepsilon = 2^{-1}$, we have a φ_1 in $\mathsf{HOMEO}(I^n; \partial I^n \text{ fixed})$ and a rectangular subdivision $\mathcal{P}_1$ that satisfy $\mathsf{P}(\Phi_{1\#}\mu, \Psi_{0\#}\lambda; \mathcal{P}_1)$ as well as $\text{mesh}(\mathcal{P}_1) < 2^{-1}$. Conditions (1), (3) and (8) are satisfied for $j = 1$, thus the first step for $j = 1$ is completed. We still need to get ψ_1 and $\mathcal{P}'_1$ to verify conditions (2), (4), (5) and (7) for $j = 1$, which is the second step. To this end we begin by selecting a positive number δ_1 such that the modulus of uniform continuity of ${\Phi_1}^{-1}$ satisfies $\omega({\Phi_1}^{-1} \colon \delta_1) < 2^{-1}$. We may assume $\delta_1 < 2^{-1}$, also. Then, in the lemma, use $\Phi_{1\#}\mu$ for μ, $\nu = \Psi_{0\#}\lambda$, $\varepsilon = \delta_1$ and $\mathcal{P} = \mathcal{P}_1$. Then there is a ψ_1 and a rectangular subdivision $\mathcal{P}'_1$ that satisfies $\mathsf{P}(\Phi_{1\#}\mu, \Psi_{1\#}\lambda; \mathcal{P}'_1)$ as well as $\text{mesh}(\mathcal{P}'_1) < 2^{-1}$, whence condition (2) is satisfied for $j = 1$. Also conditions (4), (5), (6) and (7) are satisfied for $j = 1$; hence all the conditions are satisfied for $j = 1$. Finally we shall indicate only the third and fourth steps since the inductive construction is clear from the steps two, three and four. For the third step, we select a positive number δ'_1 with $\delta'_1 < 2^{-2}$ such that the modulus of uniform continuity of ${\Psi_1}^{-1}$ satisfies $\omega({\Psi_1}^{-1} \colon \delta'_1) < 2^{-2}$. Then, in the lemma, let $\Phi_{1\#}\mu$ be μ, $\Psi_{1\#}\nu$ be ν, $\varepsilon = \delta'_1$ and $\mathcal{P} = \mathcal{P}'_1$. Then there is a φ_2 and a rectangular subdivision $\mathcal{P}_2$ that satisfies $\mathsf{P}(\Phi_{2\#}\mu, \Psi_{1\#}\lambda; \mathcal{P}_2)$ as well as $\text{mesh}(\mathcal{P}_2) < \delta'_1$, and $\mathcal{P}_2$ refines $\mathcal{P}'_1$. Condition (1) is satisfied for $j = 2$. Also conditions (3), (6) and (8) are satisfied for $j = 2$. We still must construct ψ_2 and $\mathcal{P}'_2$, which is the fourth step. This is achieved by applying the second of the four steps again. Then conditions (2), (4), (5) and (7) are satisfied for $j = 2$. Hence all the conditions are satisfied for $j = 2$.

It is easily seen that $\Phi_j(x) = x$ and $\Psi_j(x) = x$ whenever $x \in \partial I^n$. One must verify, as indicated in the diagram, that the sequences Φ_j and Ψ_j, $j = 1, 2, \ldots$, are convergent in $\mathsf{HOMEO}(I^n; \partial I^n \text{ fixed})$. To this end let us show that Φ_j, $j = 1, 2, \ldots,$ is a Cauchy sequence. We compute an upper bound for $\text{d}(\Phi_{j+1}(x), \Phi_j(x)) = \text{d}(\varphi_{j+1}(x'), x')$, where $x' = \Phi_j(x)$. From conditions (3) and (7) of the construction, we have that $\varphi_{j+1} | \sigma'_j$ is in $\mathsf{HOMEO}(\sigma'_j; \partial\sigma'_j \text{ fixed})$ whenever σ'_j is in $\mathcal{P}'_j$, and that $\mathcal{P}_{j+1}$ refines $\mathcal{P}'_j$. There is a σ_{j+1} in $\mathcal{P}_{j+1}$ and there is a σ'_j in $\mathcal{P}'_j$ such that

$\varphi_{j+1}(x') \in \sigma_{j+1} \subset \sigma_j' = \varphi_{j+1}{}^{-1}[\sigma_j']$, whence x' is also in σ_j'; that is,

$$\text{for each } x \text{ in } I^n,\ \{x, \varphi_{j+1}(x)\} \subset \sigma_j' \text{ for some } \sigma_j' \text{ in } \mathcal{P}_j'. \tag{3.6}$$

So $\mathrm{d}(\varphi_{j+1}(x'), x') \leq \mathrm{diam}(\sigma_j') \leq \mathrm{mesh}(\mathcal{P}_j') < 2^{-j}$; consequently we have $\| \Phi_{j+1}\Phi_j{}^{-1} \| < 2^{-j}$. Next, consider

$$\mathrm{d}(\Phi_{j+1}{}^{-1}(x), \Phi_j{}^{-1}(x)) = \mathrm{d}(\Phi_j{}^{-1}(\varphi_{j+1}{}^{-1}(x)), \Phi_j{}^{-1}(x)).$$

The statement displayed in (3.6) above and conditions (3) and (5) of the construction yield $\mathrm{d}(\Phi_{j+1}{}^{-1}(x), \Phi_j{}^{-1}(x)) \leq \mathrm{mesh}(\Phi_j{}^{-1}[\mathcal{P}_j']) < 2^{-j}$; thereby we have $\| \Phi_{j+1}{}^{-1}\Phi_j \| < 2^{-j}$. So $\rho(\Phi_{j+1}, \Phi_j) < 2^{-j+1}$ for every j. Denote by Φ the limit of this Cauchy sequence. An analogous computation will show that $\Psi_j, j = 1, 2, \ldots,$ is a Cauchy sequence. We denote its limit by Ψ.

Let us show $\Phi_{\#}\mu = \Psi_{\#}\lambda$. To this end let $k < j$ and let $\sigma_k \in \mathcal{P}_k$. From conditions (7) and (8) of the construction we have

$$\Phi_{j\#}\mu(\sigma_k) = \Phi_{k\#}\mu(\sigma_k) = \Psi_{(k-1)\#}\lambda(\sigma_k) = \Psi_{j\#}\lambda(\sigma_k),$$
$$\Phi_{j\#}\mu(\partial\sigma_k) = \Phi_{k\#}\mu(\partial\sigma_k) = \Psi_{(k-1)\#}\lambda(\partial\sigma_k) = \Psi_{j\#}\lambda(\partial\sigma_k) = 0,$$

and hence

$$\Phi_{\#}\mu(E) = \Psi_{\#}\lambda(E) \quad \text{whenever} \quad \mathrm{Int}(\sigma_k) \subset E \subset \sigma_k.$$

Let $\mathcal{F} = \bigcup_{j=1}^{\infty} \mathcal{P}_j$. From conditions (7) and (8) we infer $\Phi_{\#}\mu(E) = \Psi_{\#}\lambda(E)$ whenever E is the union of a countable subset of $\mathcal{F}$ because the equality holds for the union of finite subsets of $\mathcal{F}$. Observe that for each x and each open set V with $x \in V$ there is a σ in $\mathcal{F}$ such that $x \in \sigma \subset V$. So each open set U is the union of a countable subset of $\mathcal{F}$, whence $\Phi_{\#}\mu(U) = \Psi_{\#}\lambda(U)$. Thereby $\Phi_{\#}\mu = \Psi_{\#}\lambda$ follows.

As $\mu = (\Phi^{-1}\Psi)_{\#}\lambda$ and $\Phi^{-1}\Psi \in \mathsf{HOMEO}(I^n; \partial I^n \text{ fixed})$, the proof of the Oxtoby–Ulam theorem is completed. □

Of course there is the following topological n-cell version of the Oxtoby–Ulam theorem whose proof is left to the reader.

THEOREM 3.31. *Let X be a topological n-cell. If μ and ν are Lebesgue-like measures on X with $\mu(X) = \nu(X)$, then there is an h in* $\mathsf{HOMEO}(X; \partial X \text{ fixed})$ *such that $\mu = h_{\#}\nu$. Also, if ν is Lebesgue-like on X and there is an h in* $\mathsf{HOMEO}(X)$ *such that $\mu = h_{\#}\nu$, then μ is Lebesgue-like on X.*

Other consequences of the Oxtoby–Ulam theorem are the following three lemmas.

LEMMA 3.32. *Let X be a topological n-cell and μ be a Lebesgue-like measure on X. If F is a nowhere dense closed subset of X and δ is a positive number, then there is an h in* $\mathsf{HOMEO}(X; \partial X \text{ fixed})$ *such that $h_{\#}\mu(F) = 0$ and $\rho(h, \mathrm{id}) < \delta$.*

PROOF. There is no loss in assuming X is $[0,1]^n$. Let $\mathcal{P}$ be a rectangular subdivision of X with $\mu(\partial\sigma) = 0$ for each σ in $\mathcal{P}$. Let $X_0 = \bigcup_{\sigma\in\mathcal{P}} \partial\sigma$ and $\nu = \sum_{\sigma\in\mathcal{P}} \frac{\mu(\sigma)}{\mu(\sigma\setminus F)}\, \mu \llcorner (\sigma \setminus F)$. Obviously $\nu(F) = 0$. Observe that $\nu|\sigma$ and $\mu|\sigma$ are

Lebesgue-like measures on σ such that $\nu(\sigma) = \mu(\sigma)$ for every σ in $\mathcal{P}$. Applying the above theorem to $\mu|\sigma$ and $\nu|\sigma$ for each σ in $\mathcal{P}$, we infer from Proposition 3.9 that there is an h in $\mathsf{HOMEO}(X; X_0 \text{ fixed})$ such that $\nu = h_\# \mu$ and $\rho(h, \mathrm{id}) \leq \mathrm{mesh}(\mathcal{P})$. Such a subdivision $\mathcal{P}$ with $\mathrm{mesh}(\mathcal{P}) < \delta$ clearly exists. □

LEMMA 3.33. *With $I = [0, 1]^n$ let μ be a positive, continuous, complete, finite Borel measure on I. Then a subset E of $I \setminus \partial I$ is an absolute measurable space if and only if $h^{-1}[E] \cap \partial I = \emptyset$ and $h^{-1}[E] \in \mathfrak{M}(I, \mu)$ whenever $h \in \mathsf{HOMEO}(I; \partial I \text{ fixed})$.*

PROOF. Suppose that E is an absolute measurable space contained in $I \setminus \partial I$. Then $h^{-1}[E]$ is an absolute measurable space whenever h is in $\mathsf{HOMEO}(I)$. Hence $h^{-1}[E] \cap \partial I = \emptyset$ and $h^{-1}[E] \in \mathfrak{M}(I, \mu)$ whenever $h \in \mathsf{HOMEO}(I; \partial I \text{ fixed})$.

To prove the converse, let E be such that $h^{-1}[E]$ satisfies the conditions of the lemma. Let ν be a positive, continuous, complete, finite Borel measure on I. Then $\nu' = \nu \llcorner (I \setminus \partial I)$ and $\mu' = \mu \llcorner (I \setminus \partial I)$ are Lebesgue-like measures on I. By the Oxtoby–Ulam theorem there is a positive number k and an h in $\mathsf{HOMEO}(I; \partial I \text{ fixed})$ such that $k\,\nu' = h_\# \mu'$. As $h^{-1}[E] \in \mathfrak{M}(I, \mu)$, we have that E is $(h_\# \mu)$-measurable, whence $E \in \mathfrak{M}(I, \nu')$. Since $E \subset I \setminus \partial I$, we have $E \in \mathfrak{M}(I, \nu)$ and thereby $E \in \mathsf{univ}\,\mathfrak{M}(I)$. Since I is an absolute measurable space, we have that E is also an absolute measurable space. □

LEMMA 3.34. *For $n > 1$, let $\mathbb{B}_n$ be the unit ball $\{x \in \mathbb{R}^n : \|x\|_2 \leq 1\}$ and let $\mathbb{S}_{n-1}$ be its surface $\{x \in \mathbb{R}^n : \|x\|_2 = 1\}$. Denote by F that part of $\mathbb{S}_{n-1}$ defined by $\{x \in \mathbb{S}_{n-1} : x_1 \leq 0\}$. Let μ be a positive, continuous, complete, finite Borel measure on $\mathbb{B}_n$ such that $\mu|\mathbb{S}_{n-1}$ is a positive, continuous, complete, finite Borel measure on $\mathbb{S}_{n-1}$. Then a subset E of $\mathbb{B}_n \setminus F$ is an absolute measurable space if and only if $h^{-1}[E] \cap F = \emptyset$ and $h^{-1}[E] \in \mathfrak{M}(\mathbb{B}_n, \mu)$ whenever h is in $\mathsf{HOMEO}(\mathbb{B}_n; F \text{ fixed})$.*

PROOF. Let E be an absolute measurable space contained in $\mathbb{B}_n \setminus F$. As $h^{-1}[E]$ is an absolute measurable space whenever $h \in \mathsf{HOMEO}(\mathbb{B}_n)$, we have $h^{-1}[E] \cap F = \emptyset$ and $h^{-1}[E] \in \mathfrak{M}(\mathbb{B}_n, \mu)$ whenever h is in $\mathsf{HOMEO}(\mathbb{B}_n; F \text{ fixed})$.

For the converse, let us consider the two cases: $E \subset \mathbb{S}_{n-1} \setminus F$ and $E \cap \mathbb{S}_{n-1} = \emptyset$. In the first case, let E be a subset of $\mathbb{S}_{n-1} \setminus F$ such that $h^{-1}[E] \cap F = \emptyset$ and $h^{-1}[E] \in \mathfrak{M}(\mathbb{B}_n, \mu)$ whenever h is in $\mathsf{HOMEO}(\mathbb{B}_n; F \text{ fixed})$. Since $\mathrm{Cl}_{\mathbb{B}_n}\big(\mathbb{S}_{n-1} \setminus F\big) = \{x \in \mathbb{S}_{n-1} : x_1 \geq 0\}$ we see that $I = \mathrm{Cl}_{\mathbb{B}_n}\big(\mathbb{S}_{n-1} \setminus F\big)$ is an $(n-1)$-cell and that $\partial I = \{x \in \mathbb{S}_{n-1} : x_1 = 0\}$. Clearly, every g in $\mathsf{HOMEO}(I; \partial I \text{ fixed})$ has an extension h in $\mathsf{HOMEO}(\mathbb{B}_n; F \text{ fixed})$. For such an extension we have $g^{-1}[E] = h^{-1}[E]$. As $\mathfrak{M}(I, \mu|I) \subset \mathfrak{M}(\mathbb{B}_n, \mu)$, the preceding lemma shows that E is an absolute measurable space that is contained in $\mathbb{B}_n \setminus F$, and the first case is shown. In the second case, let E be a subset of $\mathbb{B}_n \setminus \mathbb{S}_{n-1}$ such that $h^{-1}[E] \cap F = \emptyset$ and $h^{-1}[E] \in \mathfrak{M}(\mathbb{B}_n, \mu)$ whenever h is in $\mathsf{HOMEO}(\mathbb{B}_n; F \text{ fixed})$. As $\mathsf{HOMEO}(\mathbb{B}_n; \mathbb{S}_{n-1} \text{ fixed}) \subset \mathsf{HOMEO}(\mathbb{B}_n; F \text{ fixed})$, we have that $h^{-1}[E]$ is in $\mathfrak{M}(\mathbb{B}_n, \mu)$ whenever h fixes $\mathbb{S}_{n-1}$. Hence E is an absolute measurable space that is contained in $\mathbb{B}_n \setminus \mathbb{S}_{n-1}$.

Now let E be such that $h^{-1}[E] \subset \mathbb{B}_n \setminus F$ and $h^{-1}[E] \in \mathfrak{M}(\mathbb{B}_n, \mu)$ whenever $h \in \mathsf{HOMEO}(\mathbb{B}_n; F \text{ fixed})$. Then $E \cap \mathbb{S}_{n-1}$ and $E \setminus \mathbb{S}_n$ are absolute measurable spaces, whence E is an absolute measurable space contained in $\mathbb{B}_n \setminus F$ and thereby the converse is proved. □

PROOF OF THEOREM 3.25. The proof is by induction on the dimension of the n-dimensional finitely triangulable space (such a space is homeomorphic to a subspace of an N-simplex[4] Δ_N). By definition, $|K_n|$ is the union of a finite collection $K_n = \{S_m{}^j : j = 1, 2, \ldots, j_m, m \leq n\}$, where S_m are m-simplices. The subcollection of all m-simplices in K_n whose dimensions do not exceed k is denoted by K_n^k and is called the k-dimensional subcomplex of K_n. So K_n^0 is the collection of all vertices of K_n.

We give the proof for $n = 2$; the proof of the inductive step is easily modeled after this proof. Let us apply Theorem 3.12. Define $X = |K_2|$, $X_0 = |K_2^0|$, $X_1 = |K_2^1|$, and $X_{2j} = S_2^j \setminus \partial S_2^j$ for $j = 1, 2, \ldots, j_2$. Then, by Theorem 3.22 applied to X_1, there is a positive, continuous, complete, finite Borel measure ν_1 on X_1 such that ν_1 and $\mathsf{HOMEO}(X_1; X_0 \text{ fixed})$ generate $\mathsf{univ}\,\mathfrak{M}(X_1)$. Observe that $\varphi'[\partial S_2] = \partial S_2$ for every φ' in $\mathsf{HOMEO}_1(X_1; X_0 \text{ fixed})$ whenever S_2 is a 2-simplex in K_2. As S_2 is convex, each φ' in $\mathsf{HOMEO}(X_1; X_0 \text{ fixed})$ can be extended to a homeomorphism in $\mathsf{HOMEO}(X; X_0 \text{ fixed})$, call it φ. Hence we have

$$\begin{aligned}\mathcal{G}_1(X) &= \{\varphi|X_1 : \varphi \in \mathsf{HOMEO}(X; X_0 \text{ fixed}), \varphi[X_1] = X_1\}\\ &= \mathsf{HOMEO}(X_1; X_0 \text{ fixed}).\end{aligned}$$

As $S_2^j \setminus \partial S_2^j$ was defined to be X_{2j} and each φ' in $\mathsf{HOMEO}(S_2^j; \partial S_2^j \text{ fixed})$ has an extension φ in $\mathsf{HOMEO}(X; X_0 \text{ fixed})$, we also have, for $j = 1, 2, \ldots, j_2$,

$$\begin{aligned}\mathcal{G}_{2j}(X) &= \{\varphi|X_{2j} : \varphi \in \mathsf{HOMEO}(X; X_0 \text{ fixed}), \varphi[X_{2j}] = X_{2j}\}\\ &\supset \mathcal{G}'_{2j}(X) = \{\varphi'|X_{2j} : \varphi' \in \mathsf{HOMEO}(S_2^j; \partial S_2^j \text{ fixed})\}.\end{aligned}$$

By Lemma 3.33 and the topological n-cell version of the Oxtoby–Ulam theorem there is a positive, continuous, complete, finite Borel measure ν_{2j} on X_{2j} such that ν_{2j} and $\mathcal{G}'_{2j}(X)$ generate $\mathsf{univ}\,\mathfrak{M}(X_{2j})$. Let $\mu = g_{1\#}\nu_1 + \sum_{j=1}^{j_2} g_{2j\#}\nu_{2j}$, where g_1 and g_{2j} are the obvious inclusion maps. Then, by Proposition 3.14 and Theorem 3.12, we have μ and $\mathsf{HOMEO}(X; X_0 \text{ fixed})$ generate $\mathsf{univ}\,\mathfrak{M}(X)$. We leave the proof of the inductive step to the reader. □

[4] By an *N-simplex* Δ_N we mean the convex hull (in the Euclidean vector space $\mathbb{R}^{N+1}$) of the $N+1$ points v_i whose $(i+1)$-coordinate is 1 and the remaining coordinates are all 0, $i = 0, 1, \ldots, N$. The points v_0, v_1, ..., v_N are called the *vertices of* Δ_N. A *q-face of* Δ_N is the convex hull of $q+1$ distinct vertices of Δ_N. Also, a face will be referred to as a simplex. A *simplicial complex* K is a collection of faces of Δ_N satisfying the condition that every face of a simplex in the collection is likewise in the collection. A subcollection L of a complex K is called a *subcomplex of* K whenever L is also a complex. The *space* $|K|$ of K is the subset of Δ_N consisting of those points which belong to simplexes of K. A simplicial complex K is said to be *n-dimensional* provided K contains an n-simplex but no $(n+1)$-simplex. These definitions are modeled after those in [**49**, pages 54–60]. Note that the simplexes in these definitions are compact; in definitions found in other books, say [**79**, pages 67–68], q-faces S_q do not contain the points in ∂S_q for $q > 0$, that is, they are 'open' faces. A triangulable space X is one that is homeomorphic to $|K|$ for some simplicial complex K; K is called a *triangulation of* X.

A sharper result holds for the n-sphere $\mathbb{S}_n = \{x \in \mathbb{R}^{n+1} : \|x\| = 1\}$. The methods of proofs for the 1-sphere used in Lemma 3.19 and in the succeeding paragraph can be used here also.

LEMMA 3.35. *$\mathsf{H}_n|\mathbb{S}_n$ and the group $\mathsf{HOMEO}(\mathbb{S}_n)$, where H_n is the Hausdorff n-dimensional measure on $\mathbb{R}^{n+1}$, generate $\mathrm{MEAS}^{\mathrm{pos,fin}}(\mathbb{S}_n)$, the collection of all positive, continuous, complete, finite Borel measures on $\mathbb{S}_n$. Hence*

$$\mathsf{univ}\,\mathfrak{M}(\mathbb{S}_n) = \bigcap\big\{\mathfrak{M}(\mathbb{S}_n, h_\#(\mathsf{H}_n|\mathbb{S}_n)) : h \in \mathsf{HOMEO}(\mathbb{S}_n)\big\}.$$

3.5. *n*-dimensional manifolds

We have seen in the last section that the manifold $\mathbb{S}_n$ has a measure μ such that it and $\mathsf{HOMEO}(\mathbb{S}_n)$ generate $\mathsf{univ}\,\mathfrak{M}(\mathbb{S}_n)$. We will show that this holds for all separable manifolds. A manifold is defined by local conditions. That is, a separable metrizable space M_n is an *n-dimensional manifold* if each point has a neighborhood that is homeomorphic to $\mathbb{R}^n$ or $[0,\infty)\times\mathbb{R}^{n-1}$. Points whose neighborhoods are of the second kind are called boundary points of the manifold. $\partial(M_n)$ denotes the set of boundary points. Let us show that $\mathsf{univ}\,\mathfrak{M}(X)$ is also characterized by local conditions.

PROPOSITION 3.36. *Let $\mathcal{U}$ be an open cover of a separable metrizable space X. Then $E \in \mathsf{univ}\,\mathfrak{M}(X)$ if and only if $E \cap U \in \mathsf{univ}\,\mathfrak{M}(X)$ whenever $U \in \mathcal{U}$.*

PROOF. As every open set is a universally measurable set in X we have that $E\cap U$ is also universally measurable whenever E is a universally measurable set in X. Conversely, suppose that $E \cap U$ is universally measurable set in X for every U in $\mathcal{U}$. As X is a Lindeloff space, the cover $\mathcal{U}$ has a countable subcover, whence E is universally measurable. □

If a point x of an n-manifold M_n has a neighborhood that is homeomorphic to $\mathbb{R}^n$, then there is an embedding $\varphi : [-1,1]^n \to M_n$ such that $\varphi(0) = x$, where 0 is the origin of $\mathbb{R}^n$, and such that $\varphi\big((-1,1)^n\big)$ is a neighborhood of x.

If x is in ∂M_n, then there is an embedding φ of $[0,1]\times[-1,1]^{n-1}$ into M_n such that $\varphi\big((0,0)\big) = x$, where $(0,0)$ is the origin of $\mathbb{R}\times\mathbb{R}^{n-1}$, and such that $\varphi\big[[0,1)\times(-1,1)^{n-1}\big]$ is a neighborhood of x.

THEOREM 3.37. *If M_n is an n-dimensional manifold, connected or not and with or without boundary, then there is a positive, continuous, complete, finite Borel measure μ on M_n such that μ and $\mathsf{HOMEO}(M_n)$ generate $\mathsf{univ}\,\mathfrak{M}(M_n)$.*

PROOF. Observe first that $\partial\partial M_n = \emptyset$. Let μ be such that μ and $\mu|\partial M_n$ are positive, continuous, complete, finite Borel measures on M_n and ∂M_n, respectively, and let $\mathcal{U}$ be the open cover

$$\{\mathrm{Int}_{M_n}(B) : B \text{ is an } n\text{-cell contained in } M_n\}.$$

By Proposition 3.36, a subset E of M_n is in $\mathsf{univ}\,\mathfrak{M}(M_n)$ if and only if $E \cap U$ is in $\mathsf{univ}\,\mathfrak{M}(M_n)$ for each U in $\mathcal{U}$. As M_n is an absolute measurable space, the last condition will be satisfied if and only if $E \cap U$ is an absolute measurable space for each U in $\mathcal{U}$. Lemmas 3.33 and 3.34 complete the proof. □

The proof given here is rather simple in that it uses only the definition of a manifold. In fact, the manifold need not be connected nor compact. We shall see shortly that the Oxtoby–Ulam theorem can be extended to compact, connected n-dimensional manifolds. Hence the last theorem can be proved in the compact case with the aid of this extension. This extension uses the next very nice topological theorem due to M. Brown [**21**] whose proof will not be provided here.

THEOREM 3.38 (Brown). *For $n > 1$ let $I = [0,1]^n$ and $U = I \setminus \partial I$. For each compact, connected, n-dimensional manifold M_n there is a continuous surjection $\varphi\colon I \to M_n$ such that $\varphi|U\colon U \to \varphi[U]$ is a homeomorphism and $\varphi[\partial I]$ is a nowhere dense subset of M_n that is disjoint from $\varphi[U]$.*

It is easily shown that the map given by Brown's theorem has the properties that $\varphi[\partial I] \supset \partial M_n$ and $\varphi^{-1}\varphi[\partial I] = \partial I$. But, in order to apply the Oxtoby–Ulam theorem to I, it will be necessary to have the added measure theoretic condition $\nu(\varphi[\partial I]) = 0$. Whenever this condition is satisfied we can use the homeomorphism $\psi = \varphi^{-1}|\varphi[U]$ to define the measure $\mu_1 = \psi_\# \nu_1$ where $\nu_1 = \nu|\varphi[U]$. The inclusion map $g\colon U \to I$ will give us the following implications:

$$\begin{pmatrix} \mathrm{M}(\varphi[U],\nu_1) \\ \nu_1 = \nu|\varphi[U] \end{pmatrix} \underset{\psi}{\Longrightarrow} \begin{pmatrix} \mathrm{M}(U,\mu_1) \\ \mu_1 = \psi_\#\nu_1 \end{pmatrix}$$
$$\underset{g}{\Longrightarrow} \begin{pmatrix} \mathrm{M}(I,\mu) \\ \mu = g_\#\mu_1 \end{pmatrix} \underset{\varphi}{\Longrightarrow} \begin{pmatrix} \mathrm{M}(M_n,\nu_0) \\ \nu_0 = \varphi_\#\mu \end{pmatrix} \tag{3.7}$$

where $\mu(\partial I) = 0$. Now $\nu_0|\varphi[U] = \nu|\varphi[U]$, whence $\nu_0 = \nu$ whenever $\nu(\varphi[\partial I]) = 0$. The next theorem shows that there is a map that satisfies the conditions of the Brown theorem and also the added measure theoretic condition.

THEOREM 3.39. *For $n > 1$ let $I = [0,1]^n$ and $U = I \setminus \partial I$. For each compact, connected, n-dimensional manifold M_n and for each positive, continuous, complete, finite Borel measure ν on M_n with $\nu(\partial M_n) = 0$ there is a continuous surjection $\varphi\colon I \to M_n$ such that $\varphi|U\colon U \to \varphi[U]$ is a homeomorphism, $\varphi[\partial I]$ is a nowhere dense subset of M_n that is disjoint from $\varphi[U]$, and $\nu(\varphi[\partial I]) = 0$.*

PROOF. Let B_i, $i = 1,2,\ldots,k$, be n-cells contained in M_n such that $\mathcal{U} = \{U_i = \mathrm{Int}_{M_n}(B_i)\colon i = 1,2,\ldots,k\}$ is an open cover of M_n. Let $\psi\colon I \to M_n$ be as provided by the Brown theorem above and consider the measure

$$\nu_1 = \nu \llcorner (M_n \setminus U_1) + a_1\, \nu \llcorner (U_1 \setminus \psi[\partial I]), \text{ where } a_1 = \tfrac{\nu(U_1)}{\nu(U_1 \setminus \psi[\partial I])},$$

which satisfies $\nu(M_n) = \nu_1(M_n)$. Clearly ν_1 is a continuous positive Borel measure since $B_1 \cap \psi[\partial I]$ is a nowhere dense subset of B_1. Observe that $(\nu_1 \llcorner U_1)|B_1$ and $(\nu \llcorner U_1)|B_1$ are Lebesgue-like on B_1 and that $\nu_1(U_1) = \nu(U_1)$. We infer from

Lemma 3.32 that there is an h_1 in $\mathsf{HOMEO}(M_n; \partial M_n \text{ fixed})$ such that $h_{1\#}\nu = \nu_1$. Let us proceed in the same manner for U_2 and ν_1. Define

$$\nu_2 = \nu_1 \llcorner (M_n \setminus U_2) + a_2\, \nu_1 \llcorner (U_2 \setminus \psi[\partial I]), \text{ where } a_2 = \frac{\nu_1(U_2)}{\nu_1(U_2 \setminus \psi[\partial I])},$$

which satisfies $\nu_1(M_n) = \nu_2(M_n)$ and $\nu_2\big((U_1 \cup U_2) \cap \psi[\partial I]\big) = 0$. Let h_2 be in $\mathsf{HOMEO}(M_n; \partial M_n \text{ fixed})$ such that $(h_2 h_1)_\# \nu = \nu_2$. After finitely many steps we have the homeomorphism $h = h_k h_{k-1} \cdots h_2 h_1$ such that h is in $\mathsf{HOMEO}(M_n; \partial M_n \text{ fixed})$, $h_\#\nu$ is a positive, continuous, complete, finite Borel measure on M_n, and $h_\#\nu(\psi[\partial I]) = 0$. Let $\varphi = h^{-1}\psi$. Then $\varphi\colon I \to M_n$ is a continuous surjection such that $\varphi|U\colon U \to \varphi[U]$ is a homeomorphism, $\varphi[\partial I]$ is nowhere dense, $\varphi[U] \cap \varphi[\partial I] = \emptyset$, and $\nu(\varphi[\partial I]) = 0$. Observe that $\rho(h, \mathrm{id}) \le \sum_{i=1}^{k} \rho(h_i, \mathrm{id})$. □

Now we can state and prove the extension of the Oxtoby–Ulam theorem to compact manifolds (S. Alpern and V. S. Prasad [**5**, page 195]).

THEOREM 3.40 (Alpern–Prasad). *Let λ be a positive, continuous, complete, finite Borel measure on a compact, connected, n-dimensional manifold M_n such that $\lambda(\partial M_n) = 0$. In order that a Borel measure μ on M_n be such that there is an h in $\mathsf{HOMEO}(M_n)$ with $\mu = h_\#\lambda$ it is necessary and sufficient that*

(1) *μ be a positive, continuous, complete Borel measure on M_n,*
(2) *$\mu(M_n) = \lambda(M_n)$,*
(3) *$\mu(\partial M_n) = 0$.*

Moreover, the homeomorphism h may have the property that it leaves ∂M_n fixed.

PROOF. If $\mu = h_\#\lambda$ for some h in $\mathsf{HOMEO}(M_n)$, then it is easily seen that μ satisfies the conditions that are enumerated.

Suppose that μ satisfies the conditions and let $\varphi\colon I \to M_n$ be as in Theorem 3.39 for the measure $\nu = \lambda + \mu$. From the implications (3.7) that precede this cited theorem we infer that there are positive, continuous, complete, finite Borel measures λ_0 and μ_0 on I such that $\varphi_\#\lambda_0 = \lambda$ and $\varphi_\#\mu_0 = \mu$ and such that λ_0 and μ_0 are Lebesgue like on I. There is an h in $\mathsf{HOMEO}(I; \partial I \text{ fixed})$ such that $h_\#\lambda_0 = \mu_0$. As h fixes ∂I and $\varphi^{-1}\varphi[\partial I] = \partial I$, there is a well defined map $H\colon M_n \to M_n$ such that the following diagram

$$\begin{array}{ccc} I & \xrightarrow{\varphi} & M_n \\ {\scriptstyle h}\uparrow & & \uparrow{\scriptstyle H} \\ I & \xrightarrow{\varphi} & M_n \end{array}$$

is commutative; indeed, if $p \in M_n$, then $\mathrm{card}(H\big[\{p\}\big]) = 1$ whence H is a map. Clearly, H is bijective. Moreover, H is continuous since each closed subset F of M_n satisfies $H^{-1}[F] = \varphi h^{-1} \varphi^{-1}[F] = F$. The diagram also yields the fact that H fixes $\varphi[\partial I]$, whence $H \in \mathsf{HOMEO}(M_n; \partial M_n \text{ fixed})$. Finally $H_\#\lambda = \mu$ is easily verified. □

The last theorem would have been better stated if we had used the definition of Lebesgue-like measures on manifolds given earlier in the discussion of the Oxtoby–Ulam theorem. That is, a measure μ is said to be *Lebesgue-like on a n-dimensional manifold* M_n if μ is a positive, continuous, complete, finite Borel measure on M_n with $\mu(\partial M_n) = 0$.

Let us turn to a nice application of the extension of Brown's theorem, Theorem 3.39. Suppose that F is a closed, nowhere dense subset of a compact n-dimensional manifold M_n and μ is a Lebesgue-like measure on M_n. Then for each positive number ε the collection

$$\mathcal{H}(F, \mu, \varepsilon) = \{h \in \mathsf{HOMEO}(M_n; \partial M_n \text{ fixed}) \colon h_{\#}\mu(F) < \varepsilon\}$$

is a dense open subset of the metric space $\mathsf{HOMEO}(M_n; \partial M_n \text{ fixed})$. That $\mathcal{H}(F, \mu, \varepsilon)$ is open follows from the upper semi-continuity of the real-valued function $h \mapsto h_{\#}\mu(F)$. Let us prove that it is dense in $\mathsf{HOMEO}(M_n; \partial M_n \text{ fixed})$. Let $f \in \mathsf{HOMEO}(M_n; \partial M_n \text{ fixed})$ and $\delta > 0$. We seek an h with $hf \in \mathcal{H}(F, \mu, \varepsilon)$ and $\rho(h, \mathrm{id}) < \delta$. By Theorem 3.39 there is a continuous surjection $\varphi \colon I \to M_n$, where I is a topological n-cell, such that $\varphi|(I \setminus \partial I)$ is a homeomorphism, $\varphi[I \setminus \partial I] \cap \varphi[\partial I] = \emptyset$, $\varphi[\partial I]$ is nowhere dense, and $f_{\#}\mu(\varphi[\partial I]) = 0$. Using the properties of the map φ, we have a topological n-cell J contained in $\varphi[I \setminus \partial I]$ such that $f_{\#}\mu(\partial J) = 0$ and $f_{\#}\mu(M_n \setminus J) < \varepsilon$. By Lemma 3.32 there is an h' in $\mathsf{HOMEO}(J; \partial J \text{ fixed})$ with $\rho(h', \mathrm{id}_J) < \delta$ and $h'_{\#}\big((f_{\#}\mu)|J\big)(F \cap J) = 0$. Let h be the extension of h' such that $h|(M_n \setminus J)$ is the identity map on $M_n \setminus J$. It is clear that $h \in \mathsf{HOMEO}(M_n; \partial M_n \text{ fixed})$, $\rho(h, \mathrm{id}) < \delta$, $(h_{\#}f_{\#}\mu)|(M_n \setminus J) = (f_{\#}\mu)|(M_n \setminus J)$, and $(h_{\#}f_{\#}\mu)|J = h'_{\#}\big((f_{\#}\mu)|J\big)$. A simple computation yields $(hf)_{\#}\mu(F) < \varepsilon$.

We now have

THEOREM 3.41. *If M_n is a compact, connected, n-dimensional manifold and μ is a positive, continuous, complete, finite Borel measure on M_n with $\mu(\partial M_n) = 0$ (that is, Lebesgue-like), then the set*

$$\{h \in \mathsf{HOMEO}(M_n; \partial M_n \text{ fixed}) \colon h_{\#}\mu(F) = 0\}$$

is a dense G_δ subset of $\mathsf{HOMEO}(M_n; \partial M_n \text{ fixed})$ *whenever F is a closed, nowhere dense subset of M_n.*

See Exercise 3.12 on page 97 for a noncompact manifold setting.

3.6. The Hilbert cube

An obvious question that results from the finite dimensional considerations of the previous section is: What can be said about the Hilbert cube $[0, 1]^{\mathbb{N}}$? The answer is easily seen due to the extension of the Oxtoby–Ulam theorem to the Hilbert cube which was proved by Oxtoby and Prasad in [**121**]. We shall denote $[0, 1]^{\mathbb{N}}$ by Q. They have proved a stronger version of the following theorem.

THEOREM 3.42 (Oxtoby–Prasad). *Let μ and ν be positive, continuous, complete Borel probability measures on Q. Then there exists an h in* $\mathsf{HOMEO}(Q)$ *such that* $h_{\#}\mu = \nu$.

We shall give their proof of the stronger version shortly. But first let us state the theorem that connects the σ-algebra $\mathsf{univ}\,\mathfrak{M}(Q)$ and the group $\mathsf{HOMEO}(Q)$.

THEOREM 3.43. *Each positive, continuous, complete, finite Borel measure μ on Q and the group* $\mathsf{HOMEO}(Q)$ *generate* $\mathrm{MEAS}^{\mathrm{pos,fin}}(Q)$ *(the collection of all positive, continuous, complete, finite Borel measures on Q) and hence they also generate* $\mathsf{univ}\,\mathfrak{M}(Q)$.

The proof follows immediately from the preceding theorem.

3.6.1. Definitions and notations. Several definitions and notations are used in the proof. We shall collect them in this subsection.

It will be convenient to select a suitable metric d on $Q = [0, 1]^{\mathbb{N}}$, namely,

$$\mathrm{d}(x, y) = \textstyle\sum_{i=1}^{\infty} |x_i - y_i|/2^i.$$

By a *rectangular set* we mean a set $R = \mathsf{X}_{i=1}^{\infty}[a_i, b_i]$ where $0 \leq a_i < b_i \leq 1$ for all i, and $[a_i, b_i] = [0, 1]$ for all but finitely many values of i. Clearly, a rectangular set is the closure of a basic open set of Q.

We shall have need to refer to certain subsets of a rectangular set R. For a fixed i, the set $\{x \in R\colon x_i = c\}$ with $a_i < c < b_i$ is called a *section of* R, and the sets $\{x \in R\colon x_i = a_i\}$ and $\{x \in R\colon x_i = b_i\}$ are called *faces of* R. Each rectangular set R has countably many faces, the union of which is a dense subset of R. Indeed, for $x \in R$ and $\varepsilon > 0$, let n be such that

$$x \in \mathsf{X}_{i=1}^{n}[a'_i, b'_i] \times \mathsf{X}_{i=n+1}^{\infty}[0, 1] \subset \{y \in Q\colon \mathrm{d}(x, y) < \varepsilon\}.$$

The face of R given by $\{y \in R\colon y_{n+1} = 0\}$ contains the point $z = (x_1, x_2, \dots, x_n, 0, 0, \dots)$ which is in both R and the ε-neighborhood of x. The *pseudo-boundary of* R, denoted by δR, is the union of all the faces of R. Of course, δR is not the same as the topological boundary $\mathrm{Bd}_Q(R)$ which is the union of finitely many faces of R, namely, the faces $\{x \in R\colon x_i = a_i\}$ with $0 < a_i$ and the faces $\{x \in R\colon x_i = b_i\}$ with $b_i < 1$. The *pseudo-interior of* Q is the set $Q \setminus \delta Q$.

We shall use several subgroups of $\mathsf{HOMEO}(Q)$. The first one is the closed subgroup that leaves every face of Q invariant, that is,

$$\mathsf{HOMEO}_0(Q) = \bigcap\{\mathsf{HOMEO}(Q; W \text{ inv})\colon W \text{ is a face of } Q\}.$$

Given a subset B of Q we have the closed subgroup

$$\mathsf{HOMEO}_0(Q; B \text{ fixed}) = \mathsf{HOMEO}(Q; B \text{ fixed}) \cap \mathsf{HOMEO}_0(Q).$$

3.6.2. Main lemma. The main lemma reduces the general positive, continuous, complete, finite Borel measure case to those that vanish on the set δQ. For the finite-dimensional space $[0,1]^n$, where $n > 1$, there is no such analogue, hence one must necessarily assume that the measures vanish on $\partial([0,1]^n)$ in the Oxtoby–Ulam theorem.

LEMMA 3.44 (Main lemma). *If μ is a positive, continuous, complete, finite Borel measure on Q and B is the union of finitely many faces of Q with $\mu(B) = 0$, then $h_{\#}\mu(\delta Q) = 0$ for all h in a some dense G_δ subset of* HOMEO(Q; B fixed).

PROOF. Let W be a face of Q that is not contained in B. There is no loss in assuming $W = \{x \in Q \colon x_k = 0\}$. For each positive integer j, we assert that

$$E_j(W) = \{f \in \mathsf{HOMEO}(Q; B \text{ fixed}) \colon f_{\#}\mu(W) < 1/j\}$$

is open and dense in HOMEO(Q; B fixed).

The upper semi-continuity of the function $f \longmapsto f_{\#}\mu(W)$ implies that $E_j(W)$ is an open subset of HOMEO(Q; B fixed).

To see that $E_j(W)$ is dense we first show that there is an h in $E_j(W)$ with small norm. As this is obvious if $\mu(W) < \frac{1}{j}$ we shall assume $\mu(W) \geq \frac{1}{j}$. Let $0 < \varepsilon < 1$ and fix a δ such that $0 < \delta < \frac{\varepsilon}{2}$ and such that the δ-neighborhood B_δ of B satisfies $\mu(B_\delta) < \frac{1}{2j}$. Choose an n such that $k < n$, $\frac{1}{2^n} < \frac{\varepsilon}{2}$, and the two faces $\{x \in Q \colon x_n = 0\}$ and $\{x \in Q \colon x_n = 1\}$ are not contained in B.

With $J = I_k \times I_n$, let $\varphi \colon Q \to J$ be the natural projection and let $d_J\big((x_k, x_n), (y_k, y_n)\big) = \frac{|x_k - y_k|}{2^k} + \frac{|x_n - y_n|}{2^n}$ be the metric on J. Then $\varphi_{\#}\mu$ is a finite Borel measure on $I_k \times I_n$, though not necessarily continuous. Let p be such that $0 < p < 1$ and $\varphi_{\#}\mu(I_k \times \{p\}) = 0$. Select an open interval I_n' contained in I_n such that $\varphi_{\#}\mu(I_k \times I_n') < \frac{1}{2j}$ and $p \in I_n' \subset I_n$.

Let us assume that there is a continuous family $\{H_t \colon 0 \leq t \leq 1\}$ in HOMEO($I_k \times I_n$; $[\delta, 1] \times I_n$ fixed} such that H_0 is the identity and H_1 maps $\{0\} \times I_n$ into $I_k \times I_n'$. The existence of such a family will be established at the end of the proof. Obviously $\big\{H_t^{-1} \colon 0 \leq t \leq 1\big\}$ is a continuous family. Define the continuous function

$$t(x) = \min\{\operatorname{dist}(x, F), \tfrac{1}{\delta}\}, \quad x \in Q,$$

where F is the union of all faces contained in B that meet W. (Note $\operatorname{dist}(x, \emptyset) = +\infty$.) Define h as follows: for x in Q, $h(x)$ is the point y whose k-th and n-th coordinates are those of $H_{t(x)}(\varphi(x)) = (y_k, y_n)$ and $y_i = x_i$ for the remaining coordinates. Also define g as follows: for x in Q, $g(x)$ is the point y whose k-th and n-th coordinates are those of $H_{t(x)}{}^{-1}(\varphi(x)) = (z_k, z_n)$ and $y_i = x_i$ for the remaining coordinates. Note that $h(x)$ and $g(x)$ have the same i-th coordinates for every i not equal to k and n. Observe that $g(h(x)) = x$ and $h(g(x)) = x$. Hence h and g are homeomorphisms and $h^{-1} = g$.

If $x \in W$, then $\varphi(x) \in \{0\} \times I_n$. Hence $H_t(\varphi(x)) \in \{0\} \times I_n$ whenever t is in $[0,1]$ and x is in W. Consequently, $h(x) \in W$ for every x in W. If W' is the face of Q opposite W and $x \in (W')_\delta$, where $(W')_\delta$ is the δ-neighborhood of W', then $\varphi[W'] = \{1\} \times I_n$,

$t(x) = 1$ and $\mathrm{dist}_J\big(\varphi(x), \varphi[(W')_\delta]\big) < \delta$. Hence $h(x) = x$ whenever $x \in (W')_\delta$. Observe that a face that is contained in B is either W' or contained in F. Since $\varphi[B] \subset \{0,1\} \times I_n$ we have $h \in \mathsf{HOMEO}(Q; B \text{ fixed})$. It is easily shown that each section $\{x : x_i = \text{constant}\}$ of Q is invariant under h whenever i is not k and n.

Let us compute an upper bound for the norm $\|h\|$. Suppose $t(x) < 1$. Then $\mathrm{d}(x, h(x)) = \mathrm{d}(\varphi(x), \varphi(h(x)) \le \delta + \frac{\varepsilon}{2} < \varepsilon$. Suppose $t(x) = 1$. Then $\mathrm{dist}(x, F) \ge \delta$ and hence $\mathrm{d}(h(x), x) = \mathrm{d}_J(\varphi(h(x)), \varphi(x))$. So, $\|h\| \le \varepsilon$.

Since B is a fixed point set of h, it follows from the continuity of h that $h[B_\delta] \subset B_\delta$. The same is true for h^{-1}, hence $h[B_\delta] = B_\delta$.

We have $t(x) = 1$ if and only if $\delta \le \mathrm{dist}(x, F)$. For such an x, $\varphi(h(x)) \in I_k \times I'_n$ whenever $x \in W$. Hence $(\varphi h)^{-1}[I_k \times I'_n] \supset W \setminus B_\delta$. We are ready to estimate $f_\# \mu(W)$ where $f = h^{-1}$.

$$
\begin{aligned}
f_\#\mu(W) &= f_\#\mu(W \cap B_\delta) + f_\#\mu(W \setminus B_\delta) \\
&\le \mu(f^{-1}[B_\delta]) + f_\#\mu((\varphi f^{-1})^{-1}[I_k \times I'_n]) \\
&= \mu(B_\delta) + (\varphi f^{-1})_\# f_\#\mu(I_k \times I'_n) \\
&= \mu(B_\delta) + \varphi_\#\mu(I_k \times I'_n) \\
&< \tfrac{1}{2j} + \tfrac{1}{2j}.
\end{aligned}
$$

We have shown that there are arbitrarily small members of $E(j)$.

Let us show that $E(j)$ is dense in $\mathsf{HOMEO}(Q; B \text{ fixed})$. To this end, let $g \in \mathsf{HOMEO}(Q; B \text{ fixed})$ and consider $g_\#\mu$. As $g_\#\mu(B) = 0$, there is a small f in $\mathsf{HOMEO}(Q; B \text{ fixed})$ such that $f_\# g_\# \mu(W) < \frac{1}{j}$. So $\rho(fg, g)$ is small if f is small enough.

Except for the promised proof of the existence of the continuous family of homeomorphisms, the proof of the main lemma is now at hand. It only remains to intersect the countable collection of open dense sets $E_j(W)$ over all j and the countable collection of all faces W of Q that are not contained in B.

Consider the disk $D = \{(r, \vartheta) : r \in [0,1],\ \vartheta \in [-\pi, \pi)\}$ using polar coordinates. The map that sends (r, ϑ) to $\big(r, \vartheta |\vartheta|^{t/(1-t)}\big)$ for $0 < |\vartheta| \le 1$ and to (r, ϑ) for the remaining ϑ's can be easily transferred to the square $[-1,1] \times [-1,1]$ by radial projection. □

It is easy to see that there are positive, continuous, complete, finite Borel measures μ on Q such that $\mu(\delta Q) = 1$ and $\mu(Q \setminus \delta Q) = 0$. Indeed, for each k let X_k be the factor of Q that is the product of all factors of Q other than I_k. Define μ_k to be the product measure on X_k generated by the usual one-dimensional Lebesgue measure on each of the factor space of X_k and let φ_k be the obvious bijective map of X_k onto $\{x \in Q : x_k = 0\}$. Then $\varphi_\#\mu_k$ is a measure such that $\varphi_\#\mu_k(Q) = 1$ and $\varphi_\#\mu_k(Q \setminus \delta Q) = 0$. Let ψ_k be the analogous map of X_k onto $\{x \in Q : x_k = 1\}$. Then $\psi_\#\mu_k$ is a measure such that $\psi_\#\mu_k(Q) = 1$ and $\psi_\#\mu_k(Q \setminus \delta Q) = 0$. With $\nu_k = \frac{1}{2}(\varphi_\#\mu_k + \psi_\#\mu_k)$, let $\nu = \sum_{k=1}^{\infty} \frac{1}{2^k} \nu_k$. It is a simple calculation to show that $\nu(Q) = 1$ and $\nu(Q \setminus \delta Q) = 0$. As δQ is dense in Q, we have ν is also a positive measure on Q.

3.6.3. The Oxtoby–Prasad theorem. The measure λ on the Hilbert cube $Q = [0, 1]^{\mathbb{N}}$ that appears in the Oxtoby–Prasad theorem, which is the next theorem, is the product measure generated by the usual one-dimensional Lebesgue measure on each factor space I_n of Q. Clearly, $\lambda(\delta Q) = 0$.

THEOREM 3.45 (Oxtoby–Prasad). *Let μ be a continuous, positive, complete Borel measure on Q such that $\mu(Q) = 1$ and let B be the union of finitely many faces of Q with $\mu(B) = 0$. Then $\mu = h_{\#}\lambda$ for some homeomorphism h in* $\mathsf{HOMEO}(Q; B \text{ fixed})$. *If $\mu(\delta Q) = 0$, then $\mu = h_{\#}\lambda$ for some homeomorphism h in* $\mathsf{HOMEO}_0(Q; B \text{ fixed})$.

As in the proof found in [**121**], three preliminary lemmas will be proved. By the main lemma, only the last statement of the theorem needs to be proved. The first lemma is the Hilbert cube analogue of Lemma 3.26. The proof of this analogue uses the Baire category version of the proof of that lemma.

LEMMA 3.46. *Let μ be a positive, continuous, complete, finite Borel measure on Q with $\mu(\delta Q) = 0$, let B be the union of finitely many faces of Q, and let $R_1 = \{x \in Q\colon x_k \leq c\}$ and $R_2 = \{x \in Q\colon x_k \geq c\}$ be the rectangular sets in Q formed by the section $P = \{x \in Q\colon x_k = c\}$. Then for any two positive numbers α_1 and α_2 with $\alpha_1 + \alpha_2 = \mu(Q)$ there is an h in* $\mathsf{HOMEO}_0(Q; B \text{ fixed})$ *such that $h_{\#}\mu(R_1) = \alpha_1$ and $h_{\#}\mu(R_2) = \alpha_2$. For such an h, $h_{\#}\mu(\delta R_1) = h_{\#}\mu(\delta R_2) = h_{\#}\mu(R_1 \cap R_2) = 0$.*

PROOF. The proof is essentially the same as for I^n, where n is finite. The reader is referred to page 65 for the proof of the finite dimensional case. First observe that Q can be written as $Y \times I_k$ where Y is the product space formed by the interval factors of Q that are not the k-th interval factor I_k. We follow the finite dimensional proof with J replaced by Y and $[-1, 1]$ replaced by I_k. A replacement for ∂J in the finite dimensional proof must be found. To this end, let F be the union of all faces of Q contained in B that intersect P. If π denotes the natural projection of Q onto P, then $\pi^{-1}\pi[F] = F$ and $\pi[F] = P \cap F$. So we shall replace ∂J with $\pi[F]$. With these replacements, the proof of the lemma proceeds in the same manner as the Baire category proof of Lemma 3.26 given by Oxtoby and Ulam. □

A *simple subdivision of Q* is a subdivision of Q into rectangular sets defined by a finite number of sections of Q. We now generalize the last lemma to simple subdivisions of Q. The proof is a straightforward induction on the cardinality of finite families of sections of Q since a rectangular set is a copy of Q and the boundary of a rectangular set is the union of a finite number of its faces. The proof is left for the reader.

LEMMA 3.47. *Let μ be a positive, continuous, complete, finite Borel measure on Q with $\mu(\delta Q) = 0$, and let B be the union of finitely many faces of Q. If $\{R_1, \ldots, R_N\}$ is a simple subdivision of Q and $\alpha_1, \ldots, \alpha_N$ are positive numbers with $\alpha_1 + \cdots + \alpha_N = \mu(Q)$, then there is an h in* $\mathsf{HOMEO}_0(Q; B \text{ fixed})$ *such that $h_{\#}\mu(R_i) = \alpha_i$ and $h_{\#}\mu(\delta R_i) = 0$ for every i.*

The proof of the Oxtoby–Prasad theorem is now in sight. It is an inductive construction which is facilitated by the next lemma.

Lemma 3.48 (Key lemma). *Let μ and ν be positive, continuous, complete, finite Borel measures on Q with $\mu(\delta Q) = 0$, $\nu(\delta Q) = 0$ and $\mu(Q) = \nu(Q)$, let B be the union of a finite number of faces of Q and let $\mathcal{P}$ be a simple subdivision of Q such that $\mu(R) = \nu(R)$ and $\mu(\delta R) = \nu(\delta R) = 0$ for each R in $\mathcal{P}$. For each positive ε there exists a simple refinement $\mathcal{P}'$ of $\mathcal{P}$ with* $\operatorname{mesh}(\mathcal{P}') < \varepsilon$ *and there exists an h in* $\mathsf{HOMEO}_0(Q; B \text{ fixed})$ *such that h leaves R invariant and δR fixed for each R in $\mathcal{P}$ and such that $\nu(R') = h_\# \mu(R')$ and $\nu(\delta R') = h_\# \mu(\delta R') = 0$ for each R' in $\mathcal{P}'$.*

Proof. Let $\mathcal{P}'$ be a refinement of $\mathcal{P}$ with $\operatorname{mesh}(\mathcal{P}') < \varepsilon$ defined by taking additional sections of Q whose ν-measures are 0. As each R in $\mathcal{P}$ is a copy of Q we may apply the last lemma to R resulting in $R'_1, \dots, R'_N$ as members of $\mathcal{P}'$ that are contained in R, where $\alpha_i = \nu(R'_i)$ for each i, and $(B \cap R) \cup \operatorname{Bd}_Q(R)$ replaces B in the application of the lemma. The homeomorphisms that are obtained in this manner will fit together to form an h in $\mathsf{HOMEO}_0(Q; B \text{ fixed})$ that meets the requirements of the lemma. □

As in the proof of the finite dimensional Oxtoby–Ulam theorem the following definition and property will be helpful. We say that a measure μ is *Lebesgue-like on Q* if μ is a positive, continuous, complete, finite Borel measure on Q and $\mu(\delta Q) = 0$. A pair μ and ν of Lebesgue-like Borel measures on Q and a simple rectangular subdivision $\mathcal{P}$ of Q are said to satisfy the property $\mathsf{P}_Q(\mu, \nu; \mathcal{P})$ if $\mu(\sigma) = \nu(\sigma)$ and $\mu(\delta\sigma) = \nu(\delta\sigma) = 0$ whenever $\sigma \in \mathcal{P}$.

Proof of the Oxtoby–Prasad theorem. As we mentioned earlier only the last statement of the theorem requires proof. Suppose that μ and ν are Lebesgue-like Borel measures on Q and $\mathcal{P}$ is a simple rectangular subdivision of Q. We have from the key lemma above that if μ, ν and $\mathcal{P}$ satisfy $\mathsf{P}_Q(\mu, \nu; \mathcal{P})$ and if $\varepsilon > 0$, then there is a φ in $\mathsf{HOMEO}_0(Q; B \text{ fixed})$ and a simple rectangular subdivision $\mathcal{P}'$ of Q such that $\varphi_\# \mu$, ν and $\mathcal{P}'$ satisfy $\mathsf{P}_Q(\varphi_\# \mu, \nu; \mathcal{P}')$ and

(1) $\varphi|\sigma \in \mathsf{HOMEO}_0(\sigma; B \cap \sigma \text{ fixed})$ whenever $\sigma \in \mathcal{P}$,
(2) $\mathcal{P}'$ refines $\mathcal{P}$,
(3) $\operatorname{mesh}(\mathcal{P}') < \varepsilon$.

Let μ be a positive, continuous, complete, finite Borel measure on Q that satisfies $\mu(B) = 0$, $\mu(Q) = 1$ and $\mu(\delta Q) = 0$ as in the last statement of the Oxtoby–Prasad Theorem 3.45. We shall inductively construct a sequence of homeomorphisms φ_j and ψ_j $j = 1, 2, \dots$, and two sequences of simple rectangular subdivisions $\mathcal{P}_j$, $\mathcal{P}'_j$, $j = 1, 2, \dots$, of Q satisfying certain requirements specified below. Analogous to the proof of the Oxtoby–Ulam theorem, the following will help in the construction.

$$
\begin{aligned}
&\mathsf{P}_Q(\Phi_{j\#}\mu, \Psi_{(j-1)\#}\lambda; \mathcal{P}_j) \Longrightarrow \\
&\qquad \mathsf{P}_Q(\Phi_{j\#}\mu, \Psi_{j\#}\lambda; \mathcal{P}'_j) \Longrightarrow \mathsf{P}_Q(\Phi_{(j+1)\#}\mu, \Psi_{j\#}\lambda; \mathcal{P}_{j+1}) \\
&\qquad\qquad \Longrightarrow \mathsf{P}_Q(\Phi_{(j+1)\#}\mu, \Psi_{(j+1)\#}\lambda; \mathcal{P}'_{j+1}),
\end{aligned}
$$

for $j \geq 1$, where $\Psi_0 = \mathrm{id}$, and $\mathcal{P}'_0$ is undefined.

(1) $\Phi_j = \varphi_j \varphi_{j-1} \cdots \varphi_1$ and $\mathrm{mesh}(\mathcal{P}_j) < 2^{-j}$,
(2) $\Psi_j = \psi_j \psi_{j-1} \cdots \psi_1$, and $\mathrm{mesh}(\mathcal{P}'_j) < 2^{-j}$,
(3) $\varphi_j | \sigma' \in \mathsf{HOMEO}_0(\sigma'; B \cap \sigma' \text{ fixed})$ whenever $\sigma' \in \mathcal{P}'_{j-1}$,
(4) $\psi_j | \sigma \in \mathsf{HOMEO}_0(\sigma; B \cap \sigma \text{ fixed})$ whenever $\sigma \in \mathcal{P}_j$,
(5) $\mathrm{mesh}(\Phi_j{}^{-1}[\mathcal{P}'_j]) < 2^{-j}$,
(6) $\mathrm{mesh}(\Psi_{j-1}{}^{-1}[\mathcal{P}_j]) < 2^{-j}$,
(7) $\mathcal{P}'_j$ refines $\mathcal{P}_j$,
(8) $\mathcal{P}_j$ refines $\mathcal{P}'_{j-1}$.

From the assertion of the initial paragraph of the current proof with μ, $\nu = \lambda$, $\mathcal{P} = \{Q\}$ and $\varepsilon = 2^{-1}$ we have the existence of a φ_1 in $\mathsf{HOMEO}_0(Q; B \text{ fixed})$ and a simple rectangular subdivision $\mathcal{P}_1$ that satisfies $\mathsf{P}_Q(\Phi_{1\#}\mu, \Psi_{0\#}\lambda; \mathcal{P}_1)$ as well as $\mathrm{mesh}(\mathcal{P}_1) < 2^{-1}$. Condition (1) is satisfied for $j = 1$. Thereby, the induction is started. The rest of the proof follows the lines of the finite dimensional case and is left to the reader. □

3.7. Zero-dimensional spaces

Let us now turn to spaces with the smallest dimension. The two of most interest at this point are the spaces $\mathbb{N}^{\mathbb{N}}$ and $\{0,1\}^{\mathbb{N}}$. The first is topologically the space $\mathcal{N}$ equal to the set of irrational numbers between 0 and 1, and the second is topologically the classical Cantor ternary set in $\mathbb{R}$. The space $\mathcal{N}$ is not as "rigid" as the Cantor space which is compact. The space $\mathcal{N}$ has been characterized in [**4**, Satz IV] as those separable completely metrizable spaces that are zero-dimensional and nowhere locally compact (see also [**85**, Theorem 4 and Corollary 3a, pages 441–442]). Such a space X has a complete metric and a sequence of subdivisions $\mathcal{P}_n$, $n = 1, 2, \ldots$, with the properties

(1) $\mathcal{P}_n$ consists of nonempty sets that are both closed and open,
(2) $\mathcal{P}_{n+1}$ refines $\mathcal{P}_n$ and the collection $\{E' \in \mathcal{P}_{n+1} : E' \subset E\}$ is infinite for each E in $\mathcal{P}_n$,
(3) $\mathrm{mesh}(\mathcal{P}_n) < \frac{1}{2^n}$.

It is not difficult to show that two such spaces are homeomorphic. Using the above properties, J. C. Oxtoby [**119**] proved the following theorem.

THEOREM 3.49 (Oxtoby). *Let μ be a Borel measure on $\mathcal{N}$. Then there exists an h in* $\mathsf{HOMEO}(\mathcal{N})$ *such that $h_\# \mu = \lambda | \mathcal{N}$, where λ is the Lebesgue measure on $\mathbb{R}$, if and only if μ is positive, continuous and complete, and satisfies $\mu(\mathcal{N}) = 1$.*

An immediate consequence of this theorem is the next theorem whose proof is left to the reader.

THEOREM 3.50. $\mathsf{HOMEO}(\mathcal{N})$ *and $\lambda | \mathcal{N}$ generate* $\mathsf{univ}\, \mathfrak{M}(\mathcal{N})$.

3.7.1. Proof of Oxtoby's theorem. We give the proof found in [**119**]. The proof uses the next two lemmas.

LEMMA 3.51. *Let μ be a positive, continuous, complete, finite Borel measure on $\mathcal{N}$. If $\{\alpha_i : i = 1, 2, \dots\}$ is a sequence of positive real numbers such that $\sum_{i=1}^{\infty} \alpha_i = \mu(\mathcal{N})$, then there exists a subdivision $\mathcal{P} = \{U_i : i = 1, 2, \dots\}$ of $\mathcal{N}$ that consists of simultaneously closed and open sets such that $\mu(U_i) = \alpha_i$ for every i.*

PROOF. Observe that the map $x \mapsto \mu([0,x] \cap \mathcal{N})$ is a homeomorphism of $[0,1]$ onto $[0, \mu(\mathcal{N})]$ and that $[r, r'] \cap \mathcal{N}$ is a closed and open subset of $\mathcal{N}$ whenever r and r' are rational numbers in $[0,1]$ with $r < r'$. For (i,j) in $\mathbb{N} \times \mathbb{N}$, let $a(i,j) = \frac{j}{j+1} \alpha_i$. The linear ordering

$$\begin{aligned} &(i,j) < (i',j') \quad \text{if and only if} \\ &i + j < i' + j', \text{ or } i + j = i' + j' \text{ and } j < j' \end{aligned}$$

well orders the set $\mathbb{N} \times \mathbb{N}$. For each (i,j) we shall inductively select intervals $I(i,j)$, closed on the left-side and open on the right-side, with rational end points such that

$$a(i,j) < \sum_{n=1}^{j} \mu(I(i,n) \cap \mathcal{N}) < a(i,j+1)$$

for all i and j. Let r_1 be a rational number such that $I(1,1) = [0, r_1)$ satisfies the required condition. Suppose that (i,j) is the k-th pair and that $r_1, r_2, \dots, r_k$ are rational numbers such that the m-th pair (i',j') corresponds to the interval $I(i',j') = [r_{m-1}, r_m)$, $m = 1, 2, \dots, k$, which satisfy the above requirement. Let (i'',j'') be the $(k+1)$-th pair. Clearly one can select a rational number r_{k+1} such that $I(i'',j'') = [r_k, r_{k+1})$ satisfies the required condition. Observe that $\bigcup_{(i,j) \in \mathbb{N} \times \mathbb{N}} I(i,j) = [0,1)$. The sets $U_i = \mathcal{N} \cap \bigcup_{j=1}^{\infty} I(i,j)$, $i = 1, 2, \dots$, constitute a subdivision of $\mathcal{N}$ which satisfies $\mu(U_i) = \alpha_i$ for all i. □

LEMMA 3.52. *Let ρ be a complete metric for $\mathcal{N}$ and let ε be a positive number. If μ and ν are two positive, continuous, complete Borel measures on $\mathcal{N}$ and if U and V are nonempty open sets such that $\mu(U) = \nu(V) < \infty$, then there are subdivisions $\{U_i : i \in \mathbb{N}\}$ of U and $\{V_i : i \in \mathbb{N}\}$ of V that consist of nonempty simultaneously closed and open sets of diameter less than ε such that $\mu(U_i) = \nu(V_i)$ for every i.*

PROOF. Let $\{H_i : i \in \mathbb{N}\}$ be a subdivision of V into nonempty closed and open sets of diameter less than ε. As U is a homeomorphic copy of $\mathcal{N}$, the previous lemma provides a subdivision $\{G_i : i \in \mathbb{N}\}$ of U by closed and open sets such that $\mu(G_i) = \nu(H_i)$ for every i. For each i let $\{G_{i,j} : j \in \mathbb{N}\}$ be a subdivision of G_i by nonempty closed and open sets of diameter less than ε. As H_i is a homeomorphic copy of $\mathcal{N}$, the previous lemma provides a subdivision $\{H_{ij} : j \in \mathbb{N}\}$ of H_i by closed and open sets such that $\mu(G_{ij}) = \nu(H_{ij})$ for every j. Clearly $\{G_{ij} : (i,j) \in \mathbb{N} \times \mathbb{N}\}$ and $\{H_{ij} : (i,j) \in \mathbb{N} \times \mathbb{N}\}$ are subdivisions of U and V, respectively, with $\mu(G_{ij}) = \nu(H_{ij})$ whenever $(i,j) \in \mathbb{N} \times \mathbb{N}$. □

PROOF OF THEOREM 3.49. Let ρ be a complete metric for $\mathcal{N}$. By repeated applications of the last lemma we obtain subdivisions $\mathcal{U}_n = \{U_{ni} : i \in \mathbb{N}\}$ and $\mathcal{V}_n = \{V_{ni} : i \in \mathbb{N}\}$,

$n \in \mathbb{N}$, of $\mathcal{N}$ by simultaneously closed and open sets such that

(1) $\operatorname{mesh}(\mathcal{U}_n) < \frac{1}{2^n}$ and $\operatorname{mesh}(\mathcal{V}_n) < \frac{1}{2^n}$,
(2) $\mu(U_{ni}) = \nu(V_{ni})$ for every i,
(3) $\mathcal{U}_{n+1}$ refines $\mathcal{U}_n$, and $\mathcal{V}_{n+1}$ refines $\mathcal{V}_n$.

Each point x of $\mathcal{N}$ corresponds to a unique nested sequence $U_{n_k i_k}$, $k = 1, 2, \ldots$. As the metric ρ is complete and this sequence corresponds to the unique nested sequence $V_{n_k i_k}$, $k = 1, 2, \ldots$, there is a unique point $y = \varphi(x)$ determined by $V_{n_k i_k}$, $k = 1, 2, \ldots$. It is clear that the map φ is bijective and that $\varphi[U_{ni}] = V_{ni}$ and $U_{ni} = \varphi^{-1}[V_{ni}]$ for every n and every i, whence $\varphi \in \mathsf{HOMEO}(\mathcal{N})$. Consequently, $\mu(U_{ni}) = \varphi_{\#}\nu(U_{ni})$ for every n and every i. Obviously the smallest σ-algebra generated by $\bigcup_{n=1}^{\infty} \mathcal{U}_n$ is $\mathfrak{B}(\mathcal{N})$. Hence $\mu = \varphi_{\#}\nu$ and the theorem is proved. □

*3.7.2. **The Cantor space.*** The next obvious zero-dimensional space is the Cantor space $\{0,1\}^{\mathbb{N}}$. Here the question is whether some positive, continuous, complete, finite Borel measure μ on $\{0,1\}^{\mathbb{N}}$ and the group $\mathsf{HOMEO}(\{0,1\}^{\mathbb{N}})$ generate $\mathsf{univ}\,\mathfrak{M}(\{0,1\}^{\mathbb{N}})$ can be resolved by methods that were used up to this point of the book, that is, by proving an analogue of the Oxtoby–Ulam theorem for the space $\{0,1\}^{\mathbb{N}}$. This analogue is not possible as results of F. J. Navarro-Bermúdez and J. C. Oxtoby [**117**], F. J. Navarro-Bermúdez [**115, 116**], and K. J. Huang [**78**] show. In 1990, some of these results were further elaborated on by R. D. Mauldin in [**105**] and problems related to them were proposed there. More recently, E. Akin [**2, 3**] made substantial advances in the investigation of topological equivalence of measures on the Cantor space. Those who are interested in more details concerning the above mentioned results are referred to Appendix C in which a detailed discussion of them as well as the results of R. G. E. Pinch [**126**] and the very recent results of T. D. Austin [**6**] and R. Dougherty, R. D. Mauldin and A. Yingst [**47**] are presented.

Following Akin, we shall designate as *topological Cantor spaces* (or, more briefly, *Cantor spaces*) those spaces that are topologically the same as $\{0,1\}^{\mathbb{N}}$. For any separable metrizable space X, the collection $\mathfrak{CO}(X)$ consisting of all simultaneously closed and open subsets of X is a countable one. Moreover, if X is a Cantor space, then $\mathfrak{CO}(X)$ is a base for the open sets of X; consequently, it is also a base for the closed sets of X.

Let μ be a continuous, finite Borel measure on a separable metrizable space X. The *value set* of μ is the subset of $\mathbb{R}$ defined by

$$\operatorname{vs}(\mu, X) = \{\mu(U) \colon U \in \mathfrak{CO}(X)\}.$$

Clearly, $\operatorname{vs}(\mu, X)$ is a countable set that contains the values 0 and $\mu(X)$. Hence, if μ is a probability measure, then $\operatorname{vs}(\mu, X) \subset [0,1]$. For Cantor spaces X, the value sets $\operatorname{vs}(\mu, X)$ are dense subsets of $[0, \mu(X)]$. Generally, if $h \colon X \to h[X]$ is a continuous map, then

$$\operatorname{vs}(\mu, X) \supset \operatorname{vs}(h_{\#}\mu, h[X]),$$

where equality holds whenever h is a homeomorphism. Consequently, the value set of a measure is a topological invariant. But the value set does not determine the equivalence class $\{h_{\#}\mu \colon h \in \mathsf{HOMEO}(X)\}$, where μ is a probability measure on a Cantor space X. Indeed, Akin has shown that the measure μ on $\{0,1\}^{\mathbb{N}}$ determined by the Bernoulli probability measure on $\{0,1\}$ with values $\frac{1}{3}$ and $\frac{2}{3}$ and the measure ν on $\{0,1,2\}^{\mathbb{N}}$ determined by the uniform measure on $\{0,1,2\}$ are not topologically equivalent and yet $\mathrm{vs}(\mu,\{0,1\}^{\mathbb{N}}) = \mathrm{vs}(\nu, h[\{0,1\}^{\mathbb{N}}])$ for every homeomorphism h of $\{0,1\}^{\mathbb{N}}$ onto $\{0,1,2\}^{\mathbb{N}}$ such that $h_{\#}\mu = \nu$ (see Proposition 1.7 of [**2**] and also Appendix C). In [**2**], Akin finds another invariant that does determine equivalence classes. This invariant uses the order topology. Let us turn our attention to this invariant.

It is not difficult to show that $\{0,1\}^{\mathbb{N}}$ is homogeneous; that is, if x_1 and x_2 are points of $\{0,1\}^{\mathbb{N}}$, then there is an h in $\mathsf{HOMEO}(\{0,1\}^{\mathbb{N}})$ such that $h(x_1) = x_2$ and $h(x_2) = x_1$. Hence it follows that, for distinct points x_0 and x_1 of the classical Cantor ternary set, there is a self-homeomorphism h of the Cantor ternary set such that $h(x_0) = 0$ and $h(x_1) = 1$. Consequently we have[5]

PROPOSITION 3.53. *If X is a Cantor space and if x_0 and x_1 are distinct points of X, then there is a linear order $\leq$ on X such that*

(1) $x_0 \leq x \leq x_1$ *whenever* $x \in X$,
(2) *the order topology induced by $\leq$ on X is precisely the topology of X.*

A separable metrizable space X with a linear order $\leq$ that satisfies the above two conditions will be denoted by $(X,\leq)$ and will be called a *linearly ordered topological space* (or, more briefly, an *ordered space*). We recall two definitions from Appendix C.

DEFINITION 3.54. *Let $(X,\leq)$ be an ordered space and let μ be a complete, finite Borel measure on X. The function $F_\mu \colon X \to [0,\mu(X)]$ defined by*

$$F_\mu(x) = \mu([x_0,x]), \quad x \in X,$$

where x_0 is the minimal element of X in the order $\leq$, is called the cumulative distribution function *of μ. Define $\widetilde{\mathrm{vs}}(\mu,X,\leq)$, called the* special value set, *to be the set of values*

$$\widetilde{\mathrm{vs}}(\mu,X,\leq) = \{\mu([x_0,x]) \colon [x_0,x] \in \mathfrak{CO}(X)\} \cup \{0\}.$$

DEFINITION 3.55. *Let $(X_1,\leq_1)$ and $(X_2,\leq_2)$ be ordered spaces. $\varphi \colon X_1 \to X_2$ is said to be an* order preserving map *if $\varphi(a) \leq_2 \varphi(b)$ whenever $a \leq_1 b$. Such a map that is also bijective is called an* order isomorphism.

For Cantor spaces X, Akin showed in [**2**] that every linear order $\leq$ on X which results in an ordered space $(X,\leq)$ is order isomorphic to the classical Cantor ternary set endowed with the usual order.[6]

Here is a useful proposition.

[5] This is Proposition C.31 from Appendix C. Akin's extensive study of Cantor spaces with linear order that satisfy the two conditions that are enumerated in the proposition is presented in Appendix C.

[6] See Theorem C.37. Akin showed this to be a consequence of Theorem C.36. Actually a direct proof of the existence of such order isomorphisms can be made.

PROPOSITION 3.56. *Let $(X, \leq)$ be an ordered space. If $F: X \to [0,1]$ is an order preserving, right continuous function satisfying $F(x_1) = 1$, where x_1 is the maximal member of X, then there is a unique probability measure μ on X such that $F = F_\mu$.*

Akin proved [**2**, Theorem 2.12] the following interesting theorem which will prove useful in Chapter 4. The theorem uses Radon–Nikodym derivatives as a substitute for homeomorphisms. Recall that a measure μ is said to be *absolutely continuous* with respect a measure ν (denoted $\mu \ll \nu$) if $\mu(E) = 0$ whenever $\nu(E) = 0$. For convenience we shall assume μ and ν are continuous, finite Borel measures on a separable metrizable space X. If μ is absolutely continuous with respect to ν, then there is a real-valued, Borel measurable function, denoted by $\frac{d\mu}{d\nu}$ and called the *Radon–Nikodym derivative* of μ with respect to ν, that satisfies $\mu(B) = \int_B \frac{d\mu}{d\nu}\, d\nu = \int_X \chi_B \cdot \frac{d\mu}{d\nu}\, d\nu$ whenever $B \in \mathfrak{B}(X)$.

THEOREM 3.57 (Akin). *Let $(X_1, \leq_1)$ and $(X_2, \leq_2)$ be ordered Cantor spaces and let μ_1 and μ_2 be positive, continuous, complete Borel probability measures on X_1 and X_2, respectively. For each real number L with $L > 1$ there exists an order isomorphism $h: (X_1, \leq_1) \longrightarrow (X_2, \leq_2)$ such that $h_\# \mu_1 \ll \mu_2$ and the Radon–Nikodym derivative $\frac{dh_\#\mu_1}{d\mu_2}$ satisfies*

$$L^{-1} \leq \frac{dh_\#\mu_1}{d\mu_2} \leq L$$

everywhere on X_2. Consequently, $\mu_2 \ll h_\#\mu_1$ and $\frac{dh_\#\mu_1}{d\mu_2} \cdot \frac{d\mu_2}{dh_\#\mu_1} = 1$ everywhere on X_2.

The proof will rely on the following two lemmas.

LEMMA 3.58. *Let D_1 and D_2 be countable dense subsets of $[0,1]$ with $\{0,1\} \subset D_1 \cap D_2$. For each real number L with $L > 1$ there exists an order isomorphism φ of the ordered space $([0,1], \leq)$ such that $\varphi[D_1] = D_2$ and $L^{-1} \leq \frac{\varphi(x_2) - \varphi(x_1)}{x_2 - x_1} \leq L$. Hence $\mathrm{Lip}(\varphi) \leq L$ and $\mathrm{Lip}(\varphi^{-1}) \leq L$.*

PROOF. Let us begin by establishing some notation. Let $L > 1$ be fixed. In the plane $\mathbb{R}^2$, a parallelogram with vertices $A = (a_1, a_2)$, $B = (b_1, b_2)$, $C = (c_1, c_2)$, $D = (d_1, d_2)$ (oriented counterclockwise) and whose two pairs of opposite parallel sides have slopes L and L^{-1}, respectively, will be denoted by $\mathrm{PARA}(A, B, C, D)$. We will assume that the first vertex A has its coordinates to be smaller than the respective coordinates of the other vertices, hence the coordinates of the opposite vertex C has its coordinates larger than the respective coordinates of the other vertices. Clearly the slope of the diagonal $\overline{AC}$ is between L^{-1} and L. We shall denote the bounded component of the open set $\mathbb{R}^2 \setminus \mathrm{PARA}(A, B, C, D)$ by $[A, C]$ (note that B and D are determined by A and C). It is easy to see, for each P in $[A, C]$, that $[P, C] \subset [A, C]$, $[A, P] \subset [A, C]$, and $[A, P] \cap [P, C] = \emptyset$. For each x and y in $\mathbb{R}$ the line parallel to the second coordinate axis of $\mathbb{R}^2$ with the first coordinate equal to x will be denoted by L_x and the line parallel to the first coordinate axis of $\mathbb{R}^2$ with second coordinate equal to y will be denoted by L^y. Clearly, $\mathsf{L}_x \cap [A, C]$ is an open linear interval of L_x if and only of $a_1 < x < c_1$. A corresponding statement holds for L^y. It will be convenient to identify $\mathbb{R}$ and L_x and to identify $\mathbb{R}$ with L^y in the coming construction.

Finally, for points $P_1 = (x_1, y_1)$ and $P_2 = (x_2, y_2)$ in $D_1 \times D_2$, we write $P_1 < P_2$ to mean $x_1 < x_2$ and $y_1 < y_2$, and $\overline{P_1P_2}$ to be the line joining P_1 and P_2. The notational introduction is completed.

Let $a_1, a_2, a_3, \ldots$, and $b_1, b_2, b_3, \ldots$, be well orderings of D_1 and D_2, respectively, where $a_1 = b_1 = 0$ and $a_2 = b_2 = 1$; and let $L > 1$. We shall inductively construct a sequence φ_n, $n = 1, 2, \ldots$, of piecewise linear functions of the interval $[0, 1]$ onto itself that are determined by a sequence $\mathcal{P}_n = \{P_i^{(n)} : 1 \le i \le 2n\}$ of finite subsets of $D_1 \times D_2$ such that, for each n,

(n-1) $L^{-1} < \operatorname{Lip}(\varphi_n) < L$;
(n-2) $\mathcal{P}_n$ satisfies $P_i^{(n)} < P_{i+1}^{(n)}$ for each i, $P_1^{(n)} = (0,0)$, $P_{2n}^{(n)} = (1,1)$; and, the graph of φ_n is the union of $\overline{P_i^{(n)} P_{i+1}^{(n)}}$, $1 \le i < 2n$;
(n-3) with $(x_k^{(n)}, y_k^{(n)}) = P_k^{(n)} \in \mathcal{P}_n$, $k = 1, 2, \ldots, 2n$, the subsets $X_n = \{x_i^{(n)} : i \le 2n\}$ and $Y_n = \{y_j^{(n)} : j \le 2n\}$ of D_1 and D_2, respectively, satisfy $\{a_i : i \le n\} \subset X_n$ and $\{b_j : j \le n\} \subset Y_n$;
(n-4) the open planar set $[P_i^{(n)}, P_{i+1}^{(n)}]$ is defined for each i, and the set $M_n = \mathcal{P}_n \cup \bigcup_{i=1}^{2n-1} [P_i^{(n)}, P_{i+1}^{(n)}]$ is connected;
(n-5) $\mathcal{P}_n \subset \mathcal{P}_{n+1}$.

Let φ_1 be the piecewise linear function associated with the collection $\mathcal{P}_1 = \{(0,0), (1,1)\}$. Clearly the conditions (1-1) through (1-4) are satisfied and condition (1-5) is vacuously satisfied. The reader can readily see how to prove the inductive step from the example of the construction of φ_2 and $\mathcal{P}_2$. Let $i_1 = 3$. Then the line $\mathsf{L}_{a_{i_1}}$ has a nonempty open interval as the intersection with the connected set M_1. Let j_1 be the least index j such that b_j is in this intersection. Let $P' = (a_{i_1}, b_{j_1})$ and let $\mathcal{P}_1' = \{P'\} \cup \mathcal{P}_1$. Let M_1' be the connected set that corresponds to the collection $\mathcal{P}_1'$. Let j_1' be the least j such that b_j is not one of those already selected. The line $\mathsf{L}^{b_{j_1'}}$ has a nonempty open interval as the intersection with the connected set M_1'. Let i_1' be the least index i such that a_i is in this intersection. Let $P'' = (a_{i_1'}, b_{j_1'})$ and let $\mathcal{P}_2 = \{P', P''\} \cup \mathcal{P}_1$, where the collection is reindexed to meet the requirements of condition (2-2). Define φ_2 accordingly. The verifications of the conditions (2-1) through (2-5) are easily made.

The sequence φ_n, $n = 1, 2, \ldots$, converges on the dense set D_1 and has uniformly bounded Lipschitzian constants. Hence, by the Arzela–Ascoli theorem, the sequence converges uniformly to a function φ such that $L^{-1} \le \operatorname{Lip}(\varphi) \le L$. We infer from conditions (n-1) through (n-5) that $\varphi[D_1] = D_2$. □

LEMMA 3.59. *Let X be the Cantor ternary set with the usual order. Suppose that μ_1 and μ_2 are positive, continuous, complete Borel probability measures on X. If φ is an order isomorphism of* $[0, 1]$ *such that* $\varphi\big[\widetilde{\mathrm{vs}}(\mu_2, X, \le)\big] = \widetilde{\mathrm{vs}}(\mu_1, X, \le)$, *then there exists an order isomorphism $h \colon X \to X$ such that $\varphi F_{\mu_2} = F_{h_\# \mu_1}$. Consequently, if x_1 and x_2 are points of X with $x_1 < x_2$, then* $\big(\operatorname{Lip}(\varphi^{-1})\big)^{-1} \mu_2\big([x_1, x_2]\big) \le h_\# \mu_1\big([x_1, x_2]\big) \le \operatorname{Lip}(\varphi) \mu_2\big([x_1, x_2]\big)$ *whenever* $\operatorname{Lip}(\varphi^{-1}) > 0$.

PROOF. Define E to be the set of all left end points of the deleted open intervals of $\mathbb{R}$ that are used in the construction of the Cantor ternary set. Observe, for each i, that v is in $\widetilde{\text{vs}}(\mu_i, X, \leq)$ if and only if there is a unique e in E such that $F_{\mu_i}(e) = v$. First let us show that there is an order isomorphism $h\colon E \to E$ such that $\varphi F_{\mu_2}(h(e)) = F_{\mu_1}(e)$. To this end, let e be in E. Then $F_{\mu_1}(e) \in \varphi\big[\widetilde{\text{vs}}(\mu_2, X, \leq)\big]$. There is a unique e' in E such that $F_{\mu_1}(e) = \varphi F_{\mu_2}(e')$. Hence $h(e) = e'$ and h is defined on E. It follows that $F_{\mu_2}(h(e)) = \varphi^{-1}F_{\mu_1}(e)$. It remains to prove that h is an order isomorphism. Let g denote the function on $\widetilde{\text{vs}}(\mu_2, X, \leq) \setminus \{0\}$ that is the inverse of the restriction of F_{μ_2} to E. Then g is an order isomorphism (see Exercise 3.13). We have $h(e) = g\varphi^{-1}F_{\mu_1}(e)$. Clearly h is an order preserving injection. It is not difficult to show that h is surjective since F_{μ_i} is a surjective map of E onto $\widetilde{\text{vs}}(\mu_i, X, \leq) \setminus \{0\}$ for each i. From the order denseness of E in X we infer the existence of an order preserving extension of h to all of X. We shall continue to use h for this extension. As E is order dense in X, this extension is an order isomorphism of X. We now have $h_{\#}\mu_1\big([0,x]\big) = \mu_1\big(h^{-1}[[0,x]]\big) = F_{\mu_1}(x') = \varphi F_{\mu_2}(x)$, where $h^{-1}(x) = x'$. Hence $\varphi F_{\mu_2} = F_{h_{\#}\mu_1}$. The final statement of the lemma follows easily from this identity. □

Proof of Theorem 3.57. As $(X_1, \leq_1)$ and $(X_2, \leq_2)$ are each order isomorphic to the Cantor ternary set X with the usual order $\leq$, we may assume that both of them are $(X, \leq)$. Given $L > 1$, let φ and h be as provided by the above two lemmas. Then, for x_1 and x_2 in X with $x_1 < x_2$, we have $L^{-1}\,\mu_2\big([x_1,x_2]\big) \leq h_{\#}\mu_1\big([x_1,x_2]\big) \leq L\,\mu_2\big([x_1,x_2]\big)$. It follows that $\mu_2 \ll h_{\#}\mu_1$ and $h_{\#}\mu_1 \ll \mu_2$, and that the Radon–Nikodym derivative of $h_{\#}\mu_1$ with respect to μ_2 exists and satisfies the requirements of the theorem. □

We complete this section by connecting the last theorem to the collection $\mathsf{univ}\,\mathfrak{M}(\{0,1\}^{\mathbb{N}})$ of all universally measurable sets in the Cantor space $\{0,1\}^{\mathbb{N}}$.

Theorem 3.60. *Let X be a Cantor space. If μ is a positive, continuous, complete, finite Borel measure on X, then μ and $\mathsf{HOMEO}(X)$ generate $\mathsf{univ}\,\mathfrak{M}(X)$.*

Proof. Let E be a subset of X such that $h^{-1}[E]$ is μ-measurable whenever $h \in \mathsf{HOMEO}(X)$. We must show that E is ν-measurable for every ν in $\text{MEAS}^{\text{pos,fin}}(X)$, the collection of all positive, continuous, complete, finite Borel measures on X. Let ν be such a measure. There is no loss in assuming $\mu(X) = \nu(X)$. There exists an h in $\mathsf{HOMEO}(X)$ such that $\nu \ll h_{\#}\mu$. We have that $h^{-1}[E]$ is μ-measurable. Hence there exist Borel sets A and B such that $A \subset h^{-1}[E] \subset B$ and $\mu(B \setminus A) = 0$. Then $A' = h[A]$ and $B' = h[B]$ are Borel sets such that $A' \subset E \subset B'$. Now $h_{\#}\mu(B' \setminus A') = \mu(B \setminus A) = 0$. Hence $\nu(B' \setminus A') = 0$ and thereby E is ν-measurable. E is a universally measurable set in X. □

3.8. Other examples

With the aid of the Oxtoby–Ulam theorem, Marczewski proved that Lebesgue measure and $\mathsf{HOMEO}(\mathbb{R}^n)$ generate $\mathsf{univ}\,\mathfrak{M}(\mathbb{R}^n)$. An even stronger theorem holds, that is, Theorem 3.37 – actually, Marczewski's result is implied by this theorem since Lebesgue measure is not finite on $\mathbb{R}^n$; the implication is a consequence of the fact that there is a finite measure on $\mathbb{R}^n$ such that its measurable sets coincide with those of Lebesgue measure. The measures in the theorem are positive. There is a nonpositive, continuous, complete, finite Borel measure μ on $\mathbb{R}^n$ such that it and $\mathsf{HOMEO}(\mathbb{R}^n)$ generate $\mathsf{univ}\,\mathfrak{M}(\mathbb{R}^n)$.

EXAMPLE 3.61. The measure $\mu = \lambda \llcorner [0,1]^n$, where λ is Lebesgue measure on $\mathbb{R}^n$, and the group $\mathsf{HOMEO}(\mathbb{R}^n)$ generate $\mathsf{univ}\,\mathfrak{M}(\mathbb{R}^n)$. For a connected n-dimensional manifold M_n with $\partial M_n = \emptyset$ let $\varphi \colon \mathbb{R}^n \to M_n$ be a topological embedding. Then the measure $\varphi_\# \lambda$ and the group $\mathsf{HOMEO}(M_n)$ generate $\mathsf{univ}\,\mathfrak{M}(M_n)$. See Exercise 3.15.

Every example given so far has been an absolute G_δ space, though not necessarily locally connected for the connected examples. The following example will supply a non absolute G_δ space X for which there exists a positive, continuous, complete Borel measure μ on X such that it and the group $\mathsf{HOMEO}(X)$ generate $\mathsf{univ}\,\mathfrak{M}(X)$.

EXAMPLE 3.62. If D is a countable space, then there is a positive, continuous, complete, finite Borel measure μ on $X = D \times [0,1]$ such that $\mathsf{univ}\,\mathfrak{M}(X)$ is generated by μ and the group $\mathsf{HOMEO}(X)$. See Exercise 3.16.

In the above example the countable space can be the space $\mathbb{Q}$ of rational numbers, a non absolute G_δ space. The next example concerns the group of homeomorphisms of the space $X = \mathcal{N} \times [0,1]$.

EXAMPLE 3.63. Let $\pi : \mathcal{N} \times [0,1] \to \mathcal{N}$ be the natural projection and define f on $\mathcal{N}$ to be the injection $x \mapsto (x, 0)$. Note that $h[\{x\} \times [0,1]]$ is a component of $\mathcal{N} \times [0,1]$ for each h in $\mathsf{HOMEO}(\mathcal{N} \times [0,1])$ and each x in $\mathcal{N}$. Hence there is a bijection $h^* \colon \mathcal{N} \to \mathcal{N}$ and a homeomorphism g_x in $\mathsf{HOMEO}([0,1])$ such that $h(x,y) = \big(h^*(x), g_x(y)\big)$ whenever $(x,y) \in \mathcal{N} \times [0,1]$. As $h^* = \pi h f$ we have that h^* is continuous. Clearly the following diagram commutes.

$$\begin{array}{ccccc}
\mathcal{N} & \xrightarrow{f} & \mathcal{N} \times [0,1] & \xrightarrow{\pi} & \mathcal{N} \\
{\scriptstyle h^*}\uparrow & & \uparrow{\scriptstyle h} & & \uparrow{\scriptstyle h^*} \\
\mathcal{N} & \xrightarrow{f} & \mathcal{N} \times [0,1] & \xrightarrow{\pi} & \mathcal{N}
\end{array}$$

Hence h^* is in $\mathsf{HOMEO}(\mathcal{N})$. (See Exercise 3.17.)

Now denote $\mathcal{N} \times [0,1]$ by X. It is easy to show that there exists a positive, continuous, complete Borel probability measure μ on X with $\mu(\mathcal{N} \times \partial[0,1]) = 0$ such that μ is not topologically equivalent to $\lambda|X$, where λ is the Lebesgue measure on $\mathbb{R}^2$; that is, $\mu \neq h_\#\big(\lambda|X\big)$ for every h in $\mathsf{HOMEO}(X)$. (See Exercise 3.18.)

For the inclusion map φ of $\mathcal{N} \times [0,1]$ into $[0,1] \times [0,1]$ and the above μ, the measure $\varphi_\# \mu$ is homeomorphic to $\lambda|([0,1] \times [0,1])$. Also the restricted measure $\mu|(\mathcal{N} \times \mathcal{N})$ is homeomorphic to $\lambda|(\mathcal{N} \times \mathcal{N})$ since $\mathcal{N}$ is to homeomorphic $\mathcal{N} \times \mathcal{N}$.

We have used products of two spaces in the last two examples. Clearly, not every measure in $\mathrm{MEAS}(X \times Y)$ need be a product measure and the group $\mathsf{HOMEO}(X \times Y)$ can be quite large in comparison to $\mathsf{HOMEO}(X) \times \mathsf{HOMEO}(Y)$. It would be interesting if some general results about the generation of universally measurable sets in product spaces could be proven.

Here is a question that arises from the second example.

QUESTION. Let $\pi_1 : \mathcal{N} \times [0,1] \to \mathcal{N}$ and $\pi_2 : \mathcal{N} \times [0,1] \to [0,1]$ be the natural projections. Suppose that μ is a positive, continuous, complete Borel probability measure on $\mathcal{N} \times [0,1]$ such that $\pi_{1\#}\mu$ is a continuous measure. Then, by Oxtoby's theorem, there is a homeomorphism φ in $\mathsf{HOMEO}(\mathcal{N})$ such that $\varphi_\# \pi_{1\#}\mu$ is the restriction of the Lebesgue measure on $\mathbb{R}$ to the set $\mathcal{N}$. Hence there is a homeomorphism h_1 in $\mathsf{HOMEO}(\mathcal{N} \times [0,1])$ such that $\pi_{1\#}h_{1\#}\mu = \varphi_\#\pi_{1\#}\mu$. Is μ homeomorphic to $\lambda|(\mathcal{N} \times [0,1])$, where λ is the Lebesgue measure on $\mathbb{R}^2$? Failing that, is there a homeomorphism h_2 in $\mathsf{HOMEO}(\mathcal{N} \times [0,1])$ such that $\pi_{2\#}h_{2\#}\mu$ is continuous?

Every example so far has been an absolute Borel space. It is time to present an example of an absolute measurable space that is not an absolute Borel space.

EXAMPLE 3.64. Let N be a nowhere dense subset of the Hilbert cube $Q = [0,1]^{\mathbb{N}}$ that is also an uncountable absolute null space, and let X_n, $n = 1, 2, \ldots$, be a discrete collection of absolute measurable subspaces of Q whose diameters form a sequence converging to 0. Suppose $N \cap (\bigcup_{n=1}^{\infty} X_n) = \emptyset$ and $\mathrm{F}_Q(X_n) \neq \emptyset$ for each n. It can happen that $N \cap \mathrm{Cl}_Q(\bigcup_{n=1}^{\infty} X_n) \neq \emptyset$. Let $X = N \cup (\bigcup_{n=1}^{\infty} X_n)$ and let μ be a continuous, complete, finite Borel measure on X such that, for each n, the measure $\mu|X_n$ is positive on X_n. Obviously X is not an absolute Borel space. If, for each n, $\mu|X_n$ and $\mathsf{HOMEO}(X_n)$ generate $\mathrm{MEAS}^{\mathrm{pos,fin}}(X_n)$, then a measure ν in $\mathrm{MEAS}^{\mathrm{pos,fin}}(X)$ is such that $h_\#\mu = \nu$ for some h in $\mathsf{HOMEO}(X)$ if and only if $\nu|X_n$ is a positive measure on X_n and $\mu(X_n) = \nu(X_n)$ for every n. Hence μ and the group $\mathsf{HOMEO}(X)$ do not generate $\mathrm{MEAS}^{\mathrm{pos,fin}}(X)$. Nevertheless, μ and $\mathsf{HOMEO}(X)$ do generate $\mathsf{univ}\,\mathfrak{M}(X)$ since each h in $\mathsf{HOMEO}(X_n)$ has an extension in $\mathsf{HOMEO}(X)$. As for examples of X_n, one may take them to be arcs or Cantor spaces or topological copies of $\mathcal{N}$ or topological k-spheres.

3.9. Comments

In this chapter we have been concerned with the interplay between a pair $(\mu, \mathsf{HOMEO}(X))$ and the collection $\mathsf{univ}\,\mathfrak{M}(X)$ of universally measurable sets in a space X, where μ is a positive, continuous, complete, finite Borel measure on X and $\mathsf{HOMEO}(X)$ is the group of homeomorphisms. The origin of this investigation is found in the early part of the twentieth century during which the investigation of those subsets of the unit interval $[0,1]$ which are measurable for every continuous, finite, complete Borel measure on $[0,1]$ was initiated. These sets were called universally measurable or absolutely measurable. In 1937, a summary of results was made by Braun, Szpilrajn and Kuratowski in the *Annexe* to the *Fundamenta Mathematicae* [**15**]. There one finds the basic fact that Lebesgue measure λ and the group $\mathsf{HOMEO}([0,1])$ generate $\mathsf{univ}\,\mathfrak{M}([0,1])$. Of course, it is not difficult to replace $[0,1]$ with $\mathbb{R}$ in this assertion. So the question then becomes: Can one replace $\mathbb{R}$ with $\mathbb{R}^n$? For the plane, this is Problem 170 proposed by Szpilrajn in *The Scottish Book* (this book was the subject of a conference whose proceedings were edited by Mauldin [**104**]). The problem was solved by Marczewski [**96**]. Its solution is a simple consequence of the Oxtoby–Ulam theorem. (See Example 3.61.) Hence properties provided by the Oxtoby–Ulam theorem are sufficient conditions for the group $\mathsf{HOMEO}(\mathbb{R}^n)$ to generate $\operatorname{univ}\mathfrak{M}(\mathbb{R}^n)$. This motivates the study of the "action" of the group $\mathsf{HOMEO}(X)$ in the generation of $\operatorname{univ}\mathfrak{M}(X)$. As we have seen in the case of the Cantor space, absolute continuity of measures leads to nice a generalization of the Oxtoby and Ulam approach to the group $\mathsf{HOMEO}(X)$ generating $\operatorname{univ}\mathfrak{M}(X)$. This approach will be expanded on in the next chapter.

3.9.1. The unit n-cube and the Hilbert cube. The Oxtoby–Ulam theorem gives a characterization of those Borel measures on $[0,1]^n$ that are homeomorphic to the Lebesgue measure λ on $[0,1]^n$. The theorem has an interesting history. It was conjectured by Ulam in 1936 that the Oxtoby–Ulam theorem is true (see [**122**, page 886]). In the following year J. von Neumann gave an unpublished proof [**155**]. The Oxtoby and Ulam proof [**122**] appeared in 1941. They applied the Baire category theorem to a closed subgroup of the topological group $\mathsf{HOMEO}([0,1]^n)$ to prove the key lemma needed for their proof. In 1975 Goffman and Pedrick [**63**] gave a measure theoretic proof of the key lemma.

The two approaches to the proof to the key lemma mentioned earlier have different consequences in terms of extensions of the Oxtoby–Ulam theorem. The Baire category proof is found to work equally well in the infinite product $[0,1]^{\mathbb{N}}$, which is the Hilbert cube. This was carried out by Oxtoby and Prasad in [**121**]. Their result is a remarkable one in that the Hilbert cube does not have an algebraic boundary (that is, $\partial[0,1]^{\mathbb{N}} = \emptyset$) and hence, due to the main Lemma 3.44 of their proof, a much cleaner theorem results. The measure theoretic approach of Goffman and Pedrick has the interesting consequence of a different nature. Their proof of the key lemma can be couched easily in a homeotopy form (see [**62**]). Hence an algebraic topological form of the Oxtoby–Ulam theorem will result for finite product spaces $[0,1]^n$. Indeed, for each pair μ and ν in $\{\nu\colon \nu([0,1]^n) = 1,\ \mu \text{ is Lebesgue-like}\}$ there is a continuous map

$$G\colon [0,1] \longrightarrow \mathsf{HOMEO}([0,1]^n; \partial[0,1]^n \text{ fixed}) \tag{3.9}$$

such that $G(0)_{\#}\mu = \mu$, $G(1)_{\#}\mu = \nu$, and $G(t)_{\#}\mu(E) = \mu(E)$ for E in $\mathfrak{B}([0,1]^n)$. There is a sharper result that is a consequence of a fact (attributed to Alexander) that the space $\mathsf{HOMEO}([0,1]^n; \partial[0,1]^n \text{ fixed})$ is contractible, which was observed by R. Berlanga and D. B. A. Epstein [**8**, Remark, page 66]. $\mathsf{HOMEO}([0,1]^n; \partial[0,1]^n \text{ fixed})$ is contractible means there is a homotopy H of $\mathsf{HOMEO}([0,1]^n; \partial[0,1]^n \text{ fixed})$ to itself such that $H(1,\cdot)$ is the identity on $\mathsf{HOMEO}([0,1]^n; \partial[0,1]^n \text{ fixed})$ and $H(0,\cdot)$ is a fixed homeomorphism. Since the proof is easy let us give it. Let us replace $[0,1]^n$ with the closed unit ball B in the Euclidean space $\mathbb{R}^n$ that is centered at the origin and $\partial[0,1]^n$ with ∂B. For each positive t let tB be the closed ball with radius t. If h is in $\mathsf{HOMEO}(B; \partial B \text{ fixed})$, then the map h_t defined by $h_t(x) = t\,h(t^{-1}x)$, $x \in tB$, is a homeomorphism of tB onto itself such that h_t is the identity map on $\partial(tB)$. For each t in $(0,1]$, the identity map on $B \setminus tB$ will yield an extension of h_t to a homeomorphism of B onto B such that its restriction to ∂B is the identity map. We shall denote this extended map by $H(t,h)$.

It is easily seen that $H(t,\cdot)$ is a continuous map of HOMEO$(B;\partial B$ fixed$)$ into itself with $H(0,h) = \mathrm{id}_B$ and $H(1,h) = h$ for every h. Hence HOMEO$(B;\partial B$ fixed$)$ is contractible. Consequently, if μ is a Lebesgue-like probability measure, then $H(1,h)_{\#}\mu = \nu$ for some h by the Oxtoby–Ulam theorem. The above argument is a finite dimensional one.

Interestingly enough, the measure theoretic approach does not seem to yield the extension of the Oxtoby–Ulam theorem to the Hilbert cube because Lemma 3.27 is a finite dimensional result. Also, the Baire category approach to the key lemma loses the homeotopy refinement in the equation (3.9) achieved by the measure theoretic approach.

QUESTION. Is there a homotopy version of the Oxtoby–Prasad theorem? Is there an equation (3.9) version?

3.9.2. Compact, connected manifolds. Observe that $[0,1]^n$ is a compact, connected manifold with boundary. The Oxtoby–Ulam theorem concerns Lebesgue-like measures on this manifold. In a natural manner, for a compact, connected manifold M_n (with or without boundary), we defined a positive, continuous, complete, finite Borel measure μ on M_n to be *Lebesgue-like* if $\mu(\partial M_n) = 0$. Applying a topological theorem due to Brown [**21**], Alpern and Prasad show in [**5**, page 195] that the Oxtoby–Ulam theorem holds for compact, connected manifolds. Another proof can be found in T. Nishiura [**118**] which does not employ the Brown theorem, just the definition of manifolds. There are results concerning homeomorphic measures on noncompact, connected, separable manifolds due to Berlanga and Epstein [**8**].[7]As their results require the notion of "ends" of a manifold, we have not included them in this chapter. To illustrate the role of ends we give the following rather simple example. In the square $[-1,1]\times[-1,1]$ consider the points $p^- = (-1/2,0)$ and $p^+ = (1/2,0)$. The noncompact manifold $X = [-1,1]\times[-1,1]\setminus\{p^-,p^+\}$ has the end-point compactification $[-1,1]\times[-1,1]$ with ends consisting of the points p^- and p^+. Let μ be a continuous, positive, complete Borel measure on X with $\mu\big(\partial([-1,1]\times[-1,1])\big) = 0$ and $\mu(X) = \lambda(X)$, where λ is the Lebesgue measure on the plane. For this example, it is not difficult to see that $h_{\#}\mu = \lambda|X$ for some h in HOMEO(X). (The proof is not a simple application of the Oxtoby–Ulam theorem for the square $[-1,1]\times[-1,1]$ because the homeomorphism h must indirectly see the ends p^- and p^+.) See the book by Alpern and Prasad [**5**, pages 196–204] for a more detailed discussion on the Berlanga–Epstein results.

3.9.3. Nonmanifolds. Remember that the basic question is the existence of a positive, continuous, complete Borel measure μ on a space X such that it and the group HOMEO(X) generate univ $\mathfrak{M}(X)$. With the aid of Oxtoby and Prasad extension of the Oxtoby–Ulam theorem to the Hilbert cube $[0,1]^{\mathbb{N}}$, a nonmanifold, the question is easily answered in the affirmative for this nonmanifold.

[7] The manifolds in this paper are σ-compact, connected manifolds. It is easy to see that such manifolds are precisely the connected, separable manifolds.

It has been shown that the answer is also in the affirmative for other nonmanifolds such as finitely triangulable n-dimensional spaces, the "$\sin(\frac{1}{x})$ space" and the Warsaw circle even though the Oxtoby–Ulam theorem does not generalize to these spaces.

Another example, which we did not present, is a Menger manifold. A Menger manifold is a separable metrizable space such that each of its point has a neighborhood that is homeomorphic to the Menger compact, universal space of dimension n (see [**50**, page 121] for the definition of the Menger universal space). H. Kato, K. Kawamura and E. D. Tymchatyn [**81**, Theorem 3.1, Corollary 4.12] have shown that the analogue of the Oxtoby–Ulam theorem is valid for the Menger compact, universal space of dimension n and for compact Menger manifolds of dimension n, where $n > 0$. The case $n = 0$ is the Cantor space; the Oxtoby–Ulam theorem analogue is not available here. Their proof is based on the work of M. Bestvina [**11**], see [**11**, pages 15 and 98] for the basic definitions.

3.9.4. The space $\mathcal{N}$. Oxtoby investigated in [**119**] homeomorphic measures on the zero-dimensional space $\mathcal{N}$, which is homeomorphic to the product space $\mathbb{N}^{\mathbb{N}}$. He showed that the equivalence classes with respect to homeomorphisms are characterized by positive real numbers, that is, two positive, continuous, complete, finite Borel measures μ and ν are homeomorphic if and only if $\mu(\mathcal{N})$ and $\nu(\mathcal{N})$ are equal to the same positive real number. The Oxtoby theorem for the space $\mathcal{N}$ leads to the fact that Lebesgue measure on $\mathcal{N}$ and $\mathsf{HOMEO}(\mathcal{N})$ generate $\operatorname{univ}\mathfrak{M}(\mathcal{N})$.

In the same article, Oxtoby defined the notion of almost homeomorphic measures. This will be discussed in the next chapter.

3.9.5. The Cantor space. The other familiar zero-dimensional space is the Cantor space $\{0, 1\}^{\mathbb{N}}$. Here, characterizations of the equivalence classes of homeomorphic positive, continuous, complete, finite Borel measures on $\{0, 1\}^{\mathbb{N}}$ were not known in the mid-1970s. In Section 3.7.2 there are several references with partial results which show that the Oxtoby–Ulam theorem does not extend to the Cantor space $\{0, 1\}^{\mathbb{N}}$. In his 1999 paper [**2**] Akin successfully "characterized" the equivalence classes of homeomorphic positive, continuous, complete, finite Borel measures on Cantor spaces X. In that paper he showed that the value set $\mathrm{vs}(\mu, X)$ is invariant under homeomorphisms, but not conversely; consequently, the word characterized in the previous sentence is enclosed in quotation marks. As every Cantor space is homeomorphic to the classical Cantor ternary set, the usual order on the Cantor ternary set provides each positive, continuous, probability Borel measure μ with a natural cumulative distribution function. After a thorough investigation of linear orders on Cantor spaces X, he showed that the special value set $\widetilde{\mathrm{vs}}(\mu, X, \leq)$ is a linear (that is, order) isomorphism invariant. Hence if $r > 0$ and if D is a countable, dense subset of $[0, r]$, then the collection $D(r)$ of all positive, continuous, complete Borel measures on X with $\mu(X) = r$ is partitioned into linear isomorphism equivalence classes $\widetilde{D}(r)$ consisting of those μ such that $D \cup \{0, r\} = \widetilde{\mathrm{vs}}(\mu, X, \leq)$, where the linear order $\leq$ on X results in the Cantor space topology of X. Exercise 3.14. shows that each such countable set

D yields a nonempty collection $\widetilde{D}(r)$. It follows that, for a fixed r, the cardinality of the collection of all linear isomorphism classes $\widetilde{D}(r)$ is $\mathfrak{c}$.

There is a stronger result for Bernoulli measures on $X = \{0,1\}^{\mathbb{N}}$. A Bernoulli measure P on $\{0,1\}$, where $0 < r < 1$, is the unique measure that assigns the values $P(\{0\}) = r$ and $P(\{1\}) = 1 - r$. This measure produces the shift invariant Bernoulli product measure on $\{0,1\}^{\mathbb{N}}$ in the obvious way. This product measure will be denoted also by β_r. The value set $\mathrm{vs}(\beta_r, X)$ consists of the numbers of the form

$$\sum_{i=0}^{n} a_i r^i (1-r)^{n-i}, \qquad n \in \mathbb{N},$$

where the coefficients a_i are integers that satisfy $0 \leq a_i \leq \frac{n!}{i!(n-i)!}$. Hence r and $1 - r$ are in $\mathrm{vs}(\beta_r, X)$. Akin has observed (see [**2**, Proposition 1.8])

THEOREM 3.65. *For each r in $(0,1)$ there are only countably many s in $(0,1)$ such that the Bernoulli measure β_s on $\{0,1\}^{\mathbb{N}}$ is homeomorphic to the Bernoulli measure β_r on $\{0,1\}^{\mathbb{N}}$. Hence, among the collection $\{\beta_r : r \in (0,1)\}$ of Bernoulli measures on $\{0,1\}^{\mathbb{N}}$, the set of homeomorphism equivalence classes has the cardinality $\mathfrak{c}$.*

The following earlier and finer conclusions were proved by Navarro-Bermúdez [**115**, Theorem 3.3 and Theorem 3.4].

THEOREM 3.66 (Navarro-Bermúdez). *Let r and s be in $(0,1)$. If r is a rational number, then the Bernoulli measures β_r and β_s on $\{0,1\}^{\mathbb{N}}$ are homeomorphic if and only if $s = r$ or $s = 1 - r$. Also, if r is a transcendental number, then the Bernoulli measures β_r and β_s on $\{0,1\}^{\mathbb{N}}$ are homeomorphic if and only if $s = r$ or $s = 1 - r$.*

The investigation of the cases where r is an algebraic number is not very easy and far from complete. The remaining cited references in Section 3.7.2 concern Bernoulli measures β_r where r is an algebraic number in the interval $(0,1)$. Several questions along this line are raised in Mauldin [**105**].

A more detailed discussion of Borel probability measures on Cantor spaces can be found in Appendix C.

3.9.6. Measure preserving homeomorphisms. On several occasions we have referred to the book [**5**] by Alpern and Prasad on measure preserving homeomorphisms. In their book these are called automorphisms. An *automorphism* of a Borel measure space $\mathrm{M}(X,\mu)$ is a g in $\mathfrak{B}$-$\mathsf{HOMEO}(X)$ such that $\mu(A) = \mu(g[A]) = \mu(g^{-1}[A])$ for every μ-measurable set A. Hence the task is to investigate those positive, probability Borel measures μ on X that have nontrivial solutions g in $\mathsf{HOMEO}(X)$ of the equation $g_{\#}\mu = \mu$. These automorphisms play an important role in dynamics. It is the typical behavior (in the sense of Baire category) of these automorphisms that is the emphasis of the book by Alpern and Prasad. The spaces X of interest in their development are connected n-dimensional manifolds of various kinds.

There is a marvelous development in the appendix of their book of the Oxtoby–Ulam theorem for the n-cube, the extension of the Oxtoby–Ulam theorem to compact, connected n-manifold, and a mention of the related Oxtoby–Prasad theorem for the

Hilbert cube. Their development, being somewhat terse, has been expanded upon in this chapter. Another reason for our repeating the proofs is that their proofs use notations that are inconsistent with ours – notational consistency will allow easier passage through the proofs.

One of the objectives of this chapter is to investigate the existence of a measure μ on separable metrizable spaces X, among which are connected n-dimensional manifolds, such that μ and the group $\mathsf{HOMEO}(X)$ generate the σ-algebra $\mathsf{univ}\,\mathfrak{M}(X)$. This was not an objective of the Alpern–Prasad book. As the reader can see, many examples of spaces X which are commonly known in mathematics have been included.

3.9.7. Terminology and notations. The meaning of "positive Borel measure on X" is not standardized in the literature. It could mean $\mu(X) > 0$ which, of course, is not what is needed in the Oxtoby–Ulam type of theorems. To emphasize the stronger condition, the statements "locally positive" or "positive on nonempty open sets" have been used in the literature. These two correspond to the condition that the topological support of μ (that is, $\mathrm{support}(\mu)$) is the whole space X. The assumption $X = \mathrm{support}(\mu)$ is not a good one since the emphasis of our book is on absolute measurable spaces and absolute null spaces. Another condition that is often used is "nonempty countable subsets are not open." This condition is used to make the Baire category theorem "reasonably" available for investigating several singular sets. To study these singular sets it is desirable that countable subsets be nowhere dense and that the space be an uncountable absolute G_δ space. There is an uncountable compact metrizable space X such that some countable subset U is dense and open (such a space is easily constructed). The set $X \setminus U$ is an uncountable nowhere dense set, hence of the first category, and U is a set of the second category. From a singular set point of view, especially those that involve the Baire category, it is the set $X \setminus U$ which forms the interesting part of X. From the measure theoretic point of view, the interesting part of a space X is also $X \setminus U$ because U is an absolute null space – a continuous Borel measure μ on X should be called positive if $\mathrm{support}(\mu) = X \setminus U$. It is known that there are many absolute null spaces that are not countable, hence countable subsets are not sufficient for the determination of the "reasonable" part a universally measurable set in an arbitrary separable metrizable space. Indeed, a "reasonable" universally measurable set M in a space is one in which every absolute null space contained in M is nowhere dense in M. The objections to the various other approaches mentioned above for a useful definition of positive measures are avoided by the use of the closure-like operator F_X in the space X. Hence a measure μ on X is positive if and only if $\mathrm{support}(\mu) = \mathrm{F}_X(X) \neq \emptyset$. Fortunately, many of the spaces X – of course, nonempty – that have been studied in the chapter are such that $\mathrm{F}_X(X) = X \neq \emptyset$, hence positive measures exist on these spaces.

There are conflicting notations in the literature for a homeomorphism h in $\mathsf{HOMEO}(X)$ acting on a Borel measure μ in $\mathrm{MEAS}(X)$. We have used $h_{\#}\mu$, which is defined by $h_{\#}\mu(B) = \mu(h^{-1}[B])$ for $B \in \mathfrak{B}(X)$. Others have used μh to mean $\mu h(B) = \mu(h[B])$ for $B \in \mathfrak{B}(X)$. Of course, these approaches are formally very different. The required adjustments have been made in our presentations of results that used the μh notation in the literature. Another objection to the μh notation is that the meaning of μh for Borel measurable maps h is rather awkward to define. The definition of Borel measurability of a map h uses h^{-1}, hence $h_\# \mu$ is the natural choice.

3.9.8. Bounded Radon–Nikodym derivatives. In anticipation of results that will be developed in the next chapter on real-valued functions, we look at the Oxtoby–Ulam theorem from the point of view of absolute continuity of measures, which gives rise to a real-valued function called the Radon–Nikodym derivative. Consider a positive, continuous, complete, finite Borel measure μ on $[0,1]^n$ and a continuous, complete, finite Borel measure ν on $[0,1]^n$. Then $\mu + \nu$ is also a positive Borel measure and $\nu \leq \mu + \nu$. Consequently, $\nu \ll \mu + \nu$ and the Radon–Nikodym derivative $\frac{d\nu}{d(\mu+\nu)}$ may be assumed to be bounded above by 1 everywhere on $[0,1]^n$. Suppose further that μ is also a Lebesgue-like measure on $[0,1]^n$ and that $\nu(\partial[0,1]^n) = 0$. Then $\mu + \nu$ will be Lebesgue-like. There is a positive constant c such that $c\,\mu([0,1]^n) = (\mu+\nu)([0,1]^n)$. Then, by the Oxtoby–Ulam theorem, we have $c\, h_\# \mu = \mu + \nu$ for some h in $\mathsf{HOMEO}([0,1]^n)$. Hence $c\, \chi_{(0,1)^n} = \frac{d(\mu+\nu)}{d(h_\#\mu)}$. Consequently, by the chain rule for Radon–Nikodym derivatives, we have

$$0 \leq \frac{d\nu}{d(h_\#\mu)} \leq c\, \chi_{(0,1)^n}$$

for some positive real number c and for some h in $\mathsf{HOMEO}([0,1]^n)$.

Observe that the above discussion applies also to compact, connected n-dimensional manifolds, to the Hilbert space, to the space $\mathcal{N}$, and to n-dimensional Menger manifolds for $n \geq 1$.

The corresponding statement for the Cantor space $\{0,1\}^{\mathbb{N}}$ is also true.

THEOREM 3.67. *Suppose that μ is a positive, continuous, complete, finite Borel measure on $\{0,1\}^{\mathbb{N}}$ and suppose that ν is a finite, continuous, complete, Borel measure on $\{0,1\}^{\mathbb{N}}$. Then there is a positive real number c and there is an h in* $\mathsf{HOMEO}(\{0,1\}^{\mathbb{N}})$ *such that $\nu \ll h_\#\mu$ and $0 \leq \frac{d\nu}{d(h_\#\mu)} \leq c$.*

PROOF. First observe that the Radon–Nikodym derivative is not dependent on the metric choice on $\{0,1\}^{\mathbb{N}}$. So we may assume that the Cantor space is the Cantor ternary set X in $\mathbb{R}$. Let c' be such that $\mu(X) + \nu(X) = c'\mu(X)$. From Theorem 3.57 with $L = 2$, we infer that there is an h in $\mathsf{HOMEO}(X)$ such that $\mu + \nu \ll c' h_\#\mu$ and $\frac{d(\mu+\nu)}{d(h_\#\mu)} \leq 2\,c'$. With $c = 2\,c'$ the proof is completed by the chain rule for Radon–Nikodym derivatives. □

The above discussion foreshadows the notion of (ac)-generation by a measure μ and the group $\mathsf{HOMEO}(X)$, see page 111. We conclude with the observation that there are several possible "topological equivalences" of positive, continuous, complete, finite Borel measures μ and ν on X. The first one is modelled after the Oxtoby–Ulam type of theorems. The second and third ones are modelled after the Cantor space theorem.

(1) (TOP) There exists an h in $\mathsf{HOMEO}(X)$ such that $\mu = h_\# \nu$.
(2) (ORDER) $\mu(X) = \nu(X)$, and there exist positive constants c and c' and there exists an h in $\mathsf{HOMEO}(X)$ such that $\mu \leq c\, h_\# \nu$ and $\nu \leq c'\, h^{-1}{}_\# \mu$. This, of course, is equivalent to the Radon–Nikodym derivatives being bounded.
(3) (AC) $\mu(X) = \nu(X)$, and there exists an h in $\mathsf{HOMEO}(X)$ such that $\mu \ll h_\# \nu$ and $\nu \ll h^{-1}{}_\# \mu$.

Exercises

3.1. Prove Proposition 3.6 on page 55.
3.2. Prove Proposition 3.7 on page 55.
3.3. For a compact metrizable space X and a closed subset F of X verify that $\mathsf{HOMEO}(X; F \text{ fixed})$ and $\mathsf{HOMEO}(X; F \text{ inv})$ are closed subgroups of $\mathsf{HOMEO}(X)$. See page 55 for definitions.
3.4. Prove Proposition 3.9 on page 56.
3.5. Prove Proposition 3.13 on page 57.
3.6. Prove Proposition 3.14 on page 57.
3.7. Prove Theorem 3.22 on page 60.
3.8. Determine the equivalence classes $\{h_\# \mu \colon h \in \mathsf{HOMEO}(\mathsf{T})\}$ of the collection $\mathrm{MEAS}^{\mathrm{pos,fin}}(\mathsf{T})$, where T is the simple triod and μ is in $\mathrm{MEAS}^{\mathrm{pos,fin}}(\mathsf{T})$.
3.9. Use Theorem 3.12 to prove Theorem 3.23 on page 61. Hint: Let $\mu|X_0 = \mathsf{H}_1\,|X_0$ and $\mu|X_1 = f_\# \lambda|(0,1]$, where λ is the Lebesgue measure on $(0,1]$ and $f \colon (0,1] \to X_1$ is a homeomorphism. Let F be the closure in X of the set of all points of maxima and all points of minima of $X_1 = \mathrm{graph}(\sin(1/x))$ (see page 61). Show that $\mathsf{HOMEO}(W)$ can be replaced with the smaller subgroup $\mathsf{HOMEO}(W; F \text{ fixed})$ in the above theorem.
3.10. Prove that the Oxtoby–Ulam theorem (page 62) implies the topological version Theorem 3.31 on page 70.
3.11. Let $\varphi \colon X \to Y$ be a continuous surjection of a compact metrizable space X. If μ is a complete, finite Borel measure on X, then we have already seen that $\varphi_\# \mu$ is a complete, finite Borel measure on Y. Prove: If ν is a continuous, complete, finite Borel measure on Y, then there is a $\mathfrak{B}$-homeomorphism ψ of Y into X such that the measure $\psi_\# \nu$ on X satisfies $\varphi'{}_\# \psi_\# \nu = \nu$, where $\varphi' = \varphi|\psi[Y]$. Hint: The map $\psi \colon Y \to X$ is a measurable selection. With $K = \{(x,y) \in X \times Y \colon x \in \varphi^{-1}(y)\}$, apply the measurable selection part of Lemma A.48.
3.12. Carry out the exercise without recourse to the Baire category theorem. Let I be a topological n-cell contained in a separable n-dimensional manifold M_n and let μ be a positive, complete, finite Borel measure on M_n. Observe that M_n is locally compact. Prove: *There is a homeomorphism* $\varphi \colon [0,1]^n \to I$ *such that* $\nu = \varphi^{-1}{}_\#(\mu|I)$ *is a measure on* $[0,1]^n$ *with the property* $\nu|([0,1]^n \setminus \partial[0,1]^n) \leq c\,\lambda|([0,1]^n \setminus \partial[0,1]^n)$, *where* λ *is the Lebesgue measure and* c *is a positive number.*

Let $\eta > 0$ and let F be a nowhere dense, closed subset of M_n. Prove: *There exists an* h *in* $\mathsf{HOMEO}(M_n; M_n \setminus I \text{ fixed})$ *such that* $\rho(h, \mathrm{id}) < \eta$ *and*

$h_{\#}\mu(F \cap (I \setminus \partial I)) = 0$. Prove: *There exists an h in* $\mathsf{HOMEO}(M_n; \partial M_n$ fixed) *with* $\rho(h, \mathrm{id})$ *small such that* $h_{\#}\mu(F \setminus \partial M_n) = 0$.

3.13. Let μ be a positive, continuous, complete Borel probability measure on the Cantor ternary set X and E be the set of left end points of the components of $[0, 1] \setminus X$. Prove $F_\mu|(E \cup \{0\})$ is an order isomorphism between $E \cup \{0\}$ and $\widetilde{\mathrm{vs}}(X, \mu)$.

3.14. Let X be the Cantor ternary set with the usual order. Show that every countable, dense subset D of $[0, 1]$ that contains both 0 and 1 corresponds to a positive, continuous, complete Borel probability measure μ on X such that $D = \widetilde{\mathrm{vs}}(\mu, X, \leq)$. Hint: If D_1 and D_2 are countable, dense subsets of $\mathbb{R}$, then there is an order preserving homeomorphism φ of $\mathbb{R}$ such that $\varphi[D_1] = D_2$. (See [**79**, page 44] or [**2**, Lemma 2.3].)

3.15. Verify Example 3.61. Let M_n be a connected n-dimensional manifold with $\partial M_n = \emptyset$. Prove: *If x_0 and x_1 are in M_n, then $h(x_0) = x_1$ for some h in* $\mathsf{HOMEO}(M_n)$. That is, M_n is homogeneous.

3.16. Verify Example 3.62. That is, prove the assertion: If D is a countable space, then there is a positive, continuous, complete, finite Borel measure μ on $X = D \times [0, 1]$ such that $\mathsf{univ}\,\mathfrak{M}(X)$ is generated by μ and and the group $\mathsf{HOMEO}(X)$. Notice that, for such a measure μ, the measure $\pi_{\#}\mu$ is not continuous, where π is the natural projection of $D \times [0, 1]$ onto D.

3.17. As in Example 3.63 let $\pi : \mathcal{N} \times [0, 1] \to \mathcal{N}$ be the natural projection and define f on $\mathcal{N}$ to be the injection $x \mapsto (x, 0)$. For each h in $\mathsf{HOMEO}(\mathcal{N} \times [0, 1])$, show that $h^* = \pi h f$ is in $\mathsf{HOMEO}(\mathcal{N})$.

3.18. As in Example 3.63, denote $\mathcal{N} \times [0, 1]$ by X. Show that there exists a positive, continuous, complete Borel probability measure μ on X such that μ is not homeomorphic to $\lambda|X$, where λ is the Lebesgue measure on $\mathbb{R}^2$; that is, $\mu \neq h_{\#}(\lambda|X)$ for every h in $\mathsf{HOMEO}(X)$. (Hint: Use the last two exercises. Distribute a linear measure in X so that it is dense in the space X.)

4

Real-valued functions

In this chapter, attention is turned to topics in analysis such as measurability, derivatives and integrals of real-valued functions. Several connections between real-valued functions of a real variable and universally measurable sets in $\mathbb{R}$ have appeared in the literature. Four connections and their generalizations will be presented. The material developed in the earlier chapters are used in the generalizations. The fifth topic concerns the images of Lusin spaces under Borel measurable real-valued functions – the classical result that these images are absolute null spaces will be proved. A brief description of the first four connections is given next before proceeding.

The first connection is a problem posed by A. J. Goldman [**64**] about σ-algebras associated with Lebesgue measurable functions; Darst's solution [**35**] will be given. A natural extension of Darst's theorem will follow from results of earlier chapters. Indeed, it will be shown that the domain of the function can be chosen to be any absolute measurable space that is not an absolute null space.

The second addresses the question of whether conditions such as bounded variation or infinitely differentiability have connections to theorems such as Purves's theorem; namely, for such functions, are the images of universally measurable sets in $\mathbb{R}$ necessarily universally measurable sets in $\mathbb{R}$? Darst's negative resolutions of these questions will be presented.

The third and fourth connect to A. M. Bruckner, R. O. Davies and C. Goffman [**24**] and to T. Świątowski [**148**]. The proofs presented here use (already anticipated at the end of Chapter 3) the Radon–Nikodym derivative and the action of homeomorphisms on measures. More specifically, the Bruckner–Davies–Goffman theorem is a result about real-valued universally measurable functions of a real variable. The original proof relied on the order topology of $\mathbb{R}$. It is known that the order topology can be replaced by the Oxtoby–Ulam theorem, thereby relaxing the requirement that the function be a universally measurable function of one real variable to n real variables. A more substantial generalization of the Bruckner–Davies–Goffman theorem that involves absolute measurable spaces will be proved. The theorem of Świątowski is a change of variable theorem for Lebesgue measurable, extended real-valued functions defined on I^n; that is, for each such function f that is real-valued almost everywhere, there is an H in $\mathsf{HOMEO}(I^n)$ such that the Lebesgue integral of fH exists. With the aid of the Radon–Nikodym analogues of the Oxtoby–Ulam theorem, extensions of this theorem to μ-measurable functions that are defined on various spaces (some that are absolute measurable and others that are not) are proved.

4.1. A solution to Goldman's problem

In order to state the Goldman problem we have need of some notation. Fix a separable metrizable space Y. Let μ be a σ-finite Borel measure on a separable metrizable space X. For maps $f: X \to Y$, define the σ-algebras

$$G(Y;X,\mu,f) = \{E : E \subset Y \text{ and } f^{-1}[E] \text{ is } \mu\text{-measurable}\},$$
$$G(Y;X,\mu) = \bigcap\{G(Y;X,\mu,f) : f \text{ is } \mu\text{-measurable}\}.$$

Of course, $G(Y;X,\mu,f)$ is the largest σ-algebra of subsets E of Y such that $f^{-1}[E]$ is in $\mathfrak{M}(X,\mu)$, and $G(Y;X,\mu)$ contains $\mathfrak{B}(Y)$.

Goldman's problem concerned the Lebesgue measure λ on $\mathbb{R}$ and the collection $G(\mathbb{R};\mathbb{R},\lambda)$. He asked for a characterization of this collection; indeed, he conjectured that this collection was $\mathfrak{B}(\mathbb{R})$. Darst proved that the collection $G(\mathbb{R};\mathbb{R},\lambda)$ is precisely the collection $\mathsf{univ}\,\mathfrak{M}(\mathbb{R})$. The following proof is essentially that of Darst [**35**]. First observe that λ corresponds to a continuous, complete, finite Borel measure μ on $\mathbb{R}$ such that the σ-algebras $\mathfrak{M}(\mathbb{R},\lambda)$ and $\mathfrak{M}(\mathbb{R},\mu)$ coincide and the σ-ideals $\mathfrak{N}(\mathbb{R},\lambda)$ and $\mathfrak{N}(\mathbb{R},\mu)$ coincide. Hence λ can be replaced by a suitable finite measure μ. Key to Darst's argument is the following

PROPOSITION 4.1. *Let μ be a complete, finite Borel measure on X. For μ-measurable maps $f: X \to Y$, the Borel measure $f_{\#}\mu$ satisfies*

$$G(Y;X,\mu,f) \supset \mathfrak{M}(Y,f_{\#}\mu) \supset \mathsf{univ}\,\mathfrak{M}(Y),$$

whence $G(Y;X,\mu) \supset \mathsf{univ}\,\mathfrak{M}(Y)$.

PROOF. Let E be a $(f_{\#}\mu)$-measurable set. Then there are Borel sets A and B such that $A \supset E \supset B$ and $f_{\#}\mu(A \setminus B) = 0$. As $f^{-1}[A \setminus B]$ is a μ-measurable set with $\mu\big(f^{-1}[A\setminus B]\big) = 0$ and μ is a complete measure, we have that $f^{-1}[E]$ is μ-measurable, whence E is in $G(Y;X,\mu,f)$ and the first inclusion is verified. The second inclusion is obvious. □

We are now ready to give Darst's solution to Goldman's problem.

THEOREM 4.2 (Darst). *In order that a subset E of $\mathbb{R}$ be such that $f^{-1}[E]$ is Lebesgue measurable for every Lebesgue measurable function $f: \mathbb{R} \to \mathbb{R}$ it is necessary and sufficient that E be a universally measurable set in $\mathbb{R}$. Consequently $G(\mathbb{R};\mathbb{R},\lambda) = \mathsf{univ}\,\mathfrak{M}(\mathbb{R})$, where λ is the usual Lebesgue measure on $\mathbb{R}$.*

PROOF. The previous proposition shows that $G(\mathbb{R};\mathbb{R},\lambda)$ contains $\mathsf{univ}\,\mathfrak{M}(\mathbb{R})$. So suppose that E is not a universally measurable set in $\mathbb{R}$. Recall that λ and $\mathsf{HOMEO}(\mathbb{R})$ generate $\mathsf{univ}\,\mathfrak{M}(\mathbb{R})$. Hence there is an h in $\mathsf{HOMEO}(\mathbb{R})$ such that $h^{-1}[E]$ is not Lebesgue measurable. As h is a λ-measurable function on $\mathbb{R}$, we have $E \notin G(\mathbb{R};\mathbb{R},\lambda)$. □

An immediate consequence of the above theorem is that the statement also holds for real-valued functions $f: X \to \mathbb{R}$ defined on absolute measurable spaces X and nonzero, continuous, complete, finite Borel measures on X.

THEOREM 4.3. *Let X be an absolute measurable space and μ be a continuous, complete Borel measure on X with $0 < \mu(X) < \infty$. In order that a subset E of $\mathbb{R}$ be such that $f^{-1}[E]$ is μ-measurable for every μ-measurable function $f: X \to \mathbb{R}$ it is necessary and sufficient that E be a universally measurable set in $\mathbb{R}$. That is, $G(\mathbb{R}; X, \mu) =$ $\mathsf{univ}\,\mathfrak{M}(\mathbb{R})$.*

PROOF. By the proposition above we already know that $f^{-1}[E]$ is μ-measurable whenever $E \in \mathsf{univ}\,\mathfrak{M}(\mathbb{R})$.

As X is an absolute measurable space and $\text{support}(\mu) \neq \emptyset$ there is an absolute Borel space X_0 such that $\mu(X_0) = \mu(X)$ and $\mu|X_0$ is a positive, continuous, complete Borel measure on X_0. There is a $\mathfrak{B}$-homeomorphism φ of X_0 onto $\mathbb{R}$ such that $\varphi_\#(\mu|X_0)$ is a positive, continuous, complete, finite Borel measure on $\mathbb{R}$ and $\mathfrak{M}(\mathbb{R}, \varphi_\#(\mu|X_0)) = \mathfrak{M}(\mathbb{R}, \lambda)$. Let E be a subset of $\mathbb{R}$ such that $E \notin \mathsf{univ}\,\mathfrak{M}(\mathbb{R})$. As E is uncountable, there is no loss in assuming $0 \notin E$. By Darst's theorem above there is an $h_0: \mathbb{R} \to \mathbb{R}$ such that h_0 is Lebesgue measurable and ${h_0}^{-1}[E] \notin \mathfrak{M}(\mathbb{R}, \lambda)$. We see from the following commutative diagram, where $f_0 = h_0\varphi$, that the map f given by $f(x) = f_0(x)$ whenever $x \in X_0$ and $f(x) = 0$ whenever $x \in X \setminus X_0$ is μ-measurable.

$$\begin{array}{ccccc} X_0 & = & X_0 & \xrightarrow[\subset]{g} & X \\ \varphi\downarrow & & f_0\downarrow & & \downarrow f \\ \mathbb{R} & \xrightarrow{h_0} & \mathbb{R} & = & \mathbb{R} \end{array}$$

Observe $f^{-1}[E] = {f_0}^{-1}[E] \subset X_0$. Since φ is a $\mathfrak{B}$-homeomorphism and ${f_0}^{-1}[E] = \varphi^{-1}{h_0}^{-1}[E]$, we have ${f_0}^{-1}[E] \notin \mathfrak{M}(X_0, \mu|X_0)$. Hence $f^{-1}[E] \notin \mathfrak{M}(X, \mu)$ because X_0 is an absolute Borel space. We have shown $E \notin G(\mathbb{R}; X, \mu)$ and thereby $G(\mathbb{R}; X, \mu) = \mathsf{univ}\,\mathfrak{M}(\mathbb{R})$. □

Recall that if Y is an uncountable absolute Borel space then there is a $\mathfrak{B}$-homeomorphism $\psi: Y \to \mathbb{R}$. For such a map we have $\mathsf{univ}\,\mathfrak{M}(Y) = \psi^{-1}[\mathsf{univ}\,\mathfrak{M}(\mathbb{R})]$. So the last theorem implies the next one.

THEOREM 4.4. *Let X be an absolute measurable space and μ be a continuous, complete Borel measure on X with $0 < \mu(X) < \infty$. Furthermore, let Y be an uncountable absolute Borel space. In order that a subset E of Y be such that $f^{-1}[E]$ is μ-measurable for every μ-measurable map $f: X \to Y$ it is necessary and sufficient that E be a universally measurable set in Y. That is, $G(Y; X, \mu) = \mathsf{univ}\,\mathfrak{M}(Y)$.*

PROOF. Consider the commutative diagram

$$\begin{array}{ccc} X & = & X \\ f\downarrow & & \downarrow g \\ Y & \xrightarrow{\psi} & \mathbb{R} \end{array}$$

where ψ is a $\mathfrak{B}$-homeomorphism and $g = \psi f$. Clearly g is μ-measurable if and only if f is μ-measurable. Suppose that E is such that $f^{-1}[E]$ is μ-measurable for every μ-measurable map $f: X \to Y$. As $f^{-1}[E] = g^{-1}\big[\psi[E]\big]$ for every μ-measurable g, we have that $\psi[E]$ is a universally measurable set in $\mathbb{R}$. Hence E is a universally measurable set in Y. □

In the above proof we have used compositions of maps. Let us discuss various facts about compositions. Consider compositions fg of maps f and g. The following statements are well-known.

(1) If f and g are Borel measurable, then so is fg.
(2) If f is Borel measurable and g is Lebesgue measurable, then fg is Lebesgue measurable.
(3) There are Lebesgue measurable functions f and g such that fg is not Lebesgue measurable.

Recall that a map $f: X \to Y$ is universally measurable if and only if $f^{-1}[U] \in \mathsf{univ}\,\mathfrak{M}(X)$ for every open set U of Y. The collection of all such maps has been denoted by $\mathsf{univ\,MAP}(X;Y)$. The collection of all maps $f: X \to Y$ such that f is μ-measurable (where μ is a continuous, complete, finite Borel measure on X) is denoted by $\mathsf{MAP}(X,\mu;Y)$. Since $\mathfrak{B}(X) \subset \mathsf{univ}\,\mathfrak{M}(X)$, we have that every Borel measurable map defined on X is also universally measurable on X.

Proposition 4.5. *$f \in \mathsf{univ\,MAP}(X;Y)$ if and only if $f^{-1}[M] \in \mathsf{univ}\,\mathfrak{M}(X)$ whenever $M \in \mathsf{univ}\,\mathfrak{M}(Y)$.*

Proof. Proposition 2.19 yields the if part. Let M be in $\mathsf{univ}\,\mathfrak{M}(Y)$, f be in $\mathsf{univ\,MAP}(X;Y)$, and μ be a continuous, complete, finite Borel measure on X. As f is μ-measurable, by Proposition 4.1, $f^{-1}[M]$ is μ-measurable. Hence $f^{-1}[M]$ is a universally measurable set in X. □

To the above list of properties of compositions of maps we can add the following two theorems which are consequences of the above results. The proofs are left to the reader as exercises.

Theorem 4.6. *Suppose that X, Y and Z are separable metrizable spaces and that μ is a continuous, complete, finite Borel measure on X. Let $f: Y \to Z$ and $g: X \to Y$.*

(1) *If f is in $\mathsf{univ\,MAP}(Y;Z)$ and g is in $\mathsf{univ\,MAP}(X;Y)$, then fg is in $\mathsf{univ\,MAP}(X;Z)$.*
(2) *If f is in $\mathsf{univ\,MAP}(Y;Z)$ and g is in $\mathsf{MAP}(X,\mu;Y)$, then fg is in $\mathsf{MAP}(X,\mu;Z)$.*

Theorem 4.7. *Suppose that X, Y and Z are separable metrizable spaces and let $f: Y \to Z$ and $g: X \to Y$. Suppose further that Y is an absolute measurable space. In order that $fg \in \mathsf{MAP}(Y,\mu;Z)$ holds for every g in $\mathsf{MAP}(X,\mu;Y)$ and for every continuous, complete, finite Borel measure μ on X it is necessary and sufficient that f be in $\mathsf{univ\,MAP}(Y;Z)$.*

4.2. Differentiability and $\mathfrak{B}$-maps

A $\mathfrak{B}$-map (see Definition A.18) is a Borel measurable map for which the images of Borel sets are Borel sets. By Purves's theorem, not every continuous function $f\colon \mathbb{R} \to \mathbb{R}$ is a $\mathfrak{B}$-map. One would suspect also that differentiability conditions on a real-valued map of a real variable would have no connections to $\mathfrak{B}$-maps. This was shown to be so by Darst in **[40]**. This section is devoted to the presentation of his example.

4.2.1. An observation. Let $f\colon [0,1] \to \mathbb{R}$ be a $\mathfrak{B}$-map. As homeomorphisms are $\mathfrak{B}$-maps it is clear that the composition $h_1 f h_0$ is also a $\mathfrak{B}$-map whenever h_1 is a homeomorphism of $\mathbb{R}$ and h_0 is a homeomorphism of $[0,1]$. This observation will be useful in the presentation of the example.

4.2.2. An example of A. H. Stone. The following continuous function of bounded variation is attributed to Arthur H. Stone by Darst in **[38]**. On the Cantor ternary set $\mathcal{C}$, define the function g by the formula

$$g\big(\textstyle\sum_{n=1}^{\infty} a_n\, 3^{-n}\big) = \sum_{n=1}^{\infty} a_{2n}\, 9^{-n},$$

where each $a_1, a_2, \ldots,$ is 0 or 2. The natural extension of g to $[0,1]$ (by making it linear on the complementary intervals of $[0,1] \setminus \mathcal{C}$) will also be denoted by g. An elementary computation will show that

$$|g(s) - g(t)| \leq 3\,|s - t| \quad \text{whenever } s \text{ and } t \text{ are in } [0,1].$$

Indeed, suppose that s and t are in $\mathcal{C}$ with $g(s) \neq g(t)$. Define k_0 and k_1 to be the natural numbers such that

$$s_{2k_0} \neq t_{2k_0} \text{ and } s_{2k} = t_{2k} \text{ whenever } k < k_0,$$
$$s_{2k_1-1} \neq t_{2k_1-1} \text{ and } s_{2k-1} = t_{2k-1} \text{ whenever } k < k_1.$$

Then

$$|g(s) - g(t)| \leq 2 \textstyle\sum_{k \geq k_0} \frac{1}{9^k} = \frac{9}{4}\frac{1}{9^{k_0}}.$$

If $k_1 \leq k_0$, then $|s-t| \geq \frac{2}{3^{2k_1-1}} - 2\sum_{j>2k_1-1} \frac{1}{3^j} = \frac{3}{9^{k_1}} \geq \frac{4}{3}|g(s)-g(t)|$. If $k_1 > k_0$, then $|s-t| \geq \frac{2}{9^{k_0}} - 2\sum_{j>2k_0} \frac{1}{3^j} \geq \frac{1}{9^{k_0}} \geq \frac{4}{9}|g(s)-g(t)|$. Hence $|g(s)-g(t)| \leq 3\,|s-t|$ whenever s and t are in $\mathcal{C}$. It now follows that g is Lipschitzian with Lipschitz constant not exceeding 3. So g is a continuous function of bounded variation. Clearly $\operatorname{card}\big(g^{-1}[y]\big) = \mathfrak{c}$ whenever $y \in g[\mathcal{C}]$. As g is Lipschitzian, the Lebesgue measure of $g[\mathcal{C}]$ is 0, whence $g[\mathcal{C}]$ is nowhere dense in $\mathbb{R}$. Moreover, $g\big[[0,1]\big] \subset [0,1]$ and g is infinitely differentiable on the open set $U = [0,1] \setminus \mathcal{C}$.

Let K denote the nowhere dense, compact subset $g[\mathcal{C}]$ of $\mathbb{R}$. It is well known that there is an infinitely differentiable function $h\colon \mathbb{R} \to \mathbb{R}$ such that $h(x) \geq 0$ for every x and such that $h(x) > 0$ if and only if $x \notin K$. For the sake of completeness, we shall give a proof. Let $u\colon \mathbb{R} \to \mathbb{R}$ be the familiar infinitely differentiable function given by the formula $u(x) = e^{-\frac{1}{x}}$ for $x > 0$, and let $\varphi_{(a,b)}(x) = u(x-a)u(b-x)$ for $0 \le a < b \le 1$. With the aid of the Taylor formula it is easily shown that, for each k, there is a constant M_k such that $|\varphi_{(a,b)}{}^{(k)}(x)| \le M_k(x-a)^2(b-x)^2$ whenever $0 \le a < x < b \le 1$. Let (a_n, b_n), $n = 1, 2, \ldots$, be a well ordering of the collection of all bounded components of $\mathbb{R} \setminus K$. For each n, let $h_n = \varphi_{(a_n,b_n)}$. Then a simple modification of $\sum_{n=1}^{\infty} \frac{1}{2^n} h_n$ on the unbounded components of $\mathbb{R} \setminus K$ will give the desired infinitely differentiable function h. Moreover, $h^{(k)} = \sum_{n=1}^{\infty} \frac{1}{2^n} h_n{}^{(k)}$ on the bounded components of $\mathbb{R} \setminus K$ for each k.

4.2.3. The example. Let g be Stone's example and let h be a bounded, infinitely differentiable function such that $h(x) \ge 0$ for every x and such that $h(x) = 0$ if and only if $x \in K$, where $K = g[\mathcal{C}]$. Then, by the fundamental theorem of calculus, $H(y) = \int_0^y h(t)\,dt$, $y \in \mathbb{R}$, is an infinitely differentiable function, whence Lipschitzian. Also, $h^{(k)}g$ is a Lipschitzian function, and $h^{(k)}\big(g(x)\big) = 0$ whenever $x \in \mathcal{C}$.

Define f to be the composition Hg. Then f is Lipschitzian. Clearly f is infinitely differentiable on $\mathbb{R} \setminus \mathcal{C}$; moreover,

$$f^{(k+1)}(x) = h^{(k)}\big(g(x)\big)\big(g'(x)\big)^{(k+1)} \quad \text{whenever} \quad x \in \mathbb{R} \setminus \mathcal{C}.$$

Let us show, at each x in $\mathcal{C}$, that f is differentiable and $f'(x) = 0 = h\big(g(x)\big)$. Indeed, for $t \ne 0$, there is an η between $g(x)$ and $g(x+t)$ such that

$$\frac{f(x+t) - f(x)}{t} = h(\eta)\frac{g(x+t) - g(x)}{t}$$

and observe that g is Lipschitzian and that $h\big(g(x)\big) = 0$ whenever $x \in \mathcal{C}$. Since h is continuous and g is Lipschitzian, it follows that $f'(x)$ exists and is equal to 0 at each x in $\mathcal{C}$.

Note that $g'(x)$ exists at every x in $\mathbb{R} \setminus \mathcal{C}$. It will be convenient to define the bounded function

$$G(x) = \begin{cases} 3, & \text{if } x \in \mathcal{C}, \\ g'(x), & \text{if } x \notin \mathcal{C}. \end{cases}$$

For $k \ge 0$, it is easily seen that the product $(h^{(k)}g)\cdot G$ is continuous because G is bounded and is continuous whenever $x \notin \mathcal{C}$. Hence

$$f^{(1)}(x) = \big(hg\big)(x)G(x) \quad \text{and} \quad f(x) = \textstyle\int_0^x \big(hg\big)(t)G(t)\,dt$$

for every x. Moreover

$$f^{(k+1)}(x) = \big(h^{(k)}g\big)(x)\big(G(x)\big)^{k+1}$$

whenever $x \notin \mathcal{C}$ and $k \ge 0$.

Let us show that $f^{(2)}(x)$ exists and equals 0 whenever $x \in \mathcal{C}$. To this end, observe that, for every x,

$$\begin{aligned} f^{(1)}(x) &= G(x)h\big(g(x)\big) \\ &= G(x) \int_0^{g(x)} h^{(1)}(t)\,dt \\ &= G(x) \int_0^x h^{(1)}\big(g(t)\big)G(t)\,dt. \end{aligned}$$

Since $G(x')h^{(1)}\big(g(x)\big) = 0$ whenever $x \in \mathcal{C}$ and $x' \in \mathbb{R}$, the first displayed equality gives $f^{(2)}(x)$ exists and is equal to $0 = \big(G(x)\big)^2 h^{(1)}\big(g(x)\big)$ for each x in $\mathcal{C}$. It now follows that, for every x,

$$\begin{aligned} f^{(2)}(x) &= \big(G(x)\big)^2 h^{(1)}\big(g(x)\big) \\ &= \big(G(x)\big)^2 \int_0^{g(x)} h^{(2)}(t)\,dt \\ &= \big(G(x)\big)^2 \int_0^x h^{(2)}\big(g(t)\big)G(t)\,dt. \end{aligned}$$

It is equally easy to show that $f^{(k+1)} = (h^{(k)}g)\cdot G^{k+1}$ is differentiable for each k.

THEOREM 4.8 (Darst). *There are infinitely differentiable real-valued functions of a real variable that are not $\mathfrak{B}$-maps.*

PROOF. With h, H and g as in the discussion immediately preceding the statement of the theorem, we have defined an infinitely differentiable function f by the formula

$$f(x) = \int_0^{g(x)} h(t)\,dt = Hg(x).$$

The set $K = g[\mathcal{C}]$ is an uncountable set. Also, $g^{-1}[\{y\}]$ is an uncountable set for each y in K. Hence $f^{-1}[\{z\}]$ is an uncountable set for each z in $H[K]$. Clearly $H|\big(g\big[[0,1]\big]\big)$ is a homeomorphism of $g\big[[0,1]\big]$ into $\mathbb{R}$ and thereby $H[K]$ is an uncountable set. As $f[\mathcal{C}] = H[K]$ we have that $f[\mathcal{C}]$ is an uncountable set that is contained in the set of uncountable order of f. Consequently f is not a $\mathfrak{B}$-map by Purves's theorem. □

4.3. Radon–Nikodym derivative and Oxtoby–Ulam theorem

At the end of the Comment section of the last chapter the notion of absolute continuity of measures was introduced in connection with the Oxtoby–Ulam theorem. Although an analogue of the Oxtoby–Ulam theorem for the Cantor space $\{0,1\}^{\mathbb{N}}$ does not exist, there is an interesting absolute continuity property shared by the spaces $[0,1]^n$ and $\{0,1\}^{\mathbb{N}}$. This absolute continuity property will be pursued further in this section.

Recall from page 86 that a measure μ is said to be *absolutely continuous* with respect a measure ν (denoted $\mu \ll \nu$) if $\mu(E) = 0$ whenever $\nu(E) = 0$. For convenience we shall assume μ and ν are continuous, complete, finite Borel measures on a separable metrizable space X. If μ is absolutely continuous with respect to ν, then there is a real-valued, Borel measurable function, denoted by $\frac{d\mu}{d\nu}$ and called the *Radon–Nikodym derivative* of μ with respect to ν, that satisfies the identity:

$$\mu(B) = \int_B \frac{d\mu}{d\nu}\, d\nu = \int_X \chi_B \cdot \frac{d\mu}{d\nu}\, d\nu \text{ whenever } B \in \mathfrak{B}(X).$$

In the investigation of the interplay of the Radon–Nikodym derivative and the Oxtoby–Ulam theorem, we will "piece together" measures associated with the "piecing together" of homeomorphisms. Here are two straightforward exercises on absolutely continuous measures and Radon–Nikodym derivatives.

PROPOSITION 4.9. *Let X_1 and X_2 be disjoint Borel sets of a separable metrizable space X, and let $\nu = \nu_1 \llcorner X_1 + \nu_2 \llcorner X_2$, where ν_1 and ν_2 are continuous, complete, finite Borel measures on X. If μ is a continuous, complete, finite Borel measure on X such that $\mu \llcorner X_1 \ll \nu_1$ and $\mu \llcorner X_2 \ll \nu_2$, then $\mu \ll \nu$ and*

$$\frac{d(\mu \llcorner (X_1 \cup X_2))}{d\nu} = \frac{d(\mu \llcorner X_1)}{d\nu_1} \cdot \chi_{X_1} + \frac{d(\mu \llcorner X_2)}{d\nu_2} \cdot \chi_{X_2} \, .$$

PROPOSITION 4.10. *Let μ and ν be continuous, complete, finite Borel measures on a separable metrizable space X. If $h \in$ HOMEO(X), then $\mu \ll h_\# \nu$ if and only if ${h^{-1}}_\# \mu \ll \nu$.*

In Chapter 3 we saw that the Oxtoby–Ulam theorem concerned Lebesgue-like measures on topological n-cells. We have seen also that it has many generalizations. But its extensions to spaces such as open subsets of $\mathbb{R}^n$ or to other positive measures more general than Lebesgue-like ones on I^n are yet to be found. In this section we will use the Radon–Nikodym derivative to extend the Oxtoby–Ulam theorem. The proofs of some of the theorems concerning real-valued functions will be facilitated by the existence of such Radon–Nikodym derivatives.

4.3.1. Homeomorphism group and Radon–Nikodym derivative. The spaces that are of interest here are those that contain open sets that are topologically equal to open sets of n-dimensional Euclidean space. Indeed, we shall concentrate on those open sets that are homeomorphic to $I^n \setminus \partial I^n$, where $I = [0, 1]$, or to $[0, 1) \times (0, 1)^{n-1}$.

As we have seen in the comment section of the previous chapter, the Oxtoby–Ulam theorem can be couched in terms of the existence of a bounded Radon–Nikodym derivative in conjunction with homeomorphisms. More generally, we have the following consequence of the Oxtoby–Ulam theorem.

LEMMA 4.11. *Let X be a separable metrizable space that contains a topological copy Y_1 of I^n such that $U_1 = Y_1 \setminus \partial Y_1$ is open in X and such that ∂Y_1 contains a topological copy Y_0 of I^{n-1} such that $U_0 = Y_0 \setminus \partial Y_0$ is open in $X \setminus U_1$. Suppose that ν is a positive, continuous, complete, finite Borel measure on X.*

(1) *If μ is a continuous, complete, finite Borel measure on X such that $\mu | Y_1$ is Lebesgue-like on Y_1, then there exists an h in* HOMEO$(X; X \setminus U_1$ fixed) *such that $\mu \llcorner Y_1 \ll h_\# \nu$ and $\frac{d(\mu \llcorner Y_1)}{d(h_\# \nu)} = \frac{\mu(U_1)}{\nu(U_1)}$ on U_1.*

(2) *If μ is a continuous, complete, finite Borel measure on X, then $\mu \llcorner U_1$ is absolutely continuous with respect to $h_\# \nu$ and the Radon–Nikodym derivative $\frac{d(\mu \llcorner U_1)}{d(h_\# \nu)}$ is bounded for some h in* $\mathsf{HOMEO}(X; X \setminus U_1 \text{ fixed})$.
(3) *If the restricted measure $\nu | U_0$ is also positive on the set U_0, then $\mu \llcorner (U_1 \cup U_0)$ is absolutely continuous with respect to $h_\# \nu$ and the Radon–Nikodym derivative $\frac{d(\mu \llcorner (U_1 \cup U_0))}{d(h_\# \nu)}$ is bounded for some h in* $\mathsf{HOMEO}(X; X \setminus (U_1 \cup U_0) \text{ fixed})$.

Proof. Statement (1) follows easily from the Oxtoby–Ulam theorem. Indeed, $\nu_0 = (\nu \llcorner U_1)|Y_1$ is Lebesgue-like on Y_1 and $\frac{d(\mu|Y_1)}{d(h'_\# \nu_0)} = \frac{\mu(U_1)}{\nu_0(U_1)}$ on U_1 for some h' in $\mathsf{HOMEO}(Y_1; \partial Y_1 \text{ fixed})$. As U_1 is open in X, there is a natural extension h of h' to a homeomorphism in $\mathsf{HOMEO}(X; X \setminus U_1 \text{ fixed})$. It follows that $\frac{d(\mu \llcorner Y_1)}{d(h_\# \nu)}$ equals $\frac{d(\mu \llcorner U_1)}{d(h_\# \nu)} \cdot \chi_{U_1}$ which is equal to $\frac{d(\mu \llcorner U_1)}{d((h_\# \nu) \llcorner U_1)} \cdot \chi_{U_1}$. For Borel sets B that are contained in U_1 we have $\big(\mu \llcorner U_1\big)(B) = \mu(B)$ and $\big((h_\# \nu) \llcorner U_1\big)(B) = \big(h_\# \nu\big)(B) = \big(h'_\# \nu_0\big)(B)$. Hence $\frac{d(\mu \llcorner U_1)}{d((h_\# \nu) \llcorner U_1)} \cdot \chi_{U_1}$ is equal to $\frac{d(\mu|Y_1)}{d(h'_\# \nu_0)} = \frac{\mu(U_1)}{\nu_0(U_1)} = \frac{\mu(U_1)}{\nu(U_1)}$ on U_1. The first statement is established.

We turn to statement (2). Let $\mu' = (\mu + \nu) \llcorner U_1$. Then $\mu'|Y_1$ is Lebesgue-like on Y_1. There is an h in $\mathsf{HOMEO}(X; X \setminus U_1 \text{ fixed})$ and a positive number c such that $\mu' \ll h_\# \nu$ and $\frac{d\mu'}{d(h_\# \nu)} = c\, \chi_{U_1}$. Since $0 \le \mu \llcorner U_1 \le \mu'$ and $0 \le \nu \llcorner U_1 \le \mu'$, we have $\mu \llcorner U_1 \ll h_\# \nu$ and $\nu \llcorner U_1 \ll h_\# \nu$ and consequently $\frac{d(\mu \llcorner U_1)}{d(h_\# \nu)} + \frac{d(\nu \llcorner U_1)}{d(h_\# \nu)} \le c$. It now follows that $\frac{d(\mu \llcorner U_1)}{d(h_\# \nu)}$ is bounded.

For statement (3), suppose further that ν is positive on U_0. There is an h'' in $\mathsf{HOMEO}(X \setminus U_1; X \setminus (U_1 \cup U_0) \text{ fixed})$ such that the measures $\mu'' = (\mu|(X \setminus U_1)) \llcorner U_0$ and $\nu'' = h''_\# (\nu|(X \setminus U_0))$ satisfy $\mu'' \ll \nu''$ and $\frac{d\mu''}{d(h''_\# \nu'')}$ is bounded on Y_0. Observe that $\mu'' = (\mu \llcorner U_0)|(X \setminus U_1)$ and $\nu'' = (\nu \llcorner U_0)|(X \setminus U_1)$ hold. As U_1 is open in X and as Y_0 is contained in ∂Y_1 and U_0 is open in $X \setminus U_1$, there is an extension h' of h'' in $\mathsf{HOMEO}(X; X \setminus (U_1 \cup U_0) \text{ fixed})$ such that $(h'_\# \nu)|(X \setminus U_1) = (h''_\# \nu)|(X \setminus U_1)$. It now follows that $\mu \llcorner U_0 \ll (h'_\# \mu) \llcorner U_0 \ll h'_\# \nu$ and $\frac{d(\mu \llcorner U_0)}{d(h'_\# \nu)}$ is bounded. There is an h in $\mathsf{HOMEO}(X; X \setminus U_1 \text{ fixed})$ (by statement (2)) such that $\mu \llcorner U_1 \ll (hh')_\# \nu$ and $\frac{d(\mu \llcorner U_1)}{d((hh')_\# \nu)}$ is bounded. From $hh' = h'$ on U_0 we see that $((hh')_\# \nu) \llcorner U_0 = (h'_\# \nu) \llcorner U_0$ and $\frac{d(\mu \llcorner U_0)}{d((hh')_\# \mu)} = \frac{d(\mu \llcorner U_0)}{d(h'_\# \mu)}$ on U_0. The proof is completed by observing that $U_1 \cap U_0 = \emptyset$ yields $\frac{d(\mu \llcorner (U_1 \cup U_0))}{d((hh')_\# \nu)} = \frac{d(\mu \llcorner U_1)}{d((hh')_\# \nu)} + \frac{d(\mu \llcorner U_0)}{d((hh')_\# \nu)}$. □

4.3.2. Applications. To apply the last lemma to open subsets of n-dimensional manifolds M we begin by preparing some groundwork. Let μ and ν be continuous, complete, finite Borel measures on M such that the support of ν is M and the support of $\nu|\partial M$ is the $(n-1)$-dimensional manifold ∂M. With the aid of the measure $\mu + \nu$ we can construct a base $\mathcal{B}$ for the open sets of M such that each V in $\mathcal{B}$ satisfies the following conditions.

(1) $\mathrm{Cl}_M(V)$ is a topological n-cell contained in X and V is equal to $\mathrm{Int}_M\big(\mathrm{Cl}_M(V)\big)$,
(2) $\mathrm{Cl}_M(V) \cap \partial M$ is either empty or a topological $(n-1)$-cell,
(3) $\mu(\mathrm{Bd}_M(V)) = 0$ and $\nu(\mathrm{Bd}_M(V)) = 0$. Note that ∂V and $\mathrm{Bd}_M(V)$ need not coincide.

LEMMA 4.12. *Let μ and ν be continuous, complete, finite Borel measures on M; and, for a finite collection V_i, $i = 1, 2, \ldots, k$, of open sets from the collection $\mathcal{B}$ described above, let $X = \bigcup_{i=1}^{k} \mathrm{Cl}_M(V_i)$. Then there is an h in* $\mathsf{HOMEO}(M; X \cap \partial M \text{ inv}) \cap \mathsf{HOMEO}(M; M \setminus X \text{ fixed})$ *such that $\mu \llcorner X \ll h_{\#}\nu$, $h_{\#}\nu(\mathrm{Bd}_M(X)) = 0$, and $\frac{d(\mu \llcorner X)}{d(h_{\#}\nu)}$ is bounded.*

PROOF. As V_i is in $\mathcal{B}$, we have $Y_i = \mathrm{Cl}_M(V_i)$ is a topological n-cell contained in M. The proof is an induction on k. For $k = 1$, either $Y_1 \cap \partial M = \emptyset$ or $Y_1 \cap \partial M$ is a topological $(n-1)$-cell. The first case follows from Lemma 4.11 (2) since $U_1 = Y_1 \setminus \partial Y_1$. For the second case we have $V_1 \setminus U_1$ is open in $Y_1 \setminus U_1$. This case follows from Lemma 4.11 (3).

Let $m > 1$ and assume the statement is true whenever $k < m$. Let X_1 and X_2 be as in the lemma with $k_1 < m$ and $k_2 < m$. Let h' be in $\mathsf{HOMEO}(M; X_1 \cap \partial M \text{ inv}) \cap \mathsf{HOMEO}(M; M \setminus X_1 \text{ fixed})$ such that $\mu \llcorner X_1 \ll h'_{\#}\nu$, $\big(h'_{\#}\nu\big)(\mathrm{Bd}_M(X_1)) = 0$, and $\frac{d(\mu \llcorner X_1)}{d(h'_{\#}\nu)}$ is bounded. Then let h'' be in $\mathsf{HOMEO}(M; X_2 \cap \partial M \text{ inv}) \cap \mathsf{HOMEO}(M; M \setminus X_2 \text{ fixed})$ such that $\mu \llcorner X_2 \ll h''h'_{\#}\nu$, $\big(h''h'_{\#}\nu\big)(\mathrm{Bd}_M(X_2)) = 0$, and $\frac{d(\mu \llcorner X_2)}{d(h''h'_{\#}\nu)}$ is bounded. Let $h = h''h'$ and $X = X_1 \cup X_2$. As $h(x) = x$ for every x in $M \setminus X$, we have $h[X] = X$ and $h[X \cap \partial M] = X \cap \partial M$. As

$$h_{\#}\nu = \nu \llcorner (M \setminus X) + (h'_{\#}\nu) \llcorner (X \setminus X_2)) + ((h''h')_{\#}\nu) \llcorner X_2,$$

we have $\mu \llcorner X = \mu \llcorner (X_1 \setminus X_2) + \mu \llcorner X_2 \ll h_{\#}\nu$ and hence $\frac{d(\mu \llcorner X)}{d(h_{\#}\nu)}$ is bounded. Only $h_{\#}\nu(\mathrm{Bd}_M(X)) = 0$ remains to be shown. To this end we have $\mathrm{Bd}_M(X) \subset \big(\mathrm{Bd}_M(X_2) \setminus \mathrm{Int}_M(X_1)\big) \cup \big(\mathrm{Bd}_M(X_1) \setminus \mathrm{Int}_M(X_2)\big)$. As $h^{-1}[\mathrm{Bd}_M(X)] \subset (h''h')^{-1}[\mathrm{Bd}_M(X_2)] \cup h'^{-1}[\mathrm{Bd}_M(X_1) \setminus \mathrm{Int}_M(X_2)]$, we have $h_{\#}\nu(\mathrm{Bd}_M(X)) = 0$. □

The next theorem is a Radon–Nikodym version of the Oxtoby–Ulam theorem for compact manifolds. This generalization permits the use of Borel measures μ which are not necessarily Lebesgue-like and to certain Borel measures ν such that $\nu|\partial M$ are finite sums of Lebesgue-like measures. The proof is an immediate consequence of the previous lemma since M is compact.

THEOREM 4.13. *Let M be a compact manifold and let ν be a positive, continuous, complete, finite Borel measure on M such that $\nu|\partial M$ is positive on ∂M whenever $\partial M \neq \emptyset$. If μ is a continuous, complete, finite Borel measure on M, then there is an h in* $\mathsf{HOMEO}(M; \partial M \text{ inv})$ *such that $\mu \ll h_{\#}\nu$ and $\frac{d\mu}{d(h_{\#}\nu)}$ is bounded.*

Let us turn to applications of Lemma 4.12 to sets that are open in $\mathbb{R}^n$ or open in $[0, \infty) \times \mathbb{R}^{n-1}$. Here one can take advantage of standard subdivisions of open sets by countably many nonoverlapping n-cubes. In fact, if μ and ν are σ-finite Borel measures on an open set X of $\mathbb{R}^n$ or of $[0, \infty) \times \mathbb{R}^{n-1}$ and if ε is a positive number, then there is a sequence $\mathcal{Y} = \{Y_i : i = 1, 2, \ldots\}$ of nonoverlapping n-cells such that

(1) $\mathcal{Y}$ covers X,
(2) $\mathcal{Y}$ is a locally finite collection in X,
(3) $\mathrm{diam}(Y_i) < \varepsilon$, $\mu(\mathrm{Bd}_X(Y_i)) = 0$ and $\nu(\mathrm{Bd}_X(Y_i)) = 0$ for every i,
(4) $\lim_{i \to \infty} \mathrm{diam}(Y_i) = 0$.

In order to apply the lemma to σ-finite measures it will be necessary to add the requirement that the measure of K be finite whenever K is a compact subset of X. Such measures are called *Radon measures*. We point out that there are σ-finite measures that are not Radon measures. Indeed, consider the measure μ on $[0,1]^2$ equal to the Hausdorff one-dimensional measure limited to the subset of $[0,1]^2$ consisting of those points whose first coordinate is a rational number.

THEOREM 4.14. *Suppose that M is either $\mathbb{R}^n$ or $[0,\infty) \times \mathbb{R}^{n-1}$. Let X be an open set in M and let μ and ν be continuous, complete, σ-finite Borel measures on M such that $\mu(K)$ and $\nu(K)$ are finite whenever K is a compact subset of X and such that ν is positive on M and $\nu|\partial M$ is positive on ∂M if ∂M is not empty. Then there exists an h in* $\mathsf{HOMEO}(M; M \setminus X \text{ fixed}) \cap \mathsf{HOMEO}(M; X \cap \partial M \text{ inv})$ *such that $\mu \llcorner X \ll h_\# \nu$ and $\frac{d(\mu \llcorner X)}{d(h_\# \nu)} \cdot \chi_K$ is bounded whenever K is a compact subset of X. Moreover, if $\varepsilon > 0$, then the distance between h and* id *can be made less than ε.*

PROOF. Let $\mathcal{Y}$ be as described in the paragraph preceding the statement of the theorem. For each m let $X_m = \bigcup_{i=1}^m Y_i$ and define h_m to be the corresponding homeomorphism given by Lemma 4.12. As $\mathcal{Y}$ is a nonoverlapping collection we have that $h_m[Y_i] = Y_i$ for every i and every m and that $h_m(x) = h_k(x)$ whenever $x \in Y_k$ and $m \geq k$. Hence h_m, $m = 1, 2, \ldots$, converges pointwise to a bijection of M. Also $\|h_k(x) - h_m(x)\| \leq \max\{\operatorname{diam}(Y_i) : i \geq \min\{k, m\}\}$, whence the sequences h_m, $m = 1, 2, \ldots$, and $h_m{}^{-1}$, $m = 1, 2, \ldots$, are uniformly convergent. Consequently there is an h in $\mathsf{HOMEO}(M; M \setminus X \text{ fixed}) \cap \mathsf{HOMEO}(M; X \cap \partial M \text{ inv})$ such that $h(x) = h_m(x)$ whenever x is in Y_m. Clearly, $\mu \llcorner X_{m+1} \ll h_\# \nu$ and $\frac{d(\mu \llcorner X_{m+1})}{d(h_\# \nu)} = \frac{d(\mu \llcorner X_{m+1})}{d(h_{m+1\#} \nu)} \cdot \chi_{X_{m+1}}$. Observe $\mu \llcorner X = \sum_{m=1}^\infty \mu \llcorner (X_m \setminus X_{m-1})$. Since $\mu \llcorner (X_{m+1} \setminus X_m) \leq \mu \llcorner X_{m+1}$, we have $\mu \llcorner X \ll h_\# \nu$ and

$$
\begin{aligned}
\frac{d(\mu \llcorner X)}{d(h_\# \nu)} &= \sum_{m=1}^\infty \frac{d(\mu \llcorner (X_m \setminus X_{m-1}))}{d(h_\# \nu)} \\
&= \sum_{m=1}^\infty \frac{d(\mu \llcorner Y_m)}{d(h_\# \nu)} \\
&= \sum_{m=1}^\infty \frac{d(\mu \llcorner Y_m)}{d(h_{m\#} \nu)} \cdot \chi_{Y_m}.
\end{aligned}
$$

Suppose that K is a compact subset of X. Then, by the conditions on $\mathcal{Y}$, there is an m_0 such that $X_m \supset K$ whenever $m \geq m_0$. Hence $\frac{d(\mu \llcorner X)}{d(h_\# \nu)}$ is bounded on X_{m_0} and thereby on K. □

In the above theorem the open subset X of the manifold $\mathbb{R}^n$ or $[0,\infty) \times \mathbb{R}^{n-1}$ is, in a sense, triangulable by a collection $\mathcal{Y}$ of n-cells that are locally finite in the set X. This permits the Radon–Nikodym derivative to be "locally bounded" in X. If one is willing to relax this locally bounded condition, then the set X can be chosen to be an open set in any n-dimensional manifold M. Such a weaker form of the theorem will be useful for "almost every" assertions as we shall see in the discussion of the Bruckner–Davies–Goffman theorem which will be developed shortly. We leave the proof of the following weaker theorem and preparatory lemma as exercises.

LEMMA 4.15. *Suppose that X is an open set in an n-dimensional manifold M and let μ and ν be continuous, complete, σ-finite Borel measures on M such that $\mu(K)$ and $\nu(K)$ are finite whenever K is a compact subset of X and such that ν is positive on M and $\nu|\partial M$ is positive on ∂M if ∂M is not empty. Then there is a sequence $\mathcal{Y} = \{Y_i : i = 1, 2, \ldots\}$ of n-cells such that*

(1) $\{\mathrm{Int}_M(Y) : Y \in \mathcal{Y}\}$ *covers* X,
(2) $\mathcal{Y}$ *is a locally finite collection in* X,
(3) *for each* i, $\mathrm{diam}(Y_i) < \varepsilon$, $\mu(\mathrm{Bd}_X(Y_i)) = 0$ *and* $\nu(\mathrm{Bd}_X(Y_i)) = 0$,
(4) $\lim_{i\to\infty} \mathrm{diam}(Y_i) = 0$.

The lemma and Theorem 4.14 imply the following theorem for open subsets of manifolds.

THEOREM 4.16. *Let X be an open subset of an n-dimensional manifold M and let μ and ν be continuous, σ-finite Borel measures on M such that $\mu(K)$ and $\nu(K)$ are finite whenever K is a compact subset of X and such that ν is positive on M and $\nu|\partial M$ is positive on ∂M if ∂M is not empty. Then there is an h in both* $\mathsf{HOMEO}(M; M \setminus X \text{ fixed})$ *and* $\mathsf{HOMEO}(M; X \cap \partial M \text{ inv})$ *such that $\mu \llcorner X \ll h_\# \nu$. Moreover, if $\varepsilon > 0$, then the distance between h and* id *can be made less than ε.*

Let us turn our attention to the Warsaw circle. The Warsaw circle can be considered as a compactification of the real line $\mathbb{R}$. Indeed, if one denotes the set of all points x of the Warsaw circle at which W is not locally connected by W_0, then $W_1 = W \setminus W_0$ is homeomorphic to $\mathbb{R}$. It is easily seen that, in the subspace W_1, there is a countable, closed set V such that each h in $\mathsf{HOMEO}(W_1; V \text{ fixed})$ has an extension H in $\mathsf{HOMEO}(W; W_0 \cup V \text{ fixed})$.

THEOREM 4.17. *Let μ and ν be continuous, complete, finite Borel measures on the Warsaw circle W such that $\nu|W_0$ and $\nu|W_1$ are positive measures on W_0 and W_1, respectively. Then there exists an h in* $\mathsf{HOMEO}(W)$ *such that $\mu \ll h_\# \nu$.*

PROOF. As W_0 is an arc, there is an h_0 in $\mathsf{HOMEO}(W_0; \partial W_0 \text{ fixed})$ such that $\mu|W_0 \ll h_{0\#}(\nu|W_0)$. It is easy to see that there is an extension h_1 of h_0 in $\mathsf{HOMEO}(W)$. So, $\mu \llcorner W_0 \ll (h_{1\#}\nu) \llcorner W_0$. It is easily seen (see Exercise 4.4) that there is an h_2 in $\mathsf{HOMEO}(W; W_0 \text{ fixed})$ such that $\mu \llcorner W_1 \ll h_{2\#}((h_{1\#}\nu) \llcorner W_1)$. Clearly we have $(h_{1\#}\nu) \llcorner W_0 = h_{2\#}((h_{1\#}\nu) \llcorner W_0)$. Hence $\mu = \mu \llcorner W_0 + \mu \llcorner W_1 \ll h_{2\#} h_{1\#} \nu$. Let $h = h_2 h_1$ to complete the proof. □

Finally we turn to the finitely triangulable spaces, the last of the many spaces that we have studied for which the Oxtoby–Ulam theorem can be used.

THEOREM 4.18. *If X is a finitely triangulable space with* $\dim X > 0$ *and if μ and ν are continuous, complete, finite Borel measures on X such that $\nu|\sigma$ is a positive measure on σ for every nondegenerate simplex σ in some triangulation K of X, then there is an h in* $\mathsf{HOMEO}(X)$ *such that $\mu \ll h_\#\nu$ and $\frac{d\mu}{d(h_\#\nu)}$ is bounded.*

PROOF. The proof is a simple induction on the dimension of X. For $\dim X = 1$ let K be a triangulation of X such that $\nu|\sigma$ is positive whenever σ is a one-dimensional simplex in K. Since X is finitely triangulable, there is an h in $\mathsf{HOMEO}(X; V \text{ fixed})$, where V is the set consisting of all vertices of X, such that $\mu \ll h_\# \nu$ and $\frac{d\mu}{d(h_\# \nu)}$ is bounded. We shall give the proof for $\dim X = 2$; the general case is easily seen from this one. Let $K^1 = \{\sigma \in K : \dim \sigma \le 1\}$ and define $X_0 = |K^1|$. There is an h_0 in $\mathsf{HOMEO}(X_0; V \text{ fixed})$ such that $\mu|X_0 \ll h_{0\#}(\nu|X_0)$ and $\frac{d(\mu|X_0)}{d(h_{0\#}(\nu|X_0))}$ is bounded on X_0. Consider one of the finitely many two-dimensional simplex σ in K. We have $\partial\sigma \subset X_0$ and that $\sigma \setminus \partial\sigma$ is open in X. Hence h_0 has an extension h_0' in $\mathsf{HOMEO}(X_0 \cup \sigma)$. It follows that h_0 has an extension h_1 in $\mathsf{HOMEO}(X)$ such that $\mu \llcorner X_0 \ll (h_{1\#}\nu) \llcorner X_0$ and $\frac{d(\mu \llcorner X_0)}{d(h_{1\#}\nu)} = \frac{d(\mu|X_0)}{d(h_{0\#}(\nu|X_0))}$ on X_0. By Lemma 4.11 we infer the existence of an h_2 in $\mathsf{HOMEO}(X; X_0 \text{ fixed})$ such that $\mu \llcorner (X \setminus X_0) \ll h_{2\#}((h_{1\#}\nu) \llcorner (X \setminus X_0))$ and $(h_{1\#}\nu) \llcorner X_0 = h_{2\#}((h_{1\#}\nu) \llcorner X_0)$. Hence $\mu = \mu \llcorner X_0 + \mu \llcorner (X \setminus X_0) \ll h_\# \nu$, where $h = h_2 h_1$. Moreover, as

$$\frac{d\mu}{d(h_\#\nu)} = \frac{d(\mu \llcorner X_0)}{d(h_{1\#}\nu)} \cdot \chi_{X_0} + \frac{d(\mu \llcorner (X \setminus X_0))}{d(h_{2\#}((h_{1\#}\nu) \llcorner (X \setminus X_0))} \cdot \chi_{(X \setminus X_0)}$$

and the second term of the right-hand side is bounded by Lemma 4.11, the Radon–Nikodym derivative $\frac{d\mu}{d(h_\#\nu)}$ is bounded. □

We must not forget the Hilbert space $[0,1]^{\mathbb{N}}$ and the space $\mathbb{N}^{\mathbb{N}}$ which is homeomorphic to $\mathcal{N}$. Also, there is the compact Menger manifold of positive dimension that was mentioned at the end of the last chapter. The proof of the corresponding theorem for these spaces (stated below) is left as an exercise for the reader (Exercise 4.5).

THEOREM 4.19. *Let X be the Hilbert cube or the space $\mathcal{N}$ or a compact Menger manifold of positive dimension. If μ and ν are continuous, complete, finite Borel measures on X such that ν is a positive measure on X, then there is an h in* $\mathsf{HOMEO}(X)$ *such that $\mu \ll h_\#\nu$ and the Radon–Nikodym derivative $\frac{d\mu}{d(h_\#\nu)}$ is bounded.*

The last of our examples is the Cantor space $\{0,1\}^{\mathbb{N}}$. For this, see Theorem 3.67.

We have found many examples of spaces X such that the notions of absolute continuity and Radon–Nikodym derivative provide connections between the group of homeomorphisms $\mathsf{HOMEO}(X)$ and the collection of all continuous, complete, finite Borel measures on X. These examples lead to the following definition, where $\mathrm{MEAS}^{\mathrm{finite}}(X) \cap \mathrm{MEAS}^{\mathrm{cont}}(X)$ is the collection of all continuous, complete, finite Borel measures on X.

DEFINITION 4.20. *Let ν be a continuous, complete, finite Borel measure on a separable metrizable space X. The measure ν and the group* $\mathsf{HOMEO}(X)$ *are said to* (ac)-generate $\mathrm{MEAS}^{\mathrm{finite}}(X) \cap \mathrm{MEAS}^{\mathrm{cont}}(X)$ *if to each continuous, complete, finite Borel measure μ on X there corresponds an h in* $\mathsf{HOMEO}(X)$ *such that $\mu \ll h_\#\nu$.*

Observe that every space X that has appeared in this application section possessed a measure ν such that it and $\mathsf{HOMEO}(X)$ (ac)-generate $\mathrm{MEAS}^{\mathrm{finite}}(X) \cap \mathrm{MEAS}^{\mathrm{cont}}(X)$. We have the following general theorem.

THEOREM 4.21. *Let X be a separable metrizable space and ν be a continuous, complete, finite Borel measure on X. If ν and* $\mathsf{HOMEO}(X)$ (ac)*-generate* $\mathrm{MEAS}^{\mathrm{finite}}(X) \cap \mathrm{MEAS}^{\mathrm{cont}}(X)$*, then ν and* $\mathsf{HOMEO}(X)$ *generate* $\mathsf{univ}\,\mathfrak{M}(X)$.

PROOF. We will apply Proposition 3.14. Let E be a subset of X such that $E \in \bigcap\{\mathfrak{M}(X, h_{\#}\nu) : h \in \mathsf{HOMEO}(X)\}$. Suppose that μ is a continuous, complete, finite Borel measure on X. There is an h in $\mathsf{HOMEO}(X)$ such that $\mu \ll h_{\#}\nu$. Since E is in $\mathfrak{M}(X, h_{\#}\nu)$, we have Borel sets A and B such that $A \subset E \subset B$ and $\big(h_{\#}\nu\big)(B \setminus A) = 0$. Hence $\mu(B \setminus A) = 0$ and thereby $E \in \mathfrak{M}(X, \mu)$. Consequently, E is a universally measurable set in X. $\square$

4.4. Zahorski spaces

In the previous section about Radon–Nikodym derivatives on various special spaces X, the measure ν has been assigned a special role due to its "positiveness" on X. Often this measure is constructed so that nice properties hold for the problem under investigation. This will happen in the discussion of the Bruckner–Davies–Goffman theorem which is given in the next section. The constructions will involve a class of absolute F_σ spaces called Zahorski spaces. (See Appendix A for more on Zahorski spaces.)

DEFINITION 4.22. *A separable metrizable space is a* Zahorski space *if it is the empty space or it is the union of a countable sequence of topological copies of the Cantor set. A subset Z of a separable metrizable space X is called a* Zahorski set *if it is a Zahorski subspace of X.*

Zahorski spaces have very natural continuous, complete, finite Borel measures associated with them. Indeed, for nonempty Zahorski spaces, positive ones are easily constructed. Such examples motivate the following definition (which is Definition A.42 in Appendix A).

DEFINITION 4.23. *Let E be a Zahorski set contained in a separable metrizable space X. A* Zahorski measure *determined by E is a continuous, complete, finite Borel measure on X such that $\mu(X \setminus E) = 0$ and $\mu(E \cap U) > 0$ whenever U is an open set in X with $E \cap U \neq \emptyset$.*

Recall the definition of the universally positive closure operator F_X from page 33. Zahorski sets and this operator are connected by

PROPOSITION 4.24. *Let X be a separable metrizable space.*

(1) *If E is a Zahorski set in X, then $\mathrm{Cl}_X(E) = \mathrm{F}_X(E)$.*
(2) *If A is an absolute measurable space contained in X, then there is a Zahorski set E in X such that $E \subset A$ and $\mathrm{Cl}_X(E) = \mathrm{F}_X(A)$.*
(3) *If E is a Zahorski set in X and B is a universally measurable set in X such that $\mathrm{F}_X(E \setminus B) \neq \emptyset$, then there is a nonempty Zahorski set E' in X with the properties $E' \cap B = \emptyset$, $E' \subset E$, and E' is dense in $\mathrm{F}_X(E \setminus B)$.*

Proof. Suppose that E is a nonempty Zahorski set in X. As every nonempty Zahorski space is not an absolute null space, we have $\mathrm{Cl}_X(E) = \mathrm{F}_X(E)$ because nonempty open subsets of a Zahorski set of X are Zahorski sets of X by Proposition A.41.

Let A be an absolute measurable subspace of X. If $\mathrm{F}_X(A) = \emptyset$, then let $E = \emptyset$. If $\mathrm{F}_X(A) \neq \emptyset$, then there is a continuous, complete, finite Borel measure μ on X such that support$(\mu) = \mathrm{F}_X(A)$. There exists a σ-compact kernel K contained in $A \cap$ support(μ) such that $\mu(X) = \mu(K)$, where K is the union of a Zahorski set and a countable set (see Proposition D.7 and Exercise D.1 in Appendix D).

Finally, let us prove statement (3). We have that $E \setminus B$ is an absolute measurable space by Propositions 2.6 and 2.5. Statement (2) completes the proof. □

Observe, for separable metrizable spaces X, that each subset E of X is associated with the closed set $\mathrm{F}_X(E)$ and the absolute null space $E \setminus \mathrm{F}_X(E)$. This leads to the following proposition whose proof is left as an exercise.

Proposition 4.25. *Let X be an absolute measurable space. If E is a universally measurable set in X such that $\mathrm{Int}_X(\mathrm{F}_X(E)) \neq \emptyset$, then there is a Zahorski set K in X such that $K \subset E$ and such that K is dense in $\mathrm{Int}_X(\mathrm{F}_X(E))$.*

Here is a nice connection between the Baire category theorem and universally null sets. Recall that universally null sets in a space X are always absolute null spaces (see Theorem 2.7).

Lemma 4.26. *Suppose that X is a separable completely metrizable space with $\mathrm{F}_X(X) = X$. If H_i, $i = 1, 2, \ldots$, is a sequence of closed sets and if Z is an absolute null space such that $X = Z \cup \bigcup_{i=1}^{\infty} H_i$, then $\bigcup_{i=1}^{\infty} \mathrm{Int}_X(H_i)$ is a dense, open subset of X.*

Proof. Observe that $Z_0 = Z \setminus \bigcup_{i=1}^{\infty} H_i$ is an absolute Borel subset of the absolute null space Z, hence a countable set. As no point of Z is isolated in X, the lemma follows from the Baire category theorem. □

Let us apply the lemma to countable covers of X by universally measurable sets.

Corollary 4.27. *Let X be a separable completely metrizable space such that $\mathrm{F}_X(X) = X$. If X_i, $i = 1, 2, \ldots$, is a sequence of universally measurable sets in X and if Z is a universally null set in X such that $X = Z \cup \bigcup_{i=1}^{\infty} X_i$, then $\bigcup_{i=1}^{\infty} \mathrm{Int}_X(\mathrm{F}_X(X_i))$ is a dense open subset of X.*

Proof. Note that $X_i \setminus \mathrm{F}_X(X_i)$ is a universally null set in X and $\mathrm{F}_X(X_i)$ is a closed set for every i. As $Z \cup \bigcup_{i=1}^{\infty} (X_i \setminus \mathrm{F}_X(X_i))$ is a universally null set in X, the corollary is an immediate consequence of the lemma. □

The next corollary is a key to the proof of the Bruckner–Davies– Goffman theorem and the proof of our generalization of it.

Corollary 4.28. *Let X be a separable completely metrizable space and let E be a Zahorski set in X such that E is dense in $\mathrm{F}_X(X)$. If Z is a universally null set in X*

and A is a universally measurable set in X such that $A \subset E$, then there is a Zahorski set F in X such that $F \subset E$, F is dense in E, $F \cap Z = \emptyset$, and

$$\chi_A | F = \chi_{\mathrm{F}_X(A)} | F.$$

PROOF. As $\mathrm{F}_X(A) = \mathrm{F}_{\mathrm{F}_X(X)}(A)$ holds because $A \subset E \subset \mathrm{F}_X(X)$, there is no loss in assuming $\mathrm{F}_X(X) = X$.

The sets $H_1 = A \setminus Z$ and $H_2 = E \setminus (A \cup Z)$ are universally measurable sets in X. We have $X = \mathrm{F}_X(H_1) \cup \mathrm{F}_X(H_2)$ because E is dense in $\mathrm{F}_X(X)$, whence the union of the disjoint open sets U and V defined by

$$U \cup V = \mathrm{Int}_X\big(\mathrm{F}_X(H_1)\big) \cup \big(\mathrm{Int}_X\big(\mathrm{F}_X(H_2)\big) \setminus \mathrm{F}_X(H_1)\big)$$

is a dense subset of X. By Proposition 2.5, A is an absolute measurable space. So, by statement (2) of Proposition 4.24, there is a Zahorski set F_1 contained in H_1 such that F_1 is dense in U. As $B = A \cup Z$ is a universally measurable set in X we infer from statement (3) of Proposition 4.24 that there is a Zahorski set E' in X such that $E' \subset E \setminus (A \cup Z)$ and E' is dense in $E \setminus (A \cup Z)$. Observe that $\mathrm{F}_X(E') = \mathrm{F}_X(H_2)$ follows from statement (1) of Proposition 4.24. Hence $F_2 = E' \cap V$ will be dense in V. Let $F = F_1 \cup F_2$. Clearly F is a Zahorski set such that $F \cap Z = \emptyset$ and $F \subset E$. As F is dense in X we have that it is also dense in E. And as $\mathrm{F}_X(A) = \mathrm{F}_X(H_1)$ we have

$$F \cap \mathrm{F}_X(A) = F_1 \subset A \quad \text{and} \quad F \setminus \mathrm{F}_X(A) = F_2 \subset X \setminus A.$$

The corollary now follows. □

Finally, we have the following connection between the homeomorphism group $\mathsf{HOMEO}(X)$ and Zahorski sets in X.

LEMMA 4.29. *Suppose that X is an absolute measurable space such that a positive, continuous, complete, finite Borel measure μ on X and $\mathsf{HOMEO}(X)$ generate $\mathrm{MEAS}^{\mathrm{pos,fin}}(X)$. If E is a Zahorski set in X such that E is contained densely in the support of μ, then there is an h in $\mathsf{HOMEO}(X)$ such that $h_\# \mu(X \setminus E) = 0$.*

PROOF. From statement (4) of Proposition A.41 there is a positive, continuous, complete, finite Borel measure ν such that $\nu(X \setminus E) = 0$ and $\nu(X) = \mu(X)$. As μ and $\mathsf{HOMEO}(X)$ generate $\mathrm{MEAS}^{\mathrm{pos,fin}}(X)$, there is an h in $\mathsf{HOMEO}(X)$ such that $h_\# \mu = \nu$ and the lemma is proved. □

As an aside let us discuss another connection between Zahorski sets in absolute measurable spaces X and continuous, complete, finite Borel measures on X. Oxtoby proved the following in [**119**, Theorems 2].[1] His proof is provided for the reader.

THEOREM 4.30. *If X is a separable completely metrizable space and if μ is a nonzero, continuous, complete, finite Borel measure on X, then there exists a G_δ set B of X*

[1] The theorem has connections to the notion of two measures being "almost homeomorphic" defined by N. Bourbaki [**14**, Section 6, Exercise 8c, page 84].

such that $\mu(X \setminus B) = 0$ and there exists a homeomorphism h of B onto $\mathcal{N}$ such that $h_{\#}(\mu|B) = c\,\lambda|\mathcal{N}$, where λ is the Lebesgue measure on $\mathbb{R}$ and $c = \mu(X)$.

PROOF. Since $\mu(X)$ is finite, there exists a countable collection U_i, $i = 1, 2, \ldots$, of open sets with $\mu(\mathrm{Bd}_X(U_i)) = 0$ that forms a basis for the open sets of the topology of X. $B_0 = \mathrm{support}(\mu) \setminus \bigcup_{i=1}^{\infty} \mathrm{Bd}_X(U_i)$ is an uncountable, zero-dimensional absolute G_δ space, and $\mu(X \setminus B_0) = 0$. (For the definition of dimension 0, see page 244.) The set B_0 is the union of two disjoint sets B and C such that B is homeomorphic to $\mathcal{N}$ and C is countable. Let h_0 be a homeomorphism of B onto $\mathcal{N}$. As $\mu(B \cap U) > 0$ whenever U is an open set such that $B \cap U \neq \emptyset$, the measure $h_{0\#}(\mu|B)$ is a positive measure on $\mathcal{N}$. Let h_1 be in $\mathsf{HOMEO}(\mathcal{N})$ such that $\big(h_1 h_0\big)_{\#}(\mu|B) = c\,\lambda|\mathcal{N}$. The proof is easily completed. □

We now have a corollary which Oxtoby derived in [**119**, Theorem 4] for the special case of absolute G_δ spaces X. (See Proposition D.7 on page 245 for a similar assertion.)

COROLLARY 4.31. *Suppose that X is an absolute measurable space and let $\mathrm{M}(X, \mu)$ be a continuous, complete, finite Borel measure space. For each μ-measurable set A of X with $0 < \mu(A)$ there is a Zahorski set E contained in A such that $\mu(A \setminus E) = 0$. Moreover, if A is an absolute Borel space contained in X, then E can be made to satisfy $\mathrm{Cl}_X(E) = \mathrm{F}_X(A)$.*

PROOF. As $X \in \mathsf{abMEAS}$, there is no loss in assuming X is a subspace of the Hilbert cube, μ is defined on $[0, 1]^{\mathbb{N}}$, and X is μ-measurable. Also, there is no loss in assuming that A is a Borel set in the Hilbert cube since each μ-measurable set contains a Borel set of equal μ-measure. Hence $A \in \mathsf{abBOR}$. We also may assume $\mathrm{Cl}_X(A) = \mathrm{F}_X(A)$ because $A \setminus \mathrm{F}_X(A)$ is countable. Consider the measure $\mu \llcorner A$. There exists a G_δ set B_0 of $[0, 1]^{\mathbb{N}}$ such that $B_0 \subset A$ and $\big(\mu \llcorner A\big)(X \setminus B_0) = 0$, as provided by the theorem. Let C be a G_δ set such that $X \setminus B_0 \subset C$ and $\big(\mu \llcorner A\big)(C) = 0$. Then $B_1 = X \setminus C$ is a σ-compact subset of A such that $\big(\mu \llcorner A\big)(X \setminus B_1) = 0$. As B_1 is zero-dimensional, and as each uncountable, compact, zero-dimensional set is equal to the union of topological copy of the Cantor set and a countable set, the set B_1 is the union of a Zahorski set E' and a countable set. Clearly, $\mu(A \setminus E') = 0$. It may happen that $A \setminus \mathrm{Cl}_X(E')$ is not empty. As $\mathrm{Cl}_X(A) = \mathrm{F}_X(A)$, there is a nonempty Zahorski set E'' contained in $A \setminus \mathrm{Cl}_X(E')$ such that $\mathrm{Cl}_X(E'') \supset A \setminus \mathrm{F}_X(E')$. Let $E = E' \cup E''$. □

4.5. Bruckner–Davies–Goffman theorem

For each continuous, complete, σ-finite Borel measure μ on X, it is well-known that a μ-measurable, real-valued function f on X is equal μ-almost everywhere to a function in the second Baire class (equivalently to the second Borel class). It was shown by Bruckner, Davies and Goffman [**24**] that universally measurable, real-valued functions f on $[0, 1]$ are connected to Baire class 1 functions by means of the group of homeomorphisms $\mathsf{HOMEO}([0, 1])$. Namely, they proved the following theorem.

THEOREM 4.32 (Bruckner–Davies–Goffman). *Let $f\colon [0,1] \to \mathbb{R}$ be a universally measurable function. Then there is an h in* HOMEO$([0,1])$ *and a Borel class* 1 *function $g\colon [0,1] \to \mathbb{R}$ such that fh and g are equal Lebesgue almost everywhere.*

They reduced the proof to an application of the Oxtoby–Ulam theorem for $[0,1]$. Their theorem will be generalized to universally measurable maps $f\colon X \to Y$ whose values are taken in absolute Borel spaces Y and whose domains are in a large class of spaces X for which a "Radon–Nikodym version of the Oxtoby–Ulam theorem" holds. Among these spaces X are absolute measurable spaces that are not absolute Borel spaces. Moreover, the measure can be any continuous, complete, σ-finite Borel measure on X.

4.5.1. Splitting the problem. The following is a characterization theorem due to Bruckner, Davies and Goffman [**24**].

THEOREM 4.33. *Suppose that X and Y are separable metrizable spaces, μ is a continuous, complete, finite Borel measure on X, and $f\colon X \to Y$ is an arbitrary map. Then there exists a Borel class* 1 *map $g\colon X \to Y$ and a homeomorphism h of X onto X such that $fh = g$ μ-almost everywhere if and only if $h_{\#}\mu(\{x\colon f(x) \neq G(x)\}) = 0$ for some Borel class* 1 *map $G\colon X \to Y$ and some homeomorphism h of X onto X.*

PROOF. Let h be a homeomorphism. Clearly G is a Borel class 1 map if and only if Gh is a Borel class 1 map. Also, E is a Borel set that contains $\{x\colon f(x) \neq G(x)\}$ if and only if $h^{-1}[E]$ is a Borel set that contains $\{t\colon fh(t) \neq g(t)\}$, where $g = Gh$. As $\mu(h^{-1}[E]) = h_{\#}\mu(E)$ for Borel sets E, we infer that $\mu(\{x\colon fh(x) \neq g(x)\}) = 0$ if and only if $h_{\#}\mu(\{x\colon f(x) \neq G(x)\}) = 0$, because $h_{\#}\mu$ is complete. □

Let us split the theorem into two parts. Notice that in the Bruckner–Davies–Goffman theorem the Lebesgue measure on the space $[0,1]$ is a positive measure and $[0,1] = \mathrm{F}_{[0,1]}([0,1])$. Hence in our splitting of the problem we shall assume that the measure μ is positive and that $X = \mathrm{F}_X(X) \neq \emptyset$.

PROPOSITION 4.34. *Suppose that X is a separable metrizable space with $X = \mathrm{F}_X(X) \neq \emptyset$. Let μ be a positive, continuous, complete, finite Borel measure on X, and let $f\colon X \to Y$ be an arbitrary map. If there is Borel class* 1 *map g such that $fh = g$ μ-almost everywhere for some homeomorphism h, then there exists a Borel class* 1 *map G and a positive, continuous, complete, finite Borel measure ν such that $f = G$ ν-almost everywhere. Hence there exists a Borel class* 1 *map G and a dense F_σ subset E of X such that $f = G$ on E.*

In the last statement of the proposition, we have a **necessary condition:** "There exists a Borel class 1 map G and a dense F_σ subset E of X such that $f = G$ on E." Notice that the condition is free of the measures μ and ν, it is a topological statement.

The next proposition gives a sufficient condition.

PROPOSITION 4.35. *Let X be a separable metrizable space such that $X = \mathrm{F}_X(X) \neq \emptyset$ and let $f\colon X \to Y$ be an arbitrary map. Suppose that μ and ν are continuous,* *complete, finite Borel measures on X. If there is a Borel class 1 map G such that $f = G$ ν-almost everywhere and if there is an h in $\mathsf{HOMEO}(X)$ such that $h_\# \mu$ is absolutely continuous with respect to ν ($h_\# \mu \ll \nu$), then there is a Borel class 1 map g such that $fh = g$ μ-almost everywhere.*

The **sufficient condition** in the proposition can be written in the Borel measure theoretic form: "There is a Borel class 1 map G such that $f = G$ ν-almost everywhere and there is an h in $\mathsf{HOMEO}(X)$ such that $h_\# \mu \ll \nu$."

A development of a class of spaces that satisfies the above necessary condition will be given in Section 4.5.2, and a class of maps and measures that satisfies the sufficient condition will be given in Section 4.5.3.

4.5.2. A necessity condition. The domain of the universally measurable function in the Bruckner–Davies–Goffman theorem is $[0, 1]$. A property of this space, which is exploited in their proof, is that the Baire category theorem holds. Another property is that G_δ universally null sets of $[0, 1]$ are necessarily sets of the first category of Baire. Also, it is an absolute measurable space with the property that an open universally null set is the empty set, whence $X = \mathrm{F}_X(X) \neq \emptyset$.

We begin with the definition of Baire space, a topological notion.

DEFINITION 4.36. *A separable metrizable space X is said to be a* Baire space[2] *if $\bigcap_{i=1}^{\infty} U_i$ is dense in X whenever U_i, $i = 1, 2, \ldots$, is a sequence of dense open sets of X.*

Taking a cue from the above discussion, we define the topological notion of a BDG space.

DEFINITION 4.37. *A separable metrizable space X is said to be a* BDG space *if X is a Baire space such that every G_δ universally null set in X is a set of the first category of Baire, and if $X = \mathrm{F}_X(X) \neq \emptyset$.*

The following propositions are easily proved.

PROPOSITION 4.38. *Let X be a* BDG *space. If X_0 is a nonempty open subspace of X, then X_0 is a* BDG *space.*

PROPOSITION 4.39. *Let X be a separable metrizable space and let X_0 be an open dense subset of X. If X_0 is a* BDG *space, then X_1 is also a* BDG *space whenever $X_0 \subset X_1 \subset X$.*

PROPOSITION 4.40. *If a separable metrizable space X is a finite union of closed* BDG *subspaces, then X is a* BDG *space.*

It will be useful to have some examples of BDG spaces.

[2] The name Baire space is often used in topology to mean the space $D^{\mathbb{N}}$, where D is an infinite discrete space (see [**51**, page 326] or [**113**, page 73] for example.) Of course we have deviated from this convention in favor of the property of the Baire category theorem.

EXAMPLE 4.41. Let E_0 be a nonempty absolute null space contained in the open interval $(0, 1)$ and let E_1 be a non absolute measurable space that is a one-dimensional Lebesgue null set in $(0, 1)$. Then $X_0 = (\{0\} \times E_0) \cup ((0, 1) \times (0, 1))$ and $X_1 = (\{0\} \times E_1) \cup ((0, 1) \times (0, 1))$ are respectively an absolute measurable space and a non absolute measurable space. It is easy to show that both are BDG spaces since $(0, 1) \times (0, 1)$ is an absolute G_δ space that is dense in X_0 and is dense in X_1.

Let us turn to some preliminary constructions that are needed in the proof of our extension of the Bruckner–Davies–Goffman theorem. Suppose that X is a nonempty BDG space and let $\mathcal{U} = \{U_i : i \in \mathbb{N}\}$ be a countable collection of mutually disjoint universally measurable sets that covers X. Then

$$\mathcal{V} = \{V_i = \operatorname{Int}_X(\mathrm{F}_X(U_i)) : i \in \mathbb{N}\}$$

is a collection of open sets whose union is dense in X. To see this, observe that $X \setminus \bigcup\{\mathrm{F}_X(U_i) : i \in \mathbb{N}\}$ is a universally null G_δ set and hence a set of the first category in X. Clearly the open sets $V_i \setminus \bigcup_{j<i} \mathrm{F}_X(U_j)$ form a mutually disjoint collection that is dense in X. Now suppose that $\mathcal{W} = \{W_n : n \in \mathbb{N}\}$ is any collection of mutually disjoint open sets and that $\mathcal{K} = \{K_n : n \in \mathbb{N}\}$ is a collection of mutually disjoint universally measurable sets that satisfy: $\mathcal{W}$ refines $\mathcal{V}$; $\mathcal{K}$ refines $\mathcal{W}$; $\bigcup \mathcal{K}$ is dense in X; and, for each n, the set K_n is $W_n \cap U_i$ for some i with $W_n \subset V_i$. (Note: properties of the operator F_X yield K_n is dense in W_n.) Clearly, such collections $\mathcal{W}$ and $\mathcal{K}$ exist.

Next let W be a nonempty open subset of X and suppose that U_i is a member of $\mathcal{U}$ such that $W \subset \mathrm{F}_X(U_i)$. Let $\mathcal{U}' = \{U'_j : j \in \mathbb{N}\}$ be any collection of mutually disjoint universally measurable sets that covers X and refines $\mathcal{U}$. Denote the collection $\{U'_j \in \mathcal{U}' : U'_j \subset U_i \cap W\}$ by $\mathcal{U}'(W, i)$ and let

$$\mathcal{V}'(W, i) = \{V'_j = \operatorname{Int}_X(\mathrm{F}_X(U'_j)) : U'_j \in \mathcal{U}'(W, i)\}.$$

Then, with $V'_0 = W \setminus \operatorname{Cl}_X\big(\bigcup \mathcal{V}'(W, i)\big)$, the union of the open collection

$$\mathcal{V}^*(W, i) = \{V'_0\} \cup \mathcal{V}'(W, i)$$

is a dense subset of W. Let $\mathcal{W}'(W, i) = \{W'_m : m \in \mathbb{N}\}$ be any collection of mutually disjoint open sets that refines $\mathcal{V}^*(W, i)$ and whose union is dense in W. If $W'_m \subset V'_0$, then let $K'_m = W'_m \cap U_i$. Otherwise, select a j such that $W'_m \subset V'_j \in \mathcal{V}'(W, i)$ and let $K'_m = W'_m \cap U'_j$. Then the collection $\mathcal{K}'(W, i) = \{K'_m : m \in \mathbb{N}\}$ of mutually disjoint universally measurable sets refines both $\mathcal{U}'(W, i)$ and $\mathcal{W}'(W, i)$, and its union is dense in $W \cap U_i$. This ends the constructions.

We turn next to maps with discrete ranges. Bruckner, Davies and Goffman, in [**24**], cleverly reduce the investigation to discrete-valued universally measurable maps. We shall slightly modify their proof for discrete-valued maps.

The following discussion will lead to two lemmas that will permit an inductive construction for the general case as well as lead to a proof of the discrete case. Let X be a nonempty BDG space and let $\varphi : D \to D$ be a map defined on a countable discrete space. Suppose that $f : X \to D$ is a universally measurable map. Then φf is also a universally measurable map. The collections $\mathcal{U} = \{(\varphi f)^{-1}[\{d\}] : d \in \varphi[D]\}$ and $\mathcal{U}' = \{f^{-1}[\{d\}] : d \in D\}$ are covers of X such that $\mathcal{U}'$ refines $\mathcal{U}$. Let $\mathcal{W}$, $\mathcal{W}'$, $\mathcal{K}$ and $\mathcal{K}'$ be as described in the construction, and let d_0 be a fixed point in D. Note that $f[K'_m]$ and $\varphi f[K_n]$ are singleton sets of D and $\varphi[D]$, respectively, whenever they are not empty. Also $f[K'_m] \subset \varphi f[K_n]$ whenever $K'_m \subset K_n$. Let $g : X \to \varphi[D]$ be defined by

$$g(x) = \begin{cases} d_0, & \text{if } x \in X \setminus \bigcup \mathcal{W} \\ \varphi f(x_n), & \text{if } x \in W_n \in \mathcal{W}, \text{ where } x_n \in K_n \end{cases} \tag{4.1}$$

and define $g' : X \to D$ analogously. Then g and g' are Borel class 1 functions such that $\mathrm{d}(g(x), g'(x)) \le \sup\{\operatorname{diam} \varphi^{-1}[\{d\}] : d \in \varphi[D]\}$. Moreover, $\varphi f = g$ on $\bigcup \mathcal{K}$, and $f = g'$ on $\bigcup \mathcal{K}'$.

We have proved the following two lemmas; the first one will start an induction and the second one will provide the inductive step.

LEMMA 4.42. *Let X be a* BDG *space and $\mathcal{V}$ be an open collection such that $\bigcup \mathcal{V}$ is dense in X. If $f : X \to D$ is a universally measurable map into a discrete space D, then there exists an open collection $\mathcal{W} = \{W_n : n \in \mathbb{N}\}$ of mutually disjoint sets that refines $\mathcal{V}$ such that $\bigcup \mathcal{W}$ is dense in X, and there exists a sequence d_n, $n = 1, 2, \ldots$, in D such that $K_n = W_n \cap f^{-1}[\{d_n\}]$ is dense in W_n for each n. The pair $\mathcal{W}$ and $\mathcal{K}$ defines a Borel class* 1 *function $g : X \to D$ such that g is constant on each W_n and on $X \setminus \bigcup \mathcal{W}$, and such that $f(x) = g(x)$ whenever $x \in \bigcup \mathcal{K}$. Moreover, $\bigcup \mathcal{K}$ is dense in X.*

LEMMA 4.43. *Let X be a* BDG *space, and D be a countable, discrete metric space with metric* d. *Suppose that $f : X \to D$ is a universally measurable map and that $\varphi : D \to D$ is an arbitrary map. Then $\mathcal{U} = \{(\varphi f)^{-1}[\{d\}] : d \in \varphi[D]\}$ and $\mathcal{U}' = \{f^{-1}[\{d\}] : d \in D\}$ are collections of mutually disjoint universally measurable sets in X. For the map φf and the collection $\mathcal{U}$, further suppose that the corresponding collections $\mathcal{W}$ and $\mathcal{K}$ and Borel class* 1 *map $g : X \to \varphi[D]$ have the properties as described in the conclusion of the previous lemma. Then there are collections $\mathcal{W}'$ and $\mathcal{K}'$ and there is a Borel class* 1 *map $g' : X \to D$ as provided by the previous lemma with the added properties:*

(1) *$\mathcal{K}'$ refines $\mathcal{K}$,*
(2) $\mathrm{d}\big(g(x), g'(x)\big) \le \sup\{\operatorname{diam} \varphi^{-1}[\{d\}] : d \in \varphi[D]\}$,
(3) *$\varphi f(x) = g(x)$ whenever $x \in \bigcup \mathcal{K}$,*
(4) *$f(x) = g'(x)$ whenever $x \in \bigcup \mathcal{K}'$,*
(5) *$\bigcup \mathcal{K}'$ is dense in X.*

Lemma 4.42 implies the countable range theorem.

THEOREM 4.44. *Suppose that X is an absolute measurable space that is a* BDG *space and let $f : X \to Y$ be a universally measurable map of X into a countable separable metrizable space Y. Then there exists a Borel class* 1 *map g and a Zahorski set Z such that Z is dense in X and $f = g$ on Z.*

PROOF. Let D be a discrete space with $\mathrm{card}(D) = \mathrm{card}(Y)$, and let $\psi: D \to Y$ be any bijection. Clearly, ψ is continuous and ψ^{-1} is Borel measurable. The map $f_0 = \psi^{-1} f$ is a universally measurable map of X into D. Let g_0 be the Borel class 1 map that is provided by Lemma 4.42 for the map f_0. Then $g = \psi g_0$ is a Borel class 1 map such that $f = g$ on $\bigcup \mathcal{K}$, which is dense in X. As X is an absolute measurable space there is a Zahorski set Z contained in $\bigcup \mathcal{K}$ that is dense in X. □

The space $\mathcal{N}$ of all irrational numbers in $(0, 1)$ will play an important role in approximations of universally measurable discrete-valued maps. As $\mathcal{N}$ is an absolute G_δ space, there is a bounded complete metric d on $\mathcal{N}$. Here is a well-known theorem (see [**85**, Corollary 1c, page 450]).

THEOREM 4.45. *If Y is an uncountable absolute Borel space, then there exists a continuous bijection $\psi: \mathcal{N} \oplus \mathbb{N} \to Y$ such that ψ^{-1} is Borel measurable, where $\mathcal{N} \oplus \mathbb{N}$ is a disjoint topological sum. Moreover, if $Y = [0, 1]^{\mathbb{N}}$, then there exists a continuous bijection $\psi: \mathcal{N} \to Y$ such that ψ^{-1} is a Borel class 1 map.*

This theorem results in a very nice factorization.

THEOREM 4.46. *Let X be a separable metrizable space and let Y be an uncountable absolute Borel space. Then a map $f: X \to Y$ is a universally measurable if and only if there is a continuous bijection $\psi: \mathcal{N} \oplus \mathbb{N} \to Y$ and a universally measurable map $F: X \to \mathcal{N} \oplus \mathbb{N}$ such that $f = \psi F$. Moreover, F is a Borel class 1 map if and only if f is a Borel class 1 map.*

Our aim is to prove the following uncountable analogue of Theorem 4.44.

THEOREM 4.47. *Suppose that X is both an absolute measurable space and a* BDG *space, and let $f: X \to Y$ be a universally measurable map of X into an absolute Borel space Y. Then there exists a Borel class 1 map g and a Zahorski set Z such that Z is dense in X and $f = g$ on Z.*

Due to the preceding theorem it is clear that only the range space $Y = \mathcal{N} \oplus \mathbb{N}$ requires a proof. We shall use a sequence of discrete-valued maps $f_n: X \to D_n$, where D_n is a subset of Y, such that f_n, $n = 1, 2, \ldots$, converges uniformly to f. Uniform convergence will require a complete metric d on $\mathcal{N} \oplus \mathbb{N}$. As $\mathcal{N} \oplus \mathbb{N}$ is an absolute G_δ space, there exists one such metric. We may assume that d is bounded by 1.

We begin by constructing continuous maps $\varphi_n: \mathcal{N} \oplus \mathbb{N} \to \mathcal{N} \oplus \mathbb{N}$ for every n in $\mathbb{N}$ such that

(1) $\varphi_n[\mathcal{N} \oplus \mathbb{N}] = D_n$,
(2) $D_{n-1} \subset D_n$,
(3) the collection $\mathcal{W}_n = \{{\varphi_n}^{-1}[\{d\}]: d \in D_n\}$ is a covering of $\mathcal{N} \oplus \mathbb{N}$ whose mesh does not exceed $\frac{1}{2^{n-1}}$,
(4) $\mathcal{W}_n$ refines $\mathcal{W}_{n-1}$,
(5) $\mathrm{card}(W \cap D_{n-1}) \leq 1$ for every W in $\mathcal{W}_n$.

To this end, observe that the collection of all simultaneously closed and open subsets of $\mathcal{N} \oplus \mathbb{N}$ forms a base for the open sets. For $n = 1$ let $D_1 = \{y_0\}$, where y_0 is a fixed point of $\mathcal{N} \oplus \mathbb{N}$, and $\varphi_1(y) = y_0$ for every y in $\mathcal{N} \oplus \mathbb{N}$. Suppose that φ_n has been constructed and consider the collection $\mathcal{W}$ of all simultaneously closed and open sets W such that $\text{card}(W \cap D_n) \leq 1$ and $\text{diam}\, W < \frac{1}{2^n}$, and such that $\mathcal{W}$ refines $\mathcal{W}_n$. Clearly $\mathcal{W}$ is a cover. As $\mathcal{N} \oplus \mathbb{N}$ is a Lindeloff space, there is a countable subcollection $\mathcal{W}_{n+1}$ of mutually disjoint, simultaneously closed and open sets that covers $\mathcal{N} \oplus \mathbb{N}$. Define φ_{n+1} on each W in $\mathcal{W}_{n+1}$ as follows: if $W \cap D_n \neq \emptyset$, then define $\varphi_{n+1}(y)$ to be the unique member of $W \cap D_n$ for every y in W; if $W \cap D_n = \emptyset$, then fix a member d of W and define $\varphi_{n+1}(y)$ to be d for every x in W. It is easily seen that φ_{n+1} satisfies the five conditions listed above.

Note that the following identities hold: $\varphi_n\varphi_n = \varphi_n, \varphi_{n+1}\varphi_n = \varphi_n$ and $\varphi_n\varphi_{n+1} = \varphi_n$. So, if $f: X \to \mathcal{N} \oplus \mathbb{N}$ is a universally measurable map, then $f_n = \varphi_n f$, $n = 1, 2, \ldots$, is a sequence of discrete-valued universally measurable maps such that $\varphi_n f_{n+1} = f_n$ for each n, and such that the sequence converges uniformly to f. Moreover, $\varphi_n|D_{n+1}$ maps the discrete space D_{n+1} onto D_n. Hence Lemma 4.43 can be applied whenever X is a BDG space.

Suppose that X is both an absolute measurable space and a BDG space and let $f: X \to \mathcal{N} \oplus \mathbb{N}$ be a universally measurable map. Consider now the sequence f_n, $n = 1, 2, \ldots$, and the sequence $\varphi_n|D_n$, $n = 1, 2, \ldots$, as provided above. Then we infer from Lemmas 4.42 and 4.43 the existence of a sequence of Borel class 1 maps g_n, $n = 1, 2, \ldots$, and a sequence of universally measurable sets K_n, $n = 1, 2, \ldots$, such that, for each n, $\text{d}(g_n(x), g_{n+1}(x)) \leq \frac{1}{2^n}$ for every x, $f_n(x) = g_n(x)$ whenever $x \in K_n$, and $K_{n+1} \subset K_n$. As g_n, $n = 1, 2, \ldots$, is a Cauchy sequence, it converges uniformly to a Borel class 1 map g. We seek a dense Zahorski set contained in X such that $f = g$ on it. One is tempted to seek it in the intersection $\bigcap_{n=1}^{\infty} K_n$. Unfortunately this intersection may not be well behaved. We now use a very clever construction due to Bruckner, Davies and Goffman that avoids this difficulty. The construction uses

Proposition 4.48. *Let X be an absolute measurable space and Y be a separable metrizable space. For a universally measurable set E in X, if $f: X \to Y$ is a universally measurable map and U is an open set such that* $\text{F}_X(U \cap E)$ *is not empty, then there is a topological copy of the Cantor set K contained in $U \cap E$ such that $f|K$ is continuous.*

Proof. By the definition of the positive closure operator F_X there exists a continuous, complete, finite Borel measure μ on X such that $\mu(U \cap E) > 0$. Since X is an absolute measurable space, there is a compact subset E_0 of $U \cap E$ such that $\mu(E_0) > 0$. As $f|E_0$ is μ-measurable on E_0, there is a topological copy K of the Cantor set contained in E_0 such that $f|E_0$ is continuous on K. □

Returning to the construction, we let V_n, $n = 1, 2, \ldots$, be a base for the topology of X. We infer from the above Proposition 4.48 the existence of a sequence C_n, $n = 1, 2, \ldots$, of mutually disjoint topological copies of the Cantor set such that $C_n \subset V_n \cap K_n$, $f|C_n$ is continuous, and C_n is nowhere dense. As $\text{d}(f(x), f_n(x)) \leq \frac{1}{2^n}$ and $g_n(x) = f_n(x)$ for x in C_n, we have $\text{d}(f(x), g_n(x)) \leq \frac{1}{2^n}$ whenever $x \in C_n$. Define g'_n to be $g'_n(x) = g_n(x)$ for x not in $\bigcup_{j=1}^{n} C_j$, and $g'_n(x) = f(x)$ for x in $\bigcup_{j=1}^{n} C_j$. As f is continuous on the compact set $\bigcup_{j=1}^{n} C_j$, we have that g'_n is a Borel class 1 map. Also, $d(g'_n(x), g'_{n+1}(x)) \leq d(g_n(x), g_{n+1}(x)) + \frac{1}{2^n}$ for every x. Hence the sequence g'_n, $n = 1, 2, \ldots$, converges uniformly to a Borel class 1 map g'. Clearly $f(x) = g'(x)$ whenever x is in $\bigcup_{n=1}^{\infty} C_n$, which is a Zahorski set that is dense in X. We have just proved the last theorem. Thereby we have proved the topological part of the generalization of the Bruckner–Davies–Goffman theorem.

THEOREM 4.49. *Let X be both an absolute measurable space and a* BDG *space and let Y be an absolute Borel space. If $f : X \to Y$ is a universally measurable map, then there exist a Borel class* 1 *map g and a Zahorski set Z that is densely contained in X such that $f = g$ on Z.*

This completes the first part of the splitting.

4.5.3. A sufficiency condition. Let us turn to the Borel measure theoretic part of the Bruckner–Davies–Goffman theorem. The domain of the universally measurable map in their theorem is $[0, 1]$. In their proof they used a theorem by W. J. Gorman III [**65**] which states that each dense Zahorski set contained in $[0, 1]$ is changed by a homeomorphism h of $[0, 1]$ into a set of Lebesgue measure 1, which was proved by Gorman without recourse to Borel probability measures (he was not aware of the measure theoretic proof given in the early 1900s).

Let us begin with a summary of Gorman's results. Gorman proved in [**66**] that if $f\colon [0, 1] \to \mathbb{R}$ is a Lebesgue measurable function with $\operatorname{card}\big(f\big[[0, 1]\big]\big) < \aleph_0$ then there is a Baire class 1 function g and a homeomorphism h such that $fh = g$ Lebesgue almost everywhere. He also proved that there is a Lebesgue measurable function f such that the property "$fh = g$ Lebesgue almost everywhere" fails for every homeomorphism h and every Baire class 1 function g. This leads to the following definition of a Gorman pair.

DEFINITION 4.50. *Let X and Y be separable metrizable spaces. Then (f, μ), where μ is a continuous, complete, finite Borel measure on X and f is a μ-measurable map from X to Y, is called a* Gorman pair *if there exist a Borel class 1 map g from X to Y, a positive Zahorski measure ν on X and an h in* $\mathsf{HOMEO}(X)$ *such that $\mu \ll h_{\#}\nu$ and $f = g$ ν-almost everywhere.*

Obviously, if (f, μ) is a Gorman pair, then there is a homeomorphism h such that fh is equal to a Borel class 1 map μ-almost everywhere. We need a class of spaces that has a rich supply of pairs (μ, ν) of continuous, complete, finite Borel measures and homeomorphisms h such that $\mu \ll h_{\#}\nu$, where the ν's are required to be positive Zahorski measures. We will define such a class. But, in anticipation of the definition, we must discuss invariant subsets of homeomorphisms.

Recall that a subset F of X is said to be invariant under a homeomorphism h is $F = h[F]$. There are spaces X and subgroups G of $\mathsf{HOMEO}(X)$ such that some nonempty closed subset F is invariant under every h in G. Let us give three examples.

EXAMPLE 4.51. For $n > 2$ let $B_n = \{x \in \mathbb{R}^n : \|x\| \leq 1\}$ and $F = \{x \in \partial B_n : x_n \leq 0\}$, where x_n is the n-th coordinate of x. Then ∂F, F, ∂B_n and B_n form a nested, closed collection of subsets of B_n, each of which are BDG spaces. Moreover, all of them are invariant with respect to the group $\mathsf{HOMEO}(B_n; F \text{ inv})$.

EXAMPLE 4.52. Let M be a compact n-dimensional manifold with nonempty boundary ∂M. Then ∂M and M form a nested, closed collection of subsets of M, each of which is a BDG space and is invariant with respect to the group $\mathsf{HOMEO}(M)$.

EXAMPLE 4.53. Let X be a finite-dimensional triangulable space with $X = \mathrm{F}_X(X)$. For a triangulation K, denote by F_j the space $\bigcup K_j$, where K_j is the j-dimensional simplicial complex of a triangulation K. Then $F_j, j = 1, 2, \ldots, n$, is a nested, closed collection of subsets of X such that each F_j is a BDG space. Moreover, if

$$\mathsf{G} = \bigcap_{j=1}^{n} \mathsf{HOMEO}(X; F_j \text{ inv}) \cap \mathsf{HOMEO}(X; F_0 \text{ fixed}),$$

where F_0 is the collection of all vertices of K, then each F_j is invariant with respect to each h in G.

The second part of the splitting uses a class of Zahorski generated measures which will be defined next. We first establish some notation.

Let X be a separable metrizable space and let $\mathcal{F}$ be a finite nested collection of nonempty closed sets such that $\mathrm{F}_X(F) = F$ for each F in $\mathcal{F}$, and $\bigcup \mathcal{F} = X$. Let $\mathsf{Z}(\mathcal{F})$ be the collection of all Zahorski measures ν on X such that $\nu(U \cap F) > 0$ whenever U is an open set and F is a member of $\mathcal{F}$ with $U \cap F \neq \emptyset$. Clearly, if X is an absolute measurable space then $\mathsf{Z}(\mathcal{F})$ is not empty. Define the group of homeomorphisms

$$\mathsf{HOMEO}(X; \mathcal{F}) = \bigcap_{F \in \mathcal{F}} \mathsf{HOMEO}(X; F \text{ inv}).$$

With the aid of this notation we define

DEFINITION 4.54. *Let $\mathcal{F}$ be a finite nested collection of nonempty closed sets of an absolute measurable space X such that $\bigcup \mathcal{F} = X$ and $\mathrm{F}_X(F) = F$ for each F in $\mathcal{F}$. A continuous, complete, finite Borel measure μ is said to be* Zahorski dominated *(more precisely, relative to $\mathcal{F}$) if there exists a subgroup G of $\mathsf{HOMEO}(X; \mathcal{F})$ such that for each positive Zahorski measure ν in $\mathsf{Z}(\mathcal{F})$ there exists an h in G that satisfies $\mu \ll h_\# \nu$. An absolute measurable space X is said to be* Zahorski generated *relative to $\mathcal{F}$ if each continuous, complete, finite Borel measure on X is Zahorski dominated.*

We have a simple characterization of a Zahorski generated space.

THEOREM 4.55. *Let X be an absolute measurable space, and $\mathcal{F}$ be a finite nested collection of nonempty closed sets such that $\bigcup \mathcal{F} = X$ and $\mathrm{F}_X(F) = F$ for each F in $\mathcal{F}$. Then X is Zahorski generated relative to $\mathcal{F}$ if and only if there is a subgroup G of $\mathsf{HOMEO}(X; \mathcal{F})$ such that for each positive, continuous, complete, finite Borel measure μ and for each ν in $\mathsf{Z}(\mathcal{F})$ there is an h in G such that $\mu \ll h_\# \nu$.*

The proof is a consequence of the inequality $\mu \leq \mu + \nu$ since $\mu + \nu$ is positive whenever ν is positive.

EXAMPLE 4.56. Let $X = [0,1] \times [0,1]$ and let F_1 be one of the four edges of ∂X. Define $\mathcal{F}_1$ be the finite collection consisting of F_1 and ∂X, and define $\mathsf{G} = \mathsf{HOMEO}(\partial X; F_1 \text{ inv})$. It is easily seen that ∂X is Zahorski generated relative to $\mathcal{F}_1$. Also X is Zahorski generated relative to $\mathcal{F} = \{\partial X, X\}$ since every homeomorphism in $\mathsf{HOMEO}(\partial X)$ has an extension in $\mathsf{HOMEO}(X)$.

Other examples of Zahorski generated spaces will be given later.

4.5.4. The extended Bruckner–Davies–Goffman theorem. We now combine the results of Sections 4.5.2 and 4.5.3.

THEOREM 4.57. *Let $f : X \to Y$ be a map of an absolute measurable space X into an absolute Borel space Y and let $\mathcal{F}$ be a finite nested collection of nonempty closed subsets such that each F in $\mathcal{F}$ is a* BDG *space and $X = \bigcup \mathcal{F}$. If X is Zahorski generated relative to $\mathcal{F}$, then every pair (f, μ), where f is universally measurable and where μ is a continuous, complete, finite Borel measure on X, is a Gorman pair. Hence there is a Borel class* 1 *map g and a homeomorphism h such that $fh = g$ μ-almost everywhere.*

PROOF. Let $\mathcal{F}$ and G be as given in Definition 4.54. Write $\mathcal{F}$ as F_i, $i = 1, 2, \ldots, n$, with $F_i \subset F_{i+1}$ for each i. There is no loss in assuming $H_i = F_i \setminus F_{i-1} \neq \emptyset$ (here $F_0 = \emptyset$). As H_i is a BDG space and $f|H_i$ is universally measurable, there is a Zahorski set Z_i that is dense in H_i and a Borel class 1 map g_i on H_i such that $f|H_i = g_i$ on Z_i. The set $Z = \bigcup_{i=1}^{n} Z_i$ is a Zahorski set. Let ν be a Zahorski measure on Z. As Z is an absolute measurable space, we may assume that ν is defined on X. Finally, since X is Zahorski generated relative to $\mathcal{F}$, there is a homeomorphism h in G such that $\mu \ll h_{\#}\nu$. Define g to be the map given by $g(x) = g_i(x)$ whenever $x \in H_i$. Clearly g is a Borel class 1 map such that $f = g$ on Z. Thereby we have shown that (f, μ) is a Gorman pair. □

REMARK 4.58. We have observed that each continuous, complete, σ-finite Borel measure μ on a separable metrizable space corresponds to a finite Borel measure μ_0 such that the μ-null sets and μ_0-null sets are the same collections. Hence $\mu \ll \mu_0$. It follows that the measures μ in the above theorem may be assumed to be σ-finite.

4.5.5. Examples. Our first application is Theorem 4.32 (Bruckner–Davies–Goffman theorem) where $Y = \mathbb{R}$ and $X = (0, 1)$. Clearly $(0, 1)$ is a BDG space and $\mathbb{R}$ is an absolute Borel space. As the Oxtoby–Ulam theorem applies to $[0, 1]$, the space $(0, 1)$ is Zahorski generated relative to $\mathcal{F} = \{(0, 1)\}$. Consequently the Bruckner–Davies–Goffman theorem follows.

There are many other examples of BDG spaces X that are Zahorski generated relative to $\mathcal{F} = \{X\}$ for appropriate subgroups G of $\mathsf{HOMEO}(X)$. The spaces in our first collection of examples uses the property that there exists positive, continuous, complete, finite Borel measures on X such that it and some subgroup G of $\mathsf{HOMEO}(X)$ generate $\mathrm{MEAS}^{\mathrm{finite}}(X)$.

COLLECTION 1: *One-dimensional manifolds*; the *Hilbert cube*; $\mathcal{N}$, which is homeomorphic to $\mathbb{N}^{\mathbb{N}}$; *Menger manifolds of positive dimension*. For these spaces the subgroup G is $\mathsf{HOMEO}(X)$.

The next examples, which use $\mathcal{F} = \{X\}$, are not in the first collection. We have seen that the Cantor space fails to have an analogue of the Oxtoby–Ulam theorem. But, by Theorem 3.57 due to Akin, the Radon–Nikodym derivative version of the Oxtoby–Ulam theorem is available (also see Theorem 3.67).

COLLECTION 2: *Any space X that is topologically equivalent to the Cantor space* $\{0,1\}^{\mathbb{N}}$. For these spaces, the subgroup G is $\mathsf{HOMEO}(X)$. *One-dimensional finitely triangulable spaces* use the subgroup $\mathsf{G} = \mathsf{HOMEO}(X; V \text{ fixed})$, where V is the set of vertices of a triangulation of the space.

We have seen that the boundary ∂M of a separable n-dimensional manifolds M plays an important role in determining a continuous, complete, finite Borel measure μ which, together with $\mathsf{HOMEO}(M)$, (ac)-generate $\mathrm{MEAS}^{\mathrm{finite}}(M)$. Thus we have

COLLECTION 3: *Separable n-dimensional manifolds M*. There are two cases for these spaces. If $\partial M = \emptyset$, then let $\mathcal{F} = \{M\}$. If $\partial M \neq \emptyset$, then let $\mathcal{F} = \{\partial M, M\}$. In both cases let $\mathsf{G} = \mathsf{HOMEO}(M)$. The verification that M is Zahorski generated by $\mathcal{F}$ follows easily from Theorem 4.16 and Proposition A.41.

COLLECTION 4: *Finitely triangulable spaces*. For a finitely triangulable space X with $X = \mathrm{F}_X(X)$, denote by F_j the space $\bigcup K_j$, where K_j is the j-dimensional simplicial complex of a triangulation K of X. Then X is Zahorski generated relative to $\mathcal{F} = \{F_j : j = 1, 2, \ldots, n\}$, where n is the dimension of X. Here, G is the subgroup $\mathsf{HOMEO}(X; \mathcal{F})$ of $\mathsf{HOMEO}(X)$. The verification that X is Zahorski generated by $\mathcal{F}$ follows easily from Theorem 4.18 and Proposition A.41.

COLLECTION 5: *Compact, connected, non locally connected spaces*. For the first space let F be the closed subset of the Warsaw circle W consisting of all points of W at which W is not locally connected, and let $\mathcal{F} = \{F, W\}$. Clearly the set F is invariant with respect to each h in the group $\mathsf{HOMEO}(W)$. It follows from Theorem 4.17 and Proposition A.41 that W is Zahorski generated relative to $\mathcal{F}$, where the subgroup G is $\mathsf{HOMEO}(W)$. The second space is $X = W \times [0,1]^n$ with $\mathcal{F} = \{F \times [0,1]^n, W \times [0,1]^n\}$. The verification of the fact that X is Zahorski generated relative to $\mathcal{F}$ is left as an exercise for the reader.

COLLECTION 6: *Absolute measurable spaces*. The above examples are completely metrizable. $X = \big(\{0\} \times E\big) \cup \big((0,1) \times (0,1)\big)$, where E is an uncountable absolute null space contained in the open interval $(0,1)$, is an absolute measurable space that is not completely metrizable. We have seen earlier (see Example 4.41) that X is a BDG space. Let $\mathcal{F} = \{X\}$. We infer from the Oxtoby–Ulam theorem that X is Zahorski generated relative to $\mathcal{F}$. Also, $X' = \big(\{0\} \times E\big) \cup \big(\mathcal{N} \times \mathcal{N}\big)$ is a BDG space that is Zahorski generated relative $\mathcal{F} = \{X'\}$ (see Example 3.64).

4.6. Change of variable

Suppose that f is a map from a space X to a space Y. By *a change of variable* of f we mean a composition fH by a homeomorphism H in $\mathsf{HOMEO}(X)$. In [**148**] Świątowski proved the following interesting theorem. (See also B. Koszela [**82**].)

THEOREM 4.59 (Świątowski). *Suppose f is a function on $[0,1]^n$ that is extended real-valued and Lebesgue measurable. If f is real-valued Lebesgue almost everywhere, then there is a change of variable fH such that fH is Lebesgue measurable and $\int_{[0,1]^n} fH\,d\lambda$ exists.*

Notice that the function f in the theorem is just Lebesgue measurable and not necessarily universally measurable. Hence it is not obvious that the composition fH is Lebesgue measurable. The proof of the theorem is a simple application of Theorem 2.49 and the Oxtoby–Ulam theorem. The theorem becomes more complicated if one adds to the Lebesgue measure the $(n-1)$-dimensional Hausdorff measure H_{n-1} restricted to the boundary ∂I^n. There is a change of variable theorem in this case also. At this point it will be convenient to prove a lemma.

LEMMA 4.60. *Let X be a separable metrizable space and ν be a continuous, complete, σ-finite Borel measure on X. Suppose that μ is a continuous, complete Borel measure on X such that $\mu \le \nu$, and suppose that H is a homeomorphism in $\mathsf{HOMEO}(X)$ such that for each point x of X there is a neighborhood U_x and a positive constant c_x such that $(H_\#\nu) \llcorner U_x \le c_x\, \mu \llcorner U_x$. If f is a ν-measurable, extended real-valued function on X that is real-valued ν-almost everywhere, then the following statements hold.*

(1) *f is μ-measurable.*
(2) *fH is ν-measurable.*
(3) *For each compact set K there is a constant B such that*

$$\int_{H^{-1}[K]} |fH|\,d\nu \le B \int_K |f|\,d\mu$$

whenever f is locally μ-integrable.

PROOF. Let g be a real-valued, Borel measurable function such that $f = g$ ν-almost everywhere. As $\mu \le \nu$, we have $f = g$ μ-almost everywhere and statement (1) follows.

Clearly $H_\#\nu \ll \mu$ holds. From statement (1) we have, for the above g, that there is a Borel set E such that $\{y : f(y) \ne g(y)\} \subset E$ and $\mu(E) = 0$. As $H^{-1}\big[\{y : f(y) \ne g(y)\}\big] = \{x : fH(x) \ne gH(x)\}$ we have $H^{-1}[E] \supset \{x : fH(x) \ne gH(x)\}$. Let $E' = H^{-1}[E]$. Then

$$\nu^*(\{x : fH(x) \ne gH(x)\}) \le \nu(E') = \nu(H^{-1}[E]) = \mu(E) = 0.$$

Hence fH is ν-measurable. Statement (2) is proved.

Observe that $\int (gH)\,d\nu = \int g\,d(H_\#\nu)$ whenever g is a nonnegative, real-valued, Borel measurable function follows from $\nu\big((gH)^{-1}[F]\big) = \big(H_\#\nu\big)(g^{-1}[F])$ for all

Borel sets F. Now let K be a compact set. Then there is a neighborhood U of K and a positive constant B such that $(H_{\#}\nu) \llcorner U \le B\,\mu \llcorner U$. Consequently,

$$\begin{aligned}\textstyle\int |f\cdot\chi_K| H\,d\nu &= \textstyle\int |g\cdot\chi_K| H\,d\nu = \int |g\cdot\chi_K|\,d(H_{\#}\nu)\\ &\le B \textstyle\int |g\cdot\chi_U|\,d\mu = B\int |f\cdot\chi_U|\,d\mu,\end{aligned}$$

and statement (3) follows. □

For complete, finite Borel measures ν on X, each ν-measurable, extended real-valued function f on X yields a finite Borel measure μ given by $\mu(B) = \int_B \frac{1}{1+|f|}\,d\nu$ whenever $B \in \mathfrak{B}(X)$. Clearly, $\mu \le \nu$. With this and the above lemma in mind we make the following definition.

DEFINITION 4.61. *Let ν be a continuous, complete, finite Borel measure on X, where X is a separable metrizable space. The measure ν is said to have the* Świątowski property *if, for each ν-measurable, extended real-valued function f that is finite valued ν-almost everywhere, there is an H in* HOMEO(X) *such that fH is ν-measurable and $\int_X fH\,d\nu$ exists.*

4.6.1. Examples. The examples of (ac)-generation found in Section 4.3, with the aid of the above lemma, will yield measures ν on spaces X that have the Świątowski property. The verifications will be left to the reader.

EXAMPLE 1: *Let X be the Hilbert cube or the space $\mathcal{N}$ or the Cantor space or a compact n-dimensional Menger manifold with $n > 0$, and let ν be a positive, continuous, complete, finite Borel measure on X. Then ν has the Świątowski property.*

EXAMPLE 2: *Let K be a triangulation of a finitely triangulable space X. If ν is a positive, continuous, finite Borel measure on the space X such that $\nu|\sigma$ is a positive measure on σ for each simplex σ in K with* $\dim\sigma \ge 1$, *then ν has the Świątowski property.*

EXAMPLE 3: *Let X be a compact manifold with or without boundary. Let ν be a positive, continuous, finite Borel measure on the space X such that $\nu|\partial X$ is a positive measure on ∂X. Then ν has the Świątowski property.*

EXAMPLE 4: *Let X be a nonempty open subset of $\mathbb{R}^n$. To each Lebesgue measurable extended real-valued function that is real-valued Lebesgue almost everywhere on X there corresponds an H in* HOMEO(X) *such that fH is Lebesgue measurable and fH is locally Lebesgue integrable.*

The verification of the next example is left as an exercise.

EXAMPLE 5: Let $X = I^n$ and let X_0 be a k-dimensional face of I^n, for example $X_0 = \{x \in I^n : x_i = 0,\ k < i \le n\}$. Let $\nu = \nu_0 + \nu_1$ where $\nu_0 = \mathsf{H}_k \llcorner X_0$ and $\nu_1 = \lambda$ is the Lebesgue measure on I^n. Then the following assertion holds: *For each ν-measurable extended real-valued function f on X that is real-valued ν-almost everywhere there is an H in* HOMEO(X) *such that $\int_X fH\,d\nu$ exists.* What can be said if ν is such that $\nu|X_0$ is Lebesgue-like on X_0 and ν_1 is Lebesgue-like on I^n?

4.7. Images of Lusin sets

In Section 2.8.1 we promised to discuss the images of Lusin sets X for Borel measurable real-valued functions f – that is, the images $f[X]$ are absolute null space. In the course of our discussion we will introduce many classes of singular sets that appear naturally in the proof.

Lusin was interested in subsets X of a separable metrizable space Y with the property that *every nowhere dense subset of Y meets X in a countable set.* Clearly Lusin was interested in the existence of uncountable subsets with this property since countable ones are easily found. It was seen in Chapter 1 that an uncountable Lusin set exists under the continuum hypothesis whenever Y is completely metrizable. Sometimes it is convenient to require that Lusin sets be uncountable; that is, a *Lusin set* is an uncountable set with the above property. In the context of singular sets, the collection of all sets with the above property, where X need not be required to be uncountable, is denoted by L_Y. This leads to the natural collection $\mathsf{COUNTABLE}$, namely those spaces X with $\mathrm{card}(X) \leq \aleph_0$. We now have $\mathsf{COUNTABLE} \subset \mathsf{L}_Y$. The next observation is that the ambient space in the definition of spaces in L_Y is not really needed. That is, if a subset of X is nowhere dense in X, then it is also nowhere dense in the ambient space Y of X. Hence one finds in the literature the property ν for separable metrizable spaces X : *every nowhere dense subset of X is countable.* We shall denote the class of all spaces that satisfy the property ν by NU. We now have $\mathsf{L}_Y \subset \mathsf{NU}$. The obvious thing to do next is to consider countable unions of spaces in NU. This class should obviously be denoted by $\sigma\,\mathsf{NU}$.[3] Every class defined above is hereditary – that is, each subspace of a space in a class is also a member of that class.

We begin our task with the continuous map case.

THEOREM 4.62. *Suppose $X \in \mathsf{NU}$. If $f\colon X \to \mathbb{R}$ is continuous, then $f[X]$ is a Lebesgue null set. Moreover, if $h \in \mathsf{HOMEO}(\mathbb{R})$, then $h^{-1}\big[f[X]\big]$ is a Lebesgue null set, whence $f[X] \in \mathsf{univ}\,\mathfrak{N}(\mathbb{R})$.*

PROOF. Let x_n, $n = 1, 2, \ldots$, be a sequence in X such that $D = \{x_n : n \in \mathbb{N}\}$ is a dense subset of X. For each positive number ε let U_n be an open set in X such that $x_n \in U_n$ and $\operatorname{diam} f[U_n] < \frac{\varepsilon}{2^n}$. Since D is dense in X we have that $E = X \setminus \bigcup_{n=1}^{\infty} U_n$ is nowhere dense in X, whence countable. As $f[X] = f[E] \cup \bigcup_{n=1}^{\infty} f[U_n]$, we have that the outer Lebesgue measure of $f[X]$ does not exceed ε. The final statement follows easily. □

Implicit in the proof are several notions that have appeared in the early literature. The first of them is that a subset X a space Z is *concentrated about a set C,*[4] that is, *every neighborhood U of C is such that* $\mathrm{card}(X \setminus U) \leq \aleph_0$. The case of a countable set C is of special interest. The collection of all separable spaces X that are concentrated about some countable set C of the ambient space Y will be denoted by CON_Y. In the event that $Y = X$ we shall just write, CON.[5] The next property, called C'' in the

[3] The reader will see in the literature the symbol L_1 used for this class. See, for example Brown and Cox [**19**].

[4] See A. S. Besicovitch [**9**].

[5] In [**19**] the symbol P is used for this class.

literature, is the following: *For every collection* $\{G(x,n)\colon x \in X, n \in \mathbb{N}\}$ *of open sets of* X *such that* $x \in G(x,n)$ *for each* x *and* n*, there necessarily exists a diagonal sequence* x_n*,* $n \in \mathbb{N}$*, of elements of* X *such that* $X = \bigcup_{n\in\mathbb{N}} G(x_n, n)$. The collection of all spaces that possess this property will be denoted by C''. It is not difficult to prove the inclusion $\mathsf{NU} \subset \mathsf{C}''$. Moreover, the class C'' is invariant with respect to continuous surjective maps. (For proofs of these two assertions see Kuratowski [**85**, Theorems 5 and 6, page 527].)

The final property is couched in terms of a metric d on a space X. The metric is used to define the δ-neighborhood of a point x in X. The collection of all such neighborhoods forms a basis for the topology of X. Observe that if X is $\mathbb{R}$ with the usual metric, then the radius of the δ-neighborhood of x is $\frac{1}{2}$ the Lebesgue measure of that neighborhood. Since we are interested in maintaining a topological approach we shall replace the metric with a continuous, complete, finite Borel measure on the ambient space Y and a basis $\mathcal{B}(Y)$ for the open sets of the topology of the space Y. It will be important that there be a rich supply of continuous, finite Borel measures μ on Y with $\mu(Y) > 0$. This will be assured if the separable metrizable space satisfies the requirement $Y = \mathrm{F}_Y(Y) \neq \emptyset$ (see page 33 for the positive closure operator F_Y). In such a space, there is a continuous, finite Borel measure μ on Y such that $\mu(V) > 0$ whenever V is an nonempty open subset of U. For subsets X of a separable metrizable space Y, where $Y = \mathrm{F}_Y(Y) \neq \emptyset$, and a basis $\mathcal{B}(Y)$ for the open sets of X, the property of interest is the following: *For each sequence* ε_n*,* $n \in \mathbb{N}$*, of positive numbers and for each continuous, complete finite Borel measure* μ *on* Y *there is a sequence* x_n*,* $n \in \mathbb{N}$*, in* X *and there is a sequence* $U(n)$*,* $n \in \mathbb{N}$*, in* $\mathcal{B}(Y)$ *such that* $x_n \in U(n)$ *and* $\mu(U(n)) < \varepsilon_n$ *for every* n *and* $X \subset \bigcup_{n\in\mathbb{N}} U(n)$. We shall denote the collection of all subsets X with this property by C_Y. Of course, this collection is dependent on the basis $\mathcal{B}(X)$, which has not been displayed in the symbol C_Y.

For $Y = \mathbb{R}$, the above property is equivalent to the notion of *strong measure zero*, the favored terminology in set theory, where $\mathcal{B}(\mathbb{R})$ is the collection of all δ-neighborhoods of the points x and only the Lebesgue measure λ is used, that is, $\lambda\big((x-\delta, x+\delta)\big) = 2\delta$. Indeed, if $X \subset [a,b]$ and h is in $\mathsf{HOMEO}(\mathbb{R})$, then $h[X]$ is a set of strong measure zero whenever X is a set of strong measure zero because $h|[a,b]$ is uniformly continuous. Note that any countable union of strong measure zero sets is a strong measure zero set. We will use the more descriptive symbol SMZ_Y for C_Y. Clearly, if $X \in \mathsf{SMZ}_Y$, then $X \in \mathsf{univ}\,\mathfrak{N}(Y) \subset \mathsf{abNULL}$.

The next theorem is Theorem 7 and the following Remark in [**85**, pages 527–528]. We shall give a proof to illustrate the use of the notions of Baire properties of sets and of functions. (See Section A.2.3 of Appendix A for the Baire property.)

THEOREM 4.63. *Suppose that* Y *is a separable metrizable space. Then* $\mathsf{C}'' \subset \mathsf{SMZ}_Y$*. Moreover, if* $X \in \mathsf{NU}$ *and if* $f\colon X \to Y$ *has the Baire property, then* $f[X] \in \mathsf{SMZ}_Y$*; consequently,* $f[X] \in \mathsf{SMZ}_Y$ *whenever* f *is Borel measurable.*

PROOF. Let ε_n, $n \in \mathbb{N}$, be a sequence of positive numbers and let μ be a continuous, finite Borel measure on Y. Then, for each y in $f[X]$, there is a sequence $G(y,n)$, $n \in \mathbb{N}$, in $\mathcal{B}(X)$ such that $y \in G(y,n)$ and $\mu(G(y,n)) < \varepsilon_n$ for every n. The first assertion is a direct consequence of the definition of property C'' applied to this sequence.

Consider the second assertion. Suppose that X is in NU. Then there is a sequence x_k, $k \in \mathbb{N}$, in X such that X is concentrated about $C = \{x_k : k \in \mathbb{N}\}$. Let $G(y,n)$ be an open set in $\mathcal{B}(Y)$. As f has the Baire property, there is an open set $V(y,n)$ in X and a first category set $P(y,n)$ of X such that $f^{-1}[G(y,n)]$ is the symmetric difference $V(y,n)\Delta P(y,n)$. Hence $P = \bigcup_{k\in\mathbb{N}} P(f(x_k), 2k)$ is a first category set and $V = \bigcup_{k\in\mathbb{N}} V(f(x_k), 2k)$ is an open set that contains the countable set C. As $X \in \mathsf{NU}$ we have $\operatorname{card}(X \setminus V) \leq \aleph_0$ and $f[X] \subset f[P \cup (X \setminus V)] \cup \bigcup_{k\in\mathbb{N}} G(f(x_k), 2k)$. Let y_{2k-1}, $k \in \mathbb{N}$, be a sequence that exhausts $f[P \cup (X \setminus V)]$, and let $y_{2k} = f(x_k)$, $k \in \mathbb{N}$. Then the sequence $U(n) = G(y_n, n)$, $n \in \mathbb{N}$, covers X and satisfies $\mu(U(n)) < \varepsilon_n$ for every n. Hence $f[X]$ is in SMZ_Y. As a Borel measurable map has the Baire property, the remainder of the second statement is trivially true. □

THEOREM 4.64. *Suppose that X is a Lusin set in a separable metrizable space Y. If $f: X \to \mathbb{R}$ is Borel measurable, then $f[X]$ is a strong measure zero set in $\mathbb{R}$, whence an absolute null space.*

PROOF. Lusin sets in a separable metrizable space are in NU. Hence the above theorem completes the proof. □

4.8. Comments

In general, the proofs provided in the chapter differ from those found in the literature. The comments will center mostly on these differences.

4.8.1. Goldman conjecture. The Goldman problem is a natural one. Goldman himself conjectured that the answer was the collection $\mathfrak{B}(\mathbb{R})$, which was not correct. The first to show that this conjecture was incorrect was Davies [**43**] who showed in 1966 that every analytic set E has the property that $f^{-1}[E]$ is Lebesgue measurable whenever f is Lebesgue measurable. Subsequently, H. G. Eggleston [**48**] showed in 1967 that every concentrated set E (a singular set, see page 128, that is also a universally measurable set in $\mathbb{R}$ but not necessarily analytic) has the property that $f^{-1}[E]$ is Lebesgue measurable whenever f is Lebesgue measurable. In 1968, Davies [**44**] extended Eggleston's result to a larger class of singular set that are also universally measurable sets. Of course it was Darst who, in [**39**], recognized that the key to the problem is the induced measure $f_\#\lambda$, that is, he proved Proposition 4.1. The extensions of Darst's theorem to Theorems 4.3 and 4.4 are natural ones since $\mathfrak{B}$-homeomorphisms preserve universally measurable sets and the extensions do not involve any topological properties or geometric properties.

A side effect of the investigation of Goldman's problem is the characterization of universally measurable maps – Proposition 4.5 is the analogue of the well-known characterization of Borel measurable maps. Hence universally measurable sets and maps are natural extensions of the notions of Borel sets and maps. A simple consequence of this proposition is following analogue of Borel sets.

Proposition 4.65. *Let X be a subspace of a separable metrizable space Y and let M be a universally measurable set in Y. Then $X \cap M$ is a universally measurable set in X.*

To see this, observe that the inclusion map of X into Y is a universally measurable map. We have the following question.

Question 4.1. For any separable metrizable space Y and any subspace X of Y let A be any universally measurable set in X. Is it true that there is a universally measurable set M in Y such that $A = X \cap M$? The answer is obviously yes if A is an absolute measurable space or if X is a universally measurable set of Y.

4.8.2. Continuous and bounded variation. Darst wrote a series of papers on continuous functions of bounded variation and Purves's theorem, culminating in his result on infinitely differentiable functions. Clearly, locally analytic functions (i.e., functions locally equal to a power series) are infinitely differentiable. In [**40**] Darst observed that locally analytic functions are $\mathfrak{B}$-maps. This is easily seen from the Purves theorem. Indeed, a function f that is analytic on a connected open subset of $\mathbb{R}$ is necessarily constant on that set whenever f has a nonempty set of points of uncountable order $U(f)$. Since each open cover of a set in $\mathbb{R}$ has a countable subcover, $U(f)$ will be countable and the Purves theorem applies.

Darst studied several classes of singular sets of $\mathbb{R}$ in [**34**] and initiated an investigation of continuous functions of bounded variation that preserved these classes. In [**36**] Darst constructed a continuous function of bounded variation that mapped a universally null set to a nonuniversally null set. The constructed function was very similar to the A. H. Stone example, which is Lipschitzian.

It is easily seen that any Lipschitzian function that is not a $\mathfrak{B}$-map yields an infinitely differentiable function that is not a $\mathfrak{B}$-map since the method used in the above proof by Darst or the method used in the proof given in the chapter will apply. Indeed, Purves's theorem together with the fact that $U(f)$ is necessarily an analytic set imply the existence of a nonempty perfect set contained in $U(f)$. From this discussion we infer that there is an infinitely differential function that is not a $\mathfrak{B}$-map if and only if there is a Lipschitzian function that is not a $\mathfrak{B}$-map.

These comments on $U(f)$ should not end without mentioning Theorem 2.10, the Darst and Grzegorek extension of Purves's theorem on $\mathfrak{B}$-maps. This theorem supports the fact that absolute measurable spaces are a very natural extension of absolute Borel spaces.

Darst's construction of his C^∞ function is related to the problem of characterizing those continuous functions f on $[0, 1]$ of bounded variation that have the property that fh is infinitely differentiable for some h in $\mathsf{HOMEO}([0, 1])$. These compositions are "inner compositions with homeomorphisms." Such functions were characterized by M. Laczkovich and D. Preiss [**87**]. Darst constructed such a continuous function of bounded variation to get his infinitely differentiable function. Another characterization of such functions can be found in C. Goffman, T. Nishiura and D. Waterman [**62**]. The construction in this chapter is a different sort of composition; hf is an "outer composition with homeomorphisms."

4.8.3. The Bruckner–Davies–Goffman theorem. This theorem was proved for a universally measurable real-valued function defined on the interval $(0, 1)$ and the Lebesgue measure λ. Their proof used the results of Gorman [**65**] concerning homeomorphisms of dense Zahorski sets in $(0, 1)$. (The classical characterization of the collection $\mathrm{MEAS}^{\mathrm{pos,fin}}\big((0,1)\big)$, which consists of all positive, continuous, complete, finite Borel measures on $(0, 1)$, by means of λ and the group $\mathsf{HOMEO}\big((0,1)\big)$ was not known to him.) Gorman showed in [**66**] that Lebesgue measurable functions f with $\mathrm{card}(f[(0,1)]) < \aleph_0$ satisfied the conclusion of the theorem of Bruckner, Davies and Goffman. He also showed that Lebesgue measurable functions with countable images need not satisfy the conclusion of the Bruckner–Davies–Goffman theorem. The argument used by Bruckner, Davies and Goffman is very subtle. It uses the usual order of the real numbers to construct certain discrete-valued approximations of a universally measurable function by Baire class 1 functions. Then Gorman's results on Zahorski sets were applied to determine the required homeomorphism h which yielded fh to be equal Lebesgue almost everywhere to a Baire class 1 function. This final step does not yield the stronger statement where Lebesgue measure is replaced by any nontrivial, continuous Borel measure on $(0, 1)$. It is the application of the classical result mentioned in the above parenthetical comment that results in the stronger theorem.

The above mentioned Gorman results concerning Lebesgue measurable functions with countable range predates the Bruckner–Davies–Goffman theorem. Gorman assumes that the function is Lebesgue measurable, a weaker condition than the universally measurable one of the later theorem. Hence the existence of his counterexample does not lead to a contradiction. Gorman used another class of sets defined by Zahorski called the M_2 class. For a set X to be in the M_2 class, each neighborhood of a point in the set X must contain a subset of X that has positive Lebesgue measure. This leads to another closure-like operation $\mathrm{F}^{\lambda}(X)$. That is, if X is a subset of $\mathbb{R}^n$, then $\mathrm{F}^{\lambda}(X)$ is the set of all points of $\mathbb{R}^n$ with the property that every neighborhood of the point contains a subset of X that has positive Lebesgue measure. Gorman proves, for $n = 1$, the following proposition, whose proof is left as an exercise.

PROPOSITION 4.66. *The closure-like operation* $\mathrm{F}^{\lambda}(X)$ *on* $\mathbb{R}^n$ *has the properties*

(1) *if* $X \subset \mathbb{R}^n$, *then* $\mathrm{F}^{\lambda}(X)$ *is a closed set;*
(2) *if* X *is Lebesgue measurable, then there is a Zahorski set* E *contained in* X *such that* E *is dense in* $\mathrm{F}^{\lambda}(X)$*;*
(3) *if* X_i, $i = 1, 2, \ldots, k$, *is a finite collection of Lebesgue measurable sets, then* $\mathrm{F}^{\lambda}(\bigcup_{i=1}^{k} X_i) = \bigcup_{i=1}^{k} \mathrm{F}^{\lambda}(X_i)$.

This proposition yields Gorman's theorem

THEOREM 4.67. *If* f *is a Lebesgue measurable function on* I^n *with* $\mathrm{card}(f[I^n]) < \aleph_0$, *then there is an* h *in* $\mathsf{HOMEO}(I^n)$ *such that* fh *is equal Lebesgue almost everywhere to some Baire class* 1 *function.*

A discussion of the Gorman results and the Bruckner–Davies–Goffman theorem also appears in Goffman, Nishiura and Waterman [**62**].

The above mentioned Gorman results lead to the definition of Gorman pairs (f, μ) on separable metrizable spaces (see page 122). The proof given in this chapter for pairs where f is a universally measurable function whose values are in the discrete space $\mathcal{N} \oplus \mathbb{N}$ is a modification of the proof in [**24**]. The characterization theorem (Theorem 4.33 on page 116) results in the splitting of the proof of the extension of the Bruckner–Davies–Goffman theorem into a topological class of BDG spaces (see page 117) and a class of Zahorski generated Borel measure spaces (see page 123). For a space X in these classes, every pair (f, μ) is a Gorman pair whenever f is a universally measurable map of X into any absolute Borel space Y and μ is a continuous, complete, finite Borel measure on X. As pointed out in the chapter, the Bruckner–Davies–Goffman theorem holds for many spaces in addition to $(0, 1)$.

Zahorski sets play an important role in the development of the chapter. These sets form a subclass of the absolute F_σ spaces. Zahorski defined and used this class of sets in his investigations of the derivative function [**158, 159**]. (A good source on the subject of the derivative function is the book by Bruckner [**22, 23**].) Their appearance in the measure theory of continuous, finite Borel measure spaces is illustrated in Corollary 4.31. Here the measure spaces are defined on absolute measurable spaces X. The proof provided here is essentially the one given for compact spaces X by B. R. Gelbaum [**59**]. Oxtoby [**119**] gave a proof based on his theorem for separable completely metrizable spaces X, which was duplicated in this chapter as Theorem 4.30. As separable completely metrizable spaces are absolute measurable spaces, the above mentioned Oxtoby's theorem is actually deducible from the compact case by simply embedding X into the Hilbert cube. Apropos to this comment, Oxtoby proved the following theorem [**119**, Theorem 3] concerning the collection of all continuous, complete Borel probability measure spaces $\mathrm{M}(X, \nu) = \big(X, \nu, \mathfrak{M}(X, \nu)\big)$, where X is an absolute G_δ space. We denote this collection by $\mathcal{O}$. We shall provide Oxtoby's proof.

Theorem 4.68. *Let* $\mathrm{M}(X, \mu)$ *be a member of* $\mathcal{O}$ *such that for each measure space* $\mathrm{M}(Y, \nu)$ *in* $\mathcal{O}$ *there exists a homeomorphism* h *of* X *into* Y *such that* $\big(h_{\#}\mu\big)|h[X] = \nu|h[X]$. *Then there exists a homeomorphism* H *of* X *into* $\mathcal{N}$ *such that* $H_{\#}\mu = \lambda$.

Proof. Observe that $\mathrm{M}(\mathcal{N}, \nu)$, where $\nu = \lambda|\mathcal{N}$ is in $\mathcal{O}$. Let h be a homeomorphism of X into $\mathcal{N}$ such that $\nu|h[X] = \big(h_{\#}\mu\big)|h[X]$. The measure $h_{\#}\mu$ is a continuous, complete Borel probability measure on $\mathcal{N}$ with $1 = h_{\#}\mu(\mathcal{N}) = h_{\#}\mu(h[X]) = \nu(h[X])$. It follows that $h[X]$ is a dense subset of $\mathcal{N}$, whence $h_{\#}\mu$ is also positive. Let h' be in $\mathsf{HOMEO}(\mathcal{N})$ such that $h'_{\#}h_{\#}\mu = \lambda$. $H = h'h$ is the desired homeomorphism. □

4.8.4. Change of variable. The change of variable theorem of Świątowski is a proposition about the integrability of a function. The reader may have noted that the Warsaw circle was not among the spaces for which our techniques applied. We have the following question.

Question 4.2. Let μ be a continuous, positive Borel measure on the Warsaw circle W such that it and $\mathsf{HOMEO}(W)$ generate $\mathsf{univ}\,\mathfrak{M}(W)$. Let f be a μ-measurable, extended real-valued function that is real-valued μ-almost everywhere. Is there an h in $\mathsf{HOMEO}(W)$ such that fh is μ-measurable and $\int_W fh\,d\mu$ exists?

There is a nonlocally compact, absolute measurable space X and a positive, continuous, complete, finite Borel measure ν on X that has the Świątowski property. Also there is a non absolute measurable space X and a positive, continuous, complete, finite Borel measure ν on X that has the Świątowski property.

4.8.5. Lusin theorems. In the discussion of the images of Lusin sets we introduced several of the many classes of singular sets with very little discussion of them. These classes have set theoretic consequences and further discussions will be delayed to Chapter 6.

There is another very famous Lusin theorem. Namely, for a μ-measurable function $f\colon [0,1] \to \mathbb{R}$, where μ is a complete, finite Borel measure, and for $\varepsilon > 0$, there is a closed set F such that $\mu([0,1] \setminus F) > \mu([0,1]) - \varepsilon$ and $f|F$ is continuous. There are other such theorems where continuity is replaced by various kinds of differentiability. If one replaces the "closed" condition by other requirements, then further measurability requirements on f must be imposed. A possible condition is that f be universally measurable. Several papers along this line have appeared; we shall list some of them rather than provide a discussion of them since they require technicalities that do not seem to fit naturally into the context of the book. The reader is referred to the following references: J. B. Brown and K. Prikry [**20**], and J. B. Brown [**16, 17**].

4.8.6. Fourier series. Investigation of everywhere convergence of Fourier series under all changes of variable was initiated by Goffman and Waterman and has been studied extensively. Of course changes of variable lead naturally to universally measurable functions. The universally measurable functions that appear are equal, except on a universally null set, to a continuous function or to a function of various types of bounded variation. The reader may find many references to this topic in Goffman, Nishiura and Waterman [**62**].

Exercises

4.1. Prove Theorems 4.6 and 4.7 on page 102.
4.2. Prove Proposition 4.9 on page 106.
4.3. Prove Proposition 4.10 on page 106.
4.4. Prove Lemma 4.15 and Theorem 4.16 on page 110.
4.4. Let μ and ν be continuous, complete, finite Borel measures on the Warsaw circle $W = W_0 \cup W_1$ such that $\nu|W_1$ is a positive measure on W_1, where W_1 consists of the points at which W is locally connected. Show that there is an h in $\mathsf{HOMEO}(W; W_0 \text{ fixed})$ such that $\mu \llcorner W_1 \ll h_\# \nu$.
4.5. Prove Theorem 4.19 on page 111.

4.6. Prove Proposition 4.25 on page 113.

4.7. Recall that $\mathcal{N}$ is the set of irrational numbers in the interval $[0, 1]$. Let X be the subspace of $\mathbb{R}^2$ given by

$$X = \big(\mathcal{N} \times \mathcal{N}\big) \cup \{(0, r) \in \mathbb{R}^2 : r \in \mathbb{Q},\ 0 \leq r \leq 1\}.$$

(a) Show that X is not an absolute G_δ space.
(b) Show that X is a BDG space (See page 117.)
(c) What can be said if the right-hand member of the union that forms the space X is replaced with an absolute null space N contained in $\{0\} \times [0, 1]$?

4.8. Recall the definition of a Baire space given on page 117.

(a) Show the existence of a separable Baire space that is an absolute Borel space but not an absolute G_δ space.
(b) Show the existence of separable Baire space that is an absolute measurable space but not an absolute Borel space.
(c) Show the existence of a separable Baire space that is not an absolute measurable space.
(d) Prove the generalization of Lemma 4.26 where the condition "completely metrizable" is replaced with the conditions "absolute Borel and Baire space."

4.9. Verify that $X = W \times [0, 1]^n$ is Zahorski generated relative to the collection $\mathcal{F} = \{W_0 \times [0, 1]^n, W \times [0, 1]^n\}$, where W is the Warsaw circle and W_0 is the set of points of W at which W is not locally connected.

4.10. Provide a verification for Example 5 on page 127.

4.11. Prove Gorman's Proposition 4.66 on page 132.

4.12. Prove Gorman's Theorem 4.67 on page 132.

4.13. Prove $\mathsf{C}_{\mathbb{R}}$ is equivalent to the notion of strong measure zero on the space $\mathbb{R}$ as asserted on page 129. See [**9**, Theorem 1].

4.14. For a separable metrizable space X, let $f : X \to \mathbb{R}$ be such that $f(x) \geq 0$ for every x in X. Find necessary and sufficient conditions for f to be such that $\int_X f d\mu$ exists (and finite) for every continuous, complete, finite Borel measure μ on X.

5

Hausdorff measure and dimension

There are two ways of looking at the dimension of a space – that is, topologically and measure theoretically.[1] The measure theoretic dimension is the Hausdorff dimension, which is a metric notion. Hence, in this chapter, it will be necessary to assume that a metric has been or will be selected whenever the Hausdorff dimension is involved. The chapter concerns the Hausdorff measure and Hausdorff dimension of universally null sets in a metric space. The recent results of O. Zindulka [**160, 161, 162, 163**] form the major part of the chapter.

There are two well-known theorems [**79**, Chapter VII], which are stated next, that influence the development of this chapter.

THEOREM 5.1. *For every separable metric space, the topological dimension does not exceed the Hausdorff dimension.*

THEOREM 5.2. *Every nonempty separable metrizable space has a metric such that the topological dimension and the Hausdorff dimension coincide.*

The first theorem will be sharpened. Indeed, it will be shown that there is a universally null subset whose Hausdorff dimension is not smaller than the topological dimension of the metric space.

5.1. Universally null sets in metric spaces

We begin with a description of the development of Zindulka's theorems on the existence of universally null sets with large Hausdorff dimensions.

Zindulka's investigation of universally null sets in metric spaces begins with compact metrizable spaces that are zero-dimensional. The cardinality of such a space is at most $\aleph_0$ or exactly $\mathfrak{c}$. The first is not very interesting from a measure theoretic point of view. The classic example of the second kind is the Cantor ternary space contained in $\mathbb{R}$. Of course, it is topologically equal to the product space $\{0, 1\}^{\mathbb{N}}$, or more generally the product space k^{ω} where k is a nondegenerate finite space with the discrete topology. The selection of metrics on these spaces is important for the study of Hausdorff measure and Hausdorff dimension. There is a one parameter family of

[1] For those who are not so familiar with topological and Hausdorff dimensions, a brief discussion of these dimensions for separable metric spaces can be found in Appendix D.

metrics $d_{(k,\alpha)}$, $0 < \alpha < 1$, on k^ω such that, for each α, the resulting metric space $\mathbb{C}(k,\alpha)$ contains a universally null set E with $\operatorname{card}(E) = \mathsf{non}\text{-}\mathbb{L}$ and with Hausdorff dimension that coincides with that of $\mathbb{C}(k,\alpha)$. This is made possible by the existence of a universally null set in $\{0,1\}^{\mathbb{N}}$ with cardinality equal to $\mathsf{non}\text{-}\mathbb{L}$. For appropriate choices of the parameter α, it is shown that there is a natural embedding of $\mathbb{C}(2,\alpha)$, where $2 = \{0,1\}$, onto a subspace $\mathcal{C}_\alpha$ of $\mathbb{R}$ such that the embedding homeomorphism and its inverse are Lipschitzian (that is, bi-Lipschitzian), thereby resulting in a bi-Lipschitzian copy of the Cantor cube $(\mathcal{C}_\alpha)^n$ in $\mathbb{R}^n$. In this way, for each s in the closed interval $[0,n]$, a universally null set in $\mathbb{R}^n$ with Hausdorff dimension equal to s is exhibited. With geometric measure theoretic tools, sharper results are shown for analytic subsets of $\mathbb{R}^n$.

The proof of the main theorem of the chapter relies on a topological dimension theoretic theorem due to Zindulka [**160**]. (This dimension theoretic theorem is fully developed in Appendix D.) It follows from this theorem that each separable metric space contains a universally null set whose Hausdorff dimension is not smaller than the topological dimension of the space.

5.2. A summary of Hausdorff dimension theory

Here is a brief survey of p-dimensional Hausdorff measure on a separable metric space X and Hausdorff dimension of its subsets.

5.2.1. Hausdorff measure.

In this section we shall assume that X is a separable metric space with the metric denoted by d.

DEFINITION 5.3. *Let E be a subset of X and let p be a real number with $0 \le p$. For $\delta > 0$, define $\mathsf{H}_p^\delta(E)$ to be the infimum of the set of numbers $\sum_{S\in G}(\operatorname{diam}(S))^p$ corresponding to all countable families G of subsets S of X such that* $\operatorname{diam}(S) \le \delta$ *and $E \subset \bigcup_{S\in G} S$.*[2] *The* p-dimensional Hausdorff outer measure *on X is*

$$\mathsf{H}_p(E) = \sup\{\,\mathsf{H}_p^\delta(E) \colon \delta > 0\,\};$$

or equivalently,

$$\mathsf{H}_p(E) = \lim_{\delta\to 0} \mathsf{H}_p^\delta(E)$$

since the limit always exists as a nonnegative extended real number. A set E is said to be H_p-measurable *if* $\mathsf{H}_p(T) = \mathsf{H}_p(T\cap E) + \mathsf{H}_p(T \setminus E)$ *whenever $T \subset X$.*

Let $\mathfrak{M}(X,\mathsf{H}_p)$ be the collection of all H_p-measurable sets. The triple $\mathrm{M}(X,\mathsf{H}_p) = \big(X,\mathfrak{M}(X,\mathsf{H}_p),\mathsf{H}_p\big)$ is a complete Borel measure space on the topological space X. The zero-dimensional Hausdorff measure is the usual counting measure on X. For $p > 0$, the measure H_p on X is continuous, that is, $\mathsf{H}_p(E) = 0$ for every singleton set E. In general, $\mathrm{M}(X,\mathsf{H}_p)$ is not σ-finite. If Y is a subset of X whose p-dimensional Hausdorff outer measure is finite, then the Hausdorff measure space $\mathrm{M}(X,\mathsf{H}_p)$ restricted to Y

[2] We use the conventions that $\operatorname{diam}(\emptyset) = 0$ and $0^0 = 1$.

is the continuous, complete, finite Borel measure space $\mathsf{M}(Y, \mathsf{H}_p)$ that is induced by the p-dimensional Hausdorff outer measure on the metric subspace Y of X. If $X = \mathbb{R}^n$, then $\mathsf{H}_n = \alpha_n \lambda_n$, where λ_n is the usual Lebesgue measure on $\mathbb{R}^n$ and α_n is a normalizing constant given by $\mathsf{H}_n([0,1]^n) = \alpha_n \lambda_n([0,1]^n)$.

The empty set may, at times, require special treatment. The reader should keep in mind the statements in the following proposition in the course of the development of this chapter, where dim denotes the topological dimension function.

PROPOSITION 5.4. *If X is a separable metric space, then, for every p,*

$$\dim \emptyset = -1 < 0 = \mathsf{H}_p(\emptyset).$$

If $0 \leq p$ and if E is a subset of a separable metric space X with $\dim E \leq 0$, *then* $\dim E \leq \mathsf{H}_p(E)$.

Here is a useful theorem concerning Lipschitzian maps. The simple proof is left to the reader.

THEOREM 5.5. *For separable metric spaces X and Y let $f\colon X \to Y$ be a Lipschitzian map with Lipschitz constant L. If $0 \leq p$ and $E \subset X$, then $\mathsf{H}_p(f[E]) \leq L^p\, \mathsf{H}_p(E)$.*

It will be convenient to define at this point the notion of a bi-Lipschitzian embedding.

DEFINITION 5.6. *Let X and Y be separable metric spaces with respective metrics d_X and d_Y. An injection $\varphi\colon X \to Y$ is called a* bi-Lipschitzian embedding *of X onto $M = \varphi[X]$ if φ is a Lipschitzian map and $(\varphi|M)^{-1}\colon M \to X$ is a Lipschitzian map.*

Clearly the Lipschitz constants of the maps φ and $(\varphi|M)^{-1}$ in the above definition are positive whenever $\mathrm{card}(X) > 1$. Hence we have the obvious theorem.

THEOREM 5.7. *Let φ be a bi-Lipschitzian embedding of X into Y. If $0 \leq p$ and $E \subset X$, then*

$$\mathsf{H}_p(E) < \infty \text{ if and only if } \mathsf{H}_p(\varphi[E]) < \infty$$

and

$$0 < \mathsf{H}_p(E) \text{ if and only if } 0 < \mathsf{H}_p(\varphi[E]).$$

In Chapter 2 we gave a development of Grzegorek's theorem (see page 20) which says that for each positive, continuous, complete, finite Borel measure μ on $\{0,1\}^{\mathbb{N}}$ there are subsets A and B of $\{0,1\}^{\mathbb{N}}$ with $\mathrm{card}(A) = \mathrm{card}(B) = \mathsf{non}\text{-}\mathbb{L}$ such that A is an absolute null space and the outer μ measure of B is positive. Grzegorek used this to solve a question posed by Darst by showing there is an absolute null space E contained in $\{0,1\}^{\mathbb{N}} \times \{0,1\}^{\mathbb{N}}$ with $\mathrm{card}(E) = \mathsf{non}\text{-}\mathbb{L}$ such that E projects naturally onto A and B as bijections. Let us use this on topological copies X and Y of $\{0,1\}^{\mathbb{N}}$ and any continuous, complete, finite Borel measure μ on Y. We have

PROPOSITION 5.8. *For metric spaces X and Y that are homeomorphic to $\{0,1\}^{\mathbb{N}}$ and a nontrivial, continuous, complete, finite Borel measure μ on Y there is a subset E*

of $X \times Y$ *with* $\operatorname{card}(E) = \text{non-}\mathbb{L}$ *and there are subsets* A *of* X *and* B *of* Y *such that* E *projects onto* A *and* B *as bijections,* E *and* A *are absolute null spaces, and* $\mu^*(B) > 0$.

5.2.2. *Hausdorff dimension.* We shall continue to assume that X is a separable metric space. It is easily seen that if $0 \leq p < q$ then $\mathsf{H}_p(E) \geq \mathsf{H}_q(E)$ for subsets E of X. Also, if $\mathsf{H}_p(E) < \infty$ then $\mathsf{H}_q(E) = 0$ whenever $p < q$. This leads to the following definition.

DEFINITION 5.9. *For subsets* E *of* X, *the* Hausdorff dimension *of* E *is the extended real number* $\dim_{\mathsf{H}} E = \sup\{\, p\colon\ \mathsf{H}_p(E) > 0\,\}$.

Of course the Hausdorff dimension is dependent on the metric d of X. The four properties of the Hausdorff dimension in the next theorem are easily proved. As stated earlier, dim denotes the topological dimension function.

THEOREM 5.10. *Let* X *be a separable metric space.*

(1) $-1 = \dim \emptyset < \dim_{\mathsf{H}} \emptyset = 0$.
(2) *If* $E \subset X$, *then* $\dim E \leq \dim_{\mathsf{H}} E$.
(3) *If* $A \subset B \subset X$, *then* $\dim_{\mathsf{H}} A \leq \dim_{\mathsf{H}} B$.
(4) *If* A_i, $i = 1, 2, \ldots$, *is a countable collection of subsets of* X, *then* $\dim_{\mathsf{H}} \bigcup_{i=1}^{\infty} A_i = \sup\{\, \dim_{\mathsf{H}} A_i\colon i = 1, 2, \ldots\,\}$.

We have the following theorem which can be summarized as "the Hausdorff dimension of a set is a bi-Lipschitzian invariant." It is a consequence of Theorem 5.5.

THEOREM 5.11. *Let* X *and* Y *be separable metric spaces. For bi-Lipschitzian embeddings* $\varphi\colon X \to Y$ *of* X *onto* $M = \varphi[X]$,

$$\dim_{\mathsf{H}} E = \dim_{\mathsf{H}} \varphi[E] \text{ whenever } E \subset X.$$

This ends our summary of topological and Hausdorff dimensions of separable metric spaces.

5.3. Cantor cubes

The Cantor ternary set has been topologically characterized among the metrizable spaces to be nonempty, compact, perfect and zero-dimensional. The expression Cantor cube is a nice way to say that a metric has been chosen so as to result in a metric product space k^{ω}, where k is a finite set with more than one member. For the purposes of Hausdorff dimension, Cantor cubes are useful in that a correct choice of a metric and a correct choice of a product measure will facilitate the computation of the precise values of p-dimensional Hausdorff measure and Hausdorff dimension. Of particular interest is the Cantor space $\{0, 1\}^{\mathbb{N}}$.[3] But it will be convenient to consider more general Cantor spaces and provide them with useful specific metrics.

[3] See Appendix C for relevant material on metric spaces that are topologically equivalent to $\{0, 1\}^{\mathbb{N}}$.

5.3.1. Measures on Cantor cubes. The compact space k^ω carries a natural, continuous, Borel, probability measure μ_k generated by the uniform Bernoulli distribution μ on the factor spaces k given by $\mu(\{x\}) = (\operatorname{card}(k))^{-1}$ for each x in k. Let n be in $\mathbb{N}$ and consider the projection $\varphi_n\colon (x_0, x_1, \dots) \longmapsto (x_0, x_1, \dots, x_{n-1}) \in k^n$. As each singleton set $\{p\}$ of k^n is open we have $\mu_k({\varphi_n}^{-1}[\{p\}]) = (\operatorname{card}(k))^{-n}$. Denote the open set ${\varphi_n}^{-1}[\{p\}]$ by U_p. Consider the metrics $\mathrm{d}_{(k,\alpha)}$ on k^ω with $0 < \alpha < 1$ that are defined in Appendix C (see page 217). For the convenience of the reader we shall repeat the definition here in a notational form consistent with the product space notation, that is, $f(m)$ is the m-th coordinate of a member f of k^ω.

For distinct f and g in k^ω define

$$\chi(f,g) = \min\{\, m \in \omega\colon f(m) \neq g(m)\,\}.$$

As $f \neq g$, we have $\chi(f,g) \in \omega$. Hence $\chi(f,g)$ is the length of the initial segment that is common to f and g. Let $0 < \alpha < 1$ and define

$$\mathrm{d}_{(k,\alpha)}(f,g) = \begin{cases} \alpha^{\chi(f,g)}, & \text{if } f \neq g, \\ 0, & \text{if } f = g. \end{cases}$$

Observe that if f and g are members of U_p, then $\mathrm{d}_{(k,\alpha)}(f,g) \leq \alpha^n$, whence $\operatorname{diam}(U_p) = \alpha^n$. Let us summarize this observation as a lemma.

LEMMA 5.12. *Let $n \in \mathbb{N}$ and $s = \frac{\ln(\operatorname{card}(k))}{|\ln \alpha|}$. If $p \in k^n$, then*

$$\mu_k(U_p) = (\operatorname{card}(k))^{-n} = \alpha^{sn} = (\operatorname{diam}(U_p))^s.$$

We need to derive another property of the metric $\mathrm{d}_{(k,\alpha)}$. Let E be a nonempty subset of k^ω. First, some notation: for $f \in k^\omega$ and $n \in \mathbb{N}$, define $f|n$ to be the point of k^n given by

$$f|n = \langle f(0), f(1), \dots, f(n-1)\rangle.$$

If $\operatorname{card}(E) = 1$, then $\operatorname{diam}(E) = 0$ and $E \subset U_{f|n}$ for every f in E and every n. For $\operatorname{card}(E) > 1$ define

$$n(E) = \min\{\, \chi(f,g)\colon f \in E, g \in E, \text{ and } f \neq g\,\}.$$

Then $\mathrm{d}_{(k,\alpha)}(f,g) \leq \alpha^{n(E)}$ whenever f and g are in E. Hence, for every f in E, we have $E \subset U_{f|n(E)}$ and $\operatorname{diam}(U_{f|n(E)}) = \operatorname{diam}(E) = \alpha^{n(E)}$. Let us also summarize this as a lemma.

LEMMA 5.13. *Assume $E \subset k^\omega$. If $\operatorname{card}(E) = 1$ and $n \in \mathbb{N}$, then $E \subset U_{f|n}$ for every f in E. If $\operatorname{card}(E) > 1$, then there exists an n in $\mathbb{N}$ such that $E \subset U_{f|n}$ and $\alpha^n = \operatorname{diam}(E) = \operatorname{diam}(U_{f|n})$ for every f in E.*

We now have a proposition on open covers. The proof is left as an exercise.

PROPOSITION 5.14. *Let $0 < s < \infty$ and $0 < \delta < \infty$, and let E_i, $i = 1, 2, \dots$, be a countable family of nonempty subsets of k^ω with $\operatorname{diam}(E_i) \leq \delta$. Then, for each* *positive number ε and for each i, there is an n_i in $\mathbb{N}$ and there is an f_i in E_i such that $E_i \subset U_{f_i|n_i}$ and $\operatorname{diam}(U_{f_i|n_i}) \le \delta$, and such that*

$$\textstyle\sum_{i=1}^{\infty}(\operatorname{diam}(U_{f_i|n_i}))^s \le \sum_{i=1}^{\infty}(\operatorname{diam}(E_i))^s + \varepsilon.$$

Of course, the ε is needed for the singleton sets of the countable family. We are now ready to show that μ_k and H_s, where $s = \frac{\ln(\operatorname{card}(k))}{|\ln\alpha|}$, coincide on the metric space $\mathbb{C}(k,\alpha)$.

LEMMA 5.15. *Let $0 < \alpha < 1$ and $s = \frac{\ln(\operatorname{card}(k))}{|\ln\alpha|}$. Then $\mu_k(B) = \mathsf{H}_s(B)$ for every Borel subset B of k^ω. Hence $\dim_{\mathsf{H}}(\mathbb{C}(k,\alpha)) = s$.*

PROOF. Let B be any Borel subset of k^ω and let $0 < \delta < \infty$ and $0 < \varepsilon$. Suppose that $E_i, i = 1, 2, \ldots$, is a cover of B by nonempty sets with $\operatorname{diam}(E_i) \le \delta$. The proposition yields sets $U_{f_i|n_i}$ with $n_i \in \mathbb{N}$ and $f_i \in E_i$ such that $E_i \subset U_{f_i|n_i}$ for every i and such that the above displayed inequality holds. We have

$$\textstyle\mu_k(B) \le \sum_{i=1}^{\infty}\mu_k(U_{f_i|n_i}) \le \sum_{i=1}^{\infty}(\operatorname{diam}(E_i))^s + \varepsilon.$$

Hence $\mu_k(B) \le \mathsf{H}_s(B)$. On the other hand, by Lemma 5.12, we have for each n in $\mathbb{N}$ the inequality

$$\textstyle\mathsf{H}_s^{\alpha^n}(k^\omega) \le \sum_{p\in k^n}(\operatorname{diam}(U_p))^s = \sum_{p\in k^n}\mu_k(U_p) = \mu_k(k^\omega).$$

Since $\alpha^n \to 0$ as $n \to \infty$, we have $\mathsf{H}_s(k^\omega) \le \mu_k(k^\omega) = 1$. The inequality $\mathsf{H}_s(B) \le \mu_k(B)$ now follows easily for Borel sets B. □

Let us turn to the metric space $\mathbb{C}(2,\alpha)$ with $0 < \alpha < \frac{1}{2}$, where $2 = \{0,1\}$. Define the real-valued function $\varphi\colon \mathbb{C}(2,\alpha) \to \mathbb{R}$ given by the absolutely convergent series $\varphi(f) = (1-\alpha)\sum_{m=0}^{\infty}\alpha^m f(m)$. Denote the image $\varphi[\mathbb{C}(2,\alpha)]$ by $\mathcal{C}_\alpha$. A straightforward computation will lead to the inequalities

$$(1-2\alpha)\,\mathrm{d}_{(2,\alpha)}(f,g) \le |\varphi(f)-\varphi(g)| \le \mathrm{d}_{(2,\alpha)}(f,g)$$

for every f and g in $\mathbb{C}(2,\alpha)$. Consequently, φ is a bi-Lipschitzian map between $\mathbb{C}(2,\alpha)$ and $\mathcal{C}_\alpha$.[4]

5.3.2. Hausdorff dimension of universally null sets. The first theorem establishes the existence of universally null sets in Cantor cubes with maximal Hausdorff dimensions.

THEOREM 5.16. *Let $k \in \omega$ with $k \ge 2$ and let $0 < \alpha < 1$. Then there exist universally null sets E in $\mathbb{C}(k,\alpha)$ with $\dim_{\mathsf{H}} E = \dim_{\mathsf{H}} \mathbb{C}(k,\alpha) = \frac{\ln(\operatorname{card}(k))}{|\ln\alpha|}$ and $\operatorname{card}(E) =$ non-$\mathbb{L}$.*

[4] Related results on Hausdorff measure and dimension of Cantor set type constructions in $[0,1]$ by C. Cabrelli, U. Molter, V. Paulauskas and R. Shonkwiler can be found in [**26**].

PROOF. Let us first construct a universally null set E_m in $\mathbb{C}(k,\alpha)$ with

$$\dim_{\mathsf{H}} E_m \geq (1 - \tfrac{1}{m}) \dim_{\mathsf{H}} \mathbb{C}(k,\alpha),$$

where $m-1$ is a positive integer. To this end, consider the product metric space $\mathbb{C}(k,\alpha^m) \times \mathbb{C}(k^{m-1},\alpha^m)$. By Proposition 5.8 there is a universally null set E' in the product metric space $\mathbb{C}(k,\alpha^m) \times \mathbb{C}(k^{m-1},\alpha^m)$ with $\operatorname{card}(E') = \text{non-}\mathbb{L}$ and the projection of E' onto B in the second factor space satisfies $\mu_{k^{m-1}}{}^*(B) > 0$. As the projection of E' onto B is a Lipschitzian map with Lipschitz constant equal to 1, we have

$$\mathsf{H}_s(E') \geq \mathsf{H}_s(B) = \mu_{k^{m-1}}{}^*(B) > 0,$$

where

$$s = \dim_{\mathsf{H}} \mathbb{C}(k^{m-1},\alpha^m) = \frac{\ln(\operatorname{card}(k^{m-1}))}{|\ln \alpha^m|}$$

$$= \frac{m-1}{m} \frac{\ln(\operatorname{card}(k))}{|\ln \alpha|} = (1 - \tfrac{1}{m}) \dim_{\mathsf{H}} \mathbb{C}(k,\alpha).$$

Hence $\dim_{\mathsf{H}}(E') \geq (1 - \frac{1}{m}) \dim_{\mathsf{H}} \mathbb{C}(k,\alpha)$. We infer from Propositions C.8 and C.9 of Appendix C that $\mathbb{C}(k,\alpha)$ is bi-Lipschitzian equivalent to the product metric space $\mathbb{C}(k,\alpha^m) \times \mathbb{C}(k^{m-1},\alpha^m)$. Hence the universally null set E_m in $\mathbb{C}(k,\alpha)$ has been constructed. To complete the proof, observe that the set $E = \bigcup_{m=1}^{\infty} E_m$ is a universally null set in $\mathbb{C}(k,\alpha)$ with $\dim_{\mathsf{H}} E = \dim_{\mathsf{H}} \mathbb{C}(k,\alpha)$. □

Let us describe an example of a metric space X that illustrates the "gap" between the Hausdorff dimension and the topological dimension of X.

EXAMPLE 5.17. Let X be the disjoint topological sum

$$\mathbb{C}(k,\alpha_1) \oplus \mathbb{C}(k,\alpha_2) \oplus [0,1]^n$$

with a metric d that satisfies the conditions that d restricted to $\mathbb{C}(k,\alpha_i)$ is $\mathrm{d}_{(k,\alpha_i)}$ for $i = 1, 2$, and that d restricted to $[0,1]^n$ is the usual Euclidean distance. Select α_1 and α_2 so that

$$n < s_1 = \frac{\ln(\operatorname{card} k)}{|\ln \alpha_1|} < s_2 = \frac{\ln(\operatorname{card} k)}{|\ln \alpha_2|}.$$

Then there exists a universally null set E in X such that

$$\dim X = n < s_1 = \dim_{\mathsf{H}} E < s_2 = \dim_{\mathsf{H}} X.$$

The verification of the last inequalities is left to the reader.

QUESTION. In the example X above there exists a universally null set E in X for which $\dim_{\mathsf{H}} E = \dim_{\mathsf{H}} X$. Is it true that there always is a universally null set E in X such that $\dim_{\mathsf{H}} E = \dim_{\mathsf{H}} X$ whenever X is a separable metric space?

Let us turn to the question of the existence of perfect closed subsets F of uncountable compact metric spaces X such that $\mathsf{H}_s(F) = 0$ whenever $s > 0$. Of course, such a subset has $\dim_{\mathsf{H}} F = 0$. Due to Grzegorek's result there is an absolute null space E

with $\operatorname{card}(E) = \mathsf{non}\text{-}\mathbb{L}$ that is contained in F. With the aid of Exercise 5.5, one can prove

PROPOSITION 5.18. *If X is a uncountable compact metric space, then there is a universally null set E in X such that* $\operatorname{card}(E) = \mathsf{non}\text{-}\mathbb{L}$ *and* $\dim_{\mathsf{H}} E = 0$. *More generally, if the metric space X is an absolute measurable space that is not an absolute null space, then there is a universally null set E in X such that* $\dim_{\mathsf{H}} E = 0$ *and* $\operatorname{card}(E) = \mathsf{non}\text{-}\mathbb{L}$.

5.3.3. Euclidean n-dimensional space. Certain Cantor cubes $\mathbb{C}(k,\alpha)$ can be bi-Lipschitzian embedded into $\mathbb{R}^n$. This will permit us to prove that $\mathbb{R}^n$ contains universally null sets E_s such that $\dim_{\mathsf{H}} E_s = s$ for each s with $0 \leq s \leq n$.

THEOREM 5.19. *For each positive integer n and for each s with $0 \leq s \leq n$ there is a universally null subset E_s of $\mathbb{R}^n$ with $\dim_{\mathsf{H}} E_s = s$ and* $\operatorname{card}(E_s) = \mathsf{non}\text{-}\mathbb{L}$.

PROOF. The case $s = 0$ follows from Proposition 5.18. Let us consider an s with $0 < s < n$. Let $\alpha = 2^{-n/s}$. Clearly $\alpha < \frac{1}{2}$. Hence $\mathbb{C}(2,\alpha)$ is bi-Lipschitzian equivalent to the subset $\mathcal{C}_\alpha$ of $\mathbb{R}$. We have that $\mathbb{C}(2^n,\alpha)$ and the product $(\mathcal{C}_\alpha)^n$ with the maximum metric are bi-Lipschitzian equivalent. It is known that the maximum metric and the Euclidean metric on $\mathbb{R}^n$ are bi-Lipschitzian equivalent. Hence there is a bi-Lipschitzian embedding of $\mathbb{C}(2^n,\alpha)$ into $\mathbb{R}^n$ endowed with the Euclidean metric. As there is a universally null set D in $\mathbb{C}(2^n,\alpha)$ with $\dim_{\mathsf{H}} D = \dim_{\mathsf{H}} \mathbb{C}(2^n,\alpha) = s$ and $\operatorname{card}(D) = \mathsf{non}\text{-}\mathbb{L}$, the existence of the required set E_s is established.

For $s = n$, define $s_i = n - \frac{1}{i}$. Let E_{s_i} be a universally null set in $\mathbb{R}^n$ such that $\dim_{\mathsf{H}} E_{s_i} = s_i$ and $\operatorname{card}(E_{s_i}) = \mathsf{non}\text{-}\mathbb{L}$. Then $E_n = \bigcup_{i=1}^{\infty} E_{s_i}$ is a universally null set in $\mathbb{R}^n$ such that $\dim_{\mathsf{H}} E_n = n$. As $\mathsf{non}\text{-}\mathbb{L} > \omega$, we have $\operatorname{card}(E_n) = \mathsf{non}\text{-}\mathbb{L}$. □

Of course the above embedded Cantor cube can be arranged to be contained in a preassigned open set. A natural question is: Can the open set be replaced by an uncountable Borel set? More specifically, is there a universally null set E contained in a Borel set B of $\mathbb{R}^n$ such that $\dim_{\mathsf{H}} E = \dim_{\mathsf{H}} B$? This question, which was posed by Zindulka in [**161**], has been answered by him in the affirmative in [**162**]. His solution uses facts from geometric measure theory which are not as elementary as those used in the above theorem. Note that a more general question has already been proposed immediately following Example 5.17. Before turning to Zindulka's solution, we shall go to a, in a sense, weaker question in the next section.

5.4. Zindulka's theorem

A classical theorem in topological dimension theory (Theorem 5.1) states that $\dim_{\mathsf{H}} X \geq \dim X$ for every separable metric space (X, d). Zindulka showed in [**162**] that every separable metric space X contains a subset E that is universally null in X with $\dim_{\mathsf{H}} E \geq \dim X$ (see Theorem 5.22 below). It follows that $\dim_{\mathsf{H}} X \geq \dim_{\mathsf{H}} E \geq \dim X$ and thereby the classical theorem has been sharpened.

The proof of Zindulka's theorem relies on a dimension theoretic theorem on the existence of a certain countable family of Lipschitzian maps for arbitrary metric spaces. This general theorem is proved in Appendix C. We state and prove the special separable metric case here since its proof is straightforward.

THEOREM 5.20 (Zindulka). *If* (X, d) *is a nonempty separable metric space, then there is a sequence of Lipschitzian functions* $h_m \colon X \to [0, 1]$, $m = 0, 1, 2, \ldots$, *such that*

$$G(r) = \bigcap_{m=0}^{\infty} h_m{}^{-1}\big[\{\, s \in [0,1] \colon s \neq r \,\}\big]$$

is a G_δ *set with* $\dim G(r) \leq 0$ *for each* r *in the open interval* $(0, 1)$. *Moreover,* $X = \bigcup_{r \in E} G(r)$ *whenever* E *is an uncountable subset of* $(0, 1)$.

PROOF. Let $\mathcal{B} = \{\, U_n \colon n \in \omega \,\}$ be a countable base for the open sets of a nonempty separable metric space X. For each pair $\langle n, j \rangle$ in $\omega \times \omega$, define the Lipschitzian function $g_{\langle n, j\rangle} \colon X \to [0, 1]$ given by the formula

$$g_{\langle n,j\rangle}(x) = 1 \wedge \big(j \operatorname{dist}\big(x, X \setminus U_n\big)\big), \quad x \in X;$$

and, for each r in the open interval $(0, 1)$, define the set

$$G(r) = \bigcap_{\langle n,j\rangle \in \omega \times \omega} g_{\langle n,j\rangle}{}^{-1}\big[\{\, s \in [0,1] \colon s \neq r \,\}\big].$$

Clearly $G(r)$ is a G_δ set. Let us show $\dim G(r) \leq 0$ for every r. To this end we first show that the collection

$$\mathcal{D}_r = \{\, g_{\langle n,j\rangle}{}^{-1}\big[(r, 1]\big] \colon \langle n,j\rangle \in \omega \times \omega \,\}$$

is a base for the open sets of X. For $x \in X$, let n be such that $x \in U_n$. There is a j such that $g_{\langle n,j\rangle}(x) > r$. Consequently, $x \in g_{\langle n,j\rangle}{}^{-1}\big[(r, 1]\big]$. Hence $\mathcal{D}_r$ is a base for the open sets of X. Next let us show that $D = G(r) \cap g_{\langle n,j\rangle}{}^{-1}\big[(r, 1]\big]$ is closed in the subspace $G(r)$. Using the fact that the distance function $\operatorname{dist}\big(\cdot, X \setminus U_n\big)$ appears in the definition of $g_{\langle n,j\rangle}$, one can easily verify $g_{\langle n,j\rangle}{}^{-1}\big[[r, 1]\big] \cap \mathrm{Cl}(U_n) \subset U_n$ and $G(r) \cap g_{\langle n,j\rangle}{}^{-1}\big[\{r\}\big] \cap U_n = \emptyset$. So

$$\begin{aligned}
\mathrm{Cl}_{G(r)}(D) &\subset G(r) \cap \mathrm{Cl}\big(g_{\langle n,j\rangle}{}^{-1}\big[(r, 1]\big] \cap U_n\big) \\
&\subset G(r) \cap g_{\langle n,j\rangle}{}^{-1}\big[[r, 1]\big] \cap \mathrm{Cl}(U_n) \\
&\subset G(r) \cap g_{\langle n,j\rangle}{}^{-1}\big[(r, 1]\big] \cap U_n = D.
\end{aligned}$$

Consequently $\dim(\mathrm{Bd}_{G(r)}(D)) = -1$ and thereby $\dim G(r) \leq 0$ follows from the characterization provided by Theorem D.3.

Let us show $X = \bigcup_{r \in E} G(r)$ whenever $\operatorname{card}(E) \geq \aleph_1$. Suppose that there is an x in X such that $x \notin G(r)$ for every r in E. From the definition of $G(r)$ there is a pair $\langle n_x, j_x \rangle$ in $\omega \times \omega$ such that x is not in $g_{\langle n_x,j_x\rangle}{}^{-1}\big[\{\, s \in [0,1] \colon s \neq r \,\}\big]$, that is $g_{\langle n_x,j_x\rangle}(x) = r$. This defines a map $\eta \colon r \mapsto \langle n, j \rangle$ of E into $\omega \times \omega$ such that $g_{\eta(r)}(x) = r$. Since E is uncountable and $\omega \times \omega$ is countable, there are two distinct r and r' in E that map to the same $\langle n,j\rangle$. This implies $g_{\langle n,j\rangle}(x) = r$ and $g_{\langle n,j\rangle}(x) = r'$, a contradiction. Thereby the required equality is established.

The proof of the theorem is completed by well ordering $\omega \times \omega$. □

The above theorem does not require the metric space X to be finite dimensional. The theorem leads to the following dimension theoretic result in which the separable metrizable spaces have their topological dimensions bounded below, unlike many theorems of dimension theory which have the dimensions bounded above.

THEOREM 5.21 (Zindulka). *If X is a separable metrizable space and if m and n are integers such that $\dim X \geq m \geq n \geq 0$, then to each metric for X there corresponds a countable family $\mathcal{F}$ of Lipschitzian maps of X into $[0,1]^n$ such that for each r in $(0,1)^n$ there is an f in $\mathcal{F}$ with $\dim f^{-1}[\{r\}] \geq m-n$.*

PROOF. Let us use the sequence h_k, $k = 0,1,2,\ldots$, of Lipschitzian functions as provided by Theorem 5.20. For each ι in ω^n – that is, $\iota = \langle \iota(0), \iota(1), \ldots, \iota(n-1)\rangle \in \omega^n$ – define the Lipschitzian function $f_\iota \colon X \to [0,1]^n$ by

$$f_\iota(x) = \langle h_{\iota(0)}(x), h_{\iota(1)}(x), \ldots, h_{\iota(n-1)}(x)\rangle,$$

and define $\mathcal{F}$ to be the countable family $\{f_\iota \colon \iota \in \omega^n\}$. Let us show that $\mathcal{F}$ satisfies the requirement of the theorem. To do this we shall use well-known theorems from topological dimension theory; the reader is referred to Theorem D.6 of Appendix D. Let $r = \langle r_1, \ldots, r_n\rangle \in (0,1)^n$. As $\dim G(r_j) \leq 0$ for $1 \leq j \leq n$ we infer from the addition theorem of dimension theory that $\dim \bigcup_{j=1}^n G(r_j) \leq n-1$; hence

$$\dim \bigcap_{j=1}^n F(r_j) \geq \dim X - \dim \bigcup_{j=1}^n G(r_j) - 1 \geq m-n,$$

where $F(t) = X \setminus G(t)$ for every t in $(0,1)$. On the other hand,

$$\begin{aligned}\bigcap_{j=1}^n F(r_j) &= \bigcap_{j=1}^n \bigcup_{k\in\omega} {h_k}^{-1}[\{r_j\}] \\ &= \bigcup_{\iota\in\omega^n} \bigcap_{j=1}^n {h_{\iota(j)}}^{-1}[\{r_j\}] = \bigcup_{\iota\in\omega^n} {f_\iota}^{-1}[\{r\}].\end{aligned}$$

As ${f_\iota}^{-1}[\{r\}]$ is closed for every ι in ω^n, by the sum theorem of dimension theory, there is an ι in ω^n such that $\dim {f_\iota}^{-1}[\{r\}] \geq m-n$ and the theorem is proved. □

The following is an application of the last theorem.

THEOREM 5.22 (Zindulka). *Let X be a separable metrizable space. For each metric for X, there exists a universally null set E in X with $\dim_{\mathsf{H}} E \geq \dim X$. Also, if $\dim X \geq 1$, then $\operatorname{card}(E) = \mathsf{non}\text{-}\mathbb{L}$ can be required of E as well.*

PROOF. We have already observed earlier that the statement of the theorem holds for spaces with $\dim X < 1$. So, let n be an integer such that $1 \leq n$, and assume $n \leq \dim X$. We infer from Theorem 5.19 that there is a universally null set D_n in

$(0, 1)^n$ such that $\dim_{\mathsf{H}} D_n = n$. Consider the countable family $\mathcal{F}$ of Lipschitzian maps provided by Theorem 5.21 with $m = n$. Define, for each f in $\mathcal{F}$, the set

$$D_n(f) = \{ r \in D_n : f^{-1}[\{r\}] \neq \emptyset \}.$$

For each r in D_n we have an f in $\mathcal{F}$ such that $\dim f^{-1}[\{r\}] \geq 0$, whence $f^{-1}[\{r\}]$ is not empty. Consequently,

$$D_n = \bigcup_{f \in \mathcal{F}} D_n(f).$$

For each f in $\mathcal{F}$ and each r in $D_n(f)$ select a point $x(f, r)$ in $f^{-1}[\{r\}]$. Define

$$E_n(f) = \{ x(f, r) : r \in D_n(f) \} \quad \text{and} \quad E_n = \bigcup_{f \in \mathcal{F}} E_n(f).$$

As $f | E_n(f) : E_n(f) \to D_n(f)$ is a continuous bijection and $D_n(f)$ is a absolute null space we have that $E_n(f)$ is also an absolute null space. So, $E_n(f)$ is a universally null set in X with $\operatorname{card}(D_n(f)) = \operatorname{card}(E_n(f))$. It follows that E_n is a universally null set in X with $\operatorname{card}(E_n) = \mathsf{non}\text{-}\mathbb{L}$. As f is a Lipschitzian map we have $\dim_{\mathsf{H}} E_n(f) \geq \dim_{\mathsf{H}} D_n(f)$ because $f[E_n(f)] = D_n(f)$. Consequently, $\dim_{\mathsf{H}} E_n \geq \dim_{\mathsf{H}} D_n \geq n$. The required set is $E = \bigcup \{ E_n : 1 \leq n \leq \dim X \}$. □

We have observed earlier the following corollary which provides a second proof of the classical Theorem 5.1 as well as a sharpening of it.

COROLLARY 5.23. *If X is a separable metric space, then there is a universally null set E in X such that* $\dim_{\mathsf{H}} X \geq \dim_{\mathsf{H}} E \geq \dim X$.

(It is shown in Appendix D that Theorem 5.1 is implied directly by Theorem 5.21 without recourse to the above corollary.)

The final theorem is an immediate consequence of Theorem 5.2.

THEOREM 5.24. *For each nonempty separable metrizable space X there is a metric for X and a universally null set E in X such that* $\dim_{\mathsf{H}} X = \dim_{\mathsf{H}} E = \dim X$.

REMARK 5.25. A very important consequence of Theorem 5.22 is that every separable metric space X with $\dim X = \infty$ contains a universally null set E in X such that $\dim_{\mathsf{H}} E = \infty$. Hence the existence of universally null sets E in X with $\dim_{\mathsf{H}} E = \dim_{\mathsf{H}} X$ is only a problem for finite topological dimensional metric spaces X. It is known that every finite dimensional separable metrizable space can be topologically embedded into $[0, 1]^{2n+1}$ (see Theorem D.5 in Appendix D). Unfortunately the embedding need not be bi-Lipschitzian, witness the case of $\dim X = 0$. In the next section we shall consider analytic metric spaces that have bi-Lipschitzian injections into some Euclidean space.

5.5. Analytic sets in $\mathbb{R}^n$

Let us return to universally null sets contained in Borel subsets of $\mathbb{R}^n$ as promised in Section 5.3.3. As we have mentioned already, Zindulka showed in [**162**] that every Borel subset B of $\mathbb{R}^n$ contains a universally null set E in B such that $\dim_{\mathsf{H}} E = \dim_{\mathsf{H}} B$;

indeed, he showed that the subset B can be any analytic subset of $\mathbb{R}^n$. Zindulka's geometric measure theoretic proof will be presented.

Let us begin with some side comments about p-dimensional Hausdorff measure H_p on $\mathbb{R}^n$. It is well-known that $0 < \mathsf{H}_n([0,1]^n) < \infty$ and $\dim_{\mathsf{H}}[0,1]^n = \dim[0,1]^n = n$. But, if E is a universally null set in $\mathbb{R}^n$, then either $\mathsf{H}_p(E) = 0$ or $\mathsf{H}_p(E) = \infty$ whenever $0 < p = \dim_{\mathsf{H}} E$. Indeed, if the contrary is assumed, then H_p restricted to E will be a continuous, finite Borel measure on E with $\big(\mathsf{H}_p \,|\, E\big)(E) > 0$ and a contradiction will appear. For analytic subsets A of $\mathbb{R}^n$ with $\mathsf{H}_p(A) = \infty$, there is the following remarkable Besicovitch–Davies–Howroyd theorem. (See Davies [**42**] and Besicovitch [**10**] for the $\mathbb{R}^n$ case, and J. D. Howroyd [**76**] for the separable metric space case.) No proof will be provided.

THEOREM 5.26 (Besicovitch–Davies–Howroyd). *Let A be an analytic set contained in $\mathbb{R}^n$ with $\mathsf{H}_p(A) = \infty$. Then there exists a compact set K contained in A such that $0 < \mathsf{H}_p(K) < \infty$. Moreover*

$$\mathsf{H}_p(A) = \sup\{\, \mathsf{H}_p(K) \colon\ K \text{ is a compact subset of } A,\ \mathsf{H}_p(K) < \infty \,\}.$$

Every analytic space is an absolute measurable space. The following question is posed.

QUESTION. The Besicovitch–Davies–Howroyd theorem leads to the following *property* BDH for subsets X of $\mathbb{R}^n$: *For each nonnegative p, $\mathsf{H}_p(X)$ is the supremum of the collection of the values $\mathsf{H}_p(K)$ where K is a compact subset of X with $\mathsf{H}_p(K) < \infty$.* What sorts of sets X have this property? For example, does every co-analytic space have property BDH? Which absolute measurable spaces possess this property? There are absolute null subspaces X of $\mathbb{R}^n$ and a p such that $\mathsf{H}_p(X) = \infty$, hence property BDH fails for such X.

5.5.1. Universally null sets in analytic spaces.

Let us begin by stating the theorem [**162**, Theorem 4.3].

THEOREM 5.27 (Zindulka). *If A is a nonempty analytic set contained in $\mathbb{R}^n$, then there is a universally null set E in $\mathbb{R}^n$ such that $E \subset A$ and $\dim_{\mathsf{H}} A = \dim_{\mathsf{H}} E$.*

As the Hausdorff dimension is a bi-Lipschitzian invariant (Theorem 5.11), we have

COROLLARY 5.28. *If A is a nonempty analytic metric space such that there is a bi-Lipschitzian injection $\varphi\colon A \to \mathbb{R}^n$, then there exists a universally null set E in A such that $\dim_{\mathsf{H}} E \geq \dim_{\mathsf{H}} A$.*

The inductive proof of Theorem 5.27 will require us to establish some geometric measure theoretic notation.

5.5.2. Geometric measure theory preliminaries.

Much of the preliminary discussion is taken from P. Mattila [**99**]. For further details, refer to Appendix D.

For $n > 1$ we consider m-dimensional linear subspaces V of $\mathbb{R}^n$ and their $(n-m)$-dimensional orthogonal complements $V^\perp$, which are linear subspaces. For convenience, we shall call V an m-dimensional plane. The collection of all m-dimensional planes in $\mathbb{R}^n$, denoted by $G(n,m)$, is called the *Grassmannian manifold.* There is a natural metric and a natural Radon probability measure $\gamma_{\mathrm{n,m}}$ on this manifold. The metric is provided by employing the natural orthogonal projections $\pi_{\mathrm{V}} : \mathbb{R}^n \to V$, $V \in G(n,m)$. The distance in $G(n,m)$ is given by

$$\mathrm{d}(V,W) = \|\pi_{\mathrm{V}} - \pi_W\|, \quad (V,W) \in G(n,m) \times G(n,m), \tag{5.1}$$

where $\|\cdot\|$ is the usual operator norm. A very nice property of the measures is

$$\gamma_{\mathrm{n,m}}(A) = \gamma_{\mathrm{n,n\text{-}m}}(\{\, V^\perp : V \in A\,\}), \quad A \subset G(n,m). \tag{5.2}$$

For $B \in \mathfrak{B}(\mathbb{R}^n)$ and $p > 0$, define MEAS(B,n,p) to be the collection of all Radon measures μ on $\mathbb{R}^n$ with the property: support$(\mu) \subset B$ and $\mu(B(x,r)) \le r^p$ whenever $x \in \mathbb{R}^n$ and $r > 0$, where $B(x,r)$ is the closed ball $\{\, z \in \mathbb{R}^n : \|z - x\| \le r\,\}$. The well-known Frostman lemma (Theorem D.32) characterizes the property $\mathsf{H}_p(B) > 0$ by the property MEAS$(B,n,p) \neq \emptyset$. The following lemma was proved by Zindulka.

LEMMA 5.29 (Zindulka). *Assume B to be a compact subset of $\mathbb{R}$ with $\mathsf{H}_p(B) > 0$. If $0 < \alpha < \frac{1}{2}$, then there exists a Lipschitzian surjection $\varphi : C \to \mathbb{C}(2,\alpha)$ for some compact subset C of B.*

PROOF. Select a μ in MEAS$(B,1,p)$. Let L be a positive number such that $L^p \sum_{i=1}^\infty (2\alpha^p)^i < \mu(B)$. We will construct a sequence $\{\, F(e) : e \in \{0,1\}^{n+1}\,\}$, $n \in \omega$, of finite collections of disjoint, closed intervals $F(e)$, as in the construction of the usual Cantor ternary set in $\mathbb{R}$, that satisfy the following conditions.

(1) For each n, $\mu(B \cap F(e)) \ge 2^{-n} L^p \sum_{i=n+1}^\infty (2\alpha^p)^i$ whenever e is in $\{0,1\}^n$, whence $B \cap F(e) \neq \emptyset$.
(2) For each n, $\mathrm{dist}_{\mathbb{R}}(F(e), F(e')) \ge L\alpha^n$ whenever e and e' are distinct members of $\{0,1\}^n$.
(3) $F(e) \supset F(e')$ whenever e is an initial segment of e'.

Let $n = 0$ and denote by $F = [a,b]$ the convex hull of B. Since μ is a continuous measure there is a point m in (a,b) such that

$$\mu([a,m]) = \tfrac{1}{2}\mu([a,b]) = \mu([m,b]).$$

Let $r_0 = L\alpha$. We have $\mu([m - r_0, m + r_0]) \le L^p\alpha^p$. Hence $\mu([a, m - r_0])$ and $\mu([m + r_0, b])$ are not smaller than $\frac{1}{2} L^p \sum_{i=2}^\infty (2\alpha^p)^i$. Let $F(0)$ be the convex hull of $B \cap [a, m - r_0]$ and $F(1)$ be the convex hull of $B \cap [m + r_0, b]$. The conditions (1) and (2) are satisfied.

Let us indicate the construction of $F(0,0)$ and $F(0,1)$. In the above construction replace F with $F(0) = [a_0, b_0]$, r_0 with $r_1 = L\alpha^2$ and replace $L^p \sum_{i=1}^\infty (2\alpha^p)^i$ with $\frac{1}{2} L^p \sum_{i=2}^\infty (2\alpha^p)^i$. Let $F(0,0)$ be the convex hull of $B \cap [a_0, m - r_1]$ and $F(0,1)$ be the convex hull of $B \cap [m + r_1, b_0]$. Then the statements enumerated above hold for $F(0,0)$ and $F(0,1)$. We leave the completion of the inductive construction to the reader.

Let $C = \bigcap_{n\in\omega}\bigcup\{F(p): p \in \{0,1\}^n\}$. Clearly $K \cap B \neq \emptyset$ whenever K is a component of C. There is a natural surjection φ of C onto $\{0,1\}^\omega$. Indeed, if $p = (p_0, p_1, \ldots) \in \{0,1\}^\omega$, then $F_{(p|n)}$, $n \in \omega$, is a nested family, where $p|n = (p_0, p_1, \ldots, p_{n-1})$ for $n \in \omega$. Hence, $x \mapsto p$ for each x in $\bigcap_{n\in\omega} F(p|n)$ defines the map φ. Note that φ is constant on each component of C. Recall from Section 5.3.1 the metric $\mathrm{d}_{(2,\alpha)}$ for the space $\mathbb{C}(2,\alpha)$. Let us show that φ is Lipschitzian. If x and y are in the same component of C, then $\mathrm{d}_{(2,\alpha)}(\varphi(x), \varphi(y)) = 0$. So suppose that x and y are in different components of C. There is a smallest n in ω such that $x \in \bigcap_{m\leq n} F(\varphi(x)|m)$ and $y \in \bigcap_{m\leq n} F(\varphi(y)|m)$. Hence $|x - y| \geq L\alpha^n = L\,\mathrm{d}_{(2,\alpha)}(\varphi(x), \varphi(y))$ and thereby φ is Lipschitzian. □

We now have the existence theorem.

THEOREM 5.30 (Zindulka). *If B is a nonempty compact subset of $\mathbb{R}$, then there exists a universally null set E in $\mathbb{R}$ such that $E \subset B$ and $\dim_{\mathsf{H}} B = \dim_{\mathsf{H}} E$.*

PROOF. The case $\dim_{\mathsf{H}} B = 0$ is easily seen. So assume $\dim_{\mathsf{H}} B > 0$. For $0 < p < \dim_{\mathsf{H}} B$, let $\alpha = 2^{-1/p}$ and let N be an absolute null space contained in $\mathbb{C}(2,\alpha)$ with $\dim_{\mathsf{H}} N = \dim_{\mathsf{H}} \mathbb{C}(2,\alpha) = p$. Let C be a compact subset of B and $\varphi: C \to \mathbb{C}(2,\alpha)$ as provided by the lemma. Let E_p be a subset of the set C such that $\mathrm{card}(E_p \cap \varphi^{-1}[\{y\}]) = 1$ for each y in N. Then $\varphi|E_p$ is a continuous bijection onto the absolute null space N, whence E_p is an absolute null space. As $\varphi|E_p$ is Lipschitzian, we have $\dim_{\mathsf{H}} E_p \geq p$. The proof is easily completed. □

5.5.3. Proof of Theorem 5.27. If A is a nonempty analytic set in $\mathbb{R}^n$ with $\dim_{\mathsf{H}} A = 0$, then the required set E is easily found. Hence we shall assume $\dim_{\mathsf{H}} A > 0$. Observe, for each p with $0 < p < \dim_{\mathsf{H}} A$, that the Besicovitch–Davies–Howroyd theorem yields a compact set B_p contained in A with $0 < \mathsf{H}_p(B_p) < \infty$. Hence it is enough to prove that there exists a universally null set E_p in $\mathbb{R}^n$ that is contained in B_p such that $\dim_{\mathsf{H}} E_p = \dim_{\mathsf{H}} B_p$.

LEMMA 5.31 (Zindulka). *If B is a compact subset of $\mathbb{R}^n$ and if p is a positive number such that $0 < \mathsf{H}_p(B) < \infty$, then there exists a universally null set E in $\mathbb{R}^n$ such that $E \subset B$ and $\dim_{\mathsf{H}} E = \dim_{\mathsf{H}} B = p$.*

PROOF. The proof is by induction on n. Let $n = 1$. As $\mathsf{H}_p(B) > 0$, we have $B \neq \emptyset$. Lemma 5.29 completes the proof.

Let us prove the inductive step. Several theorems from geometric measure theory will be used – their statements are found in Appendix D – the reader is reminded of the notations that are found in Section 5.5.2. Let B be a compact set in $\mathbb{R}^{n+1}$ with $0 < \mathsf{H}_p(B) < \infty$. We shall consider two cases; namely, $0 < p \leq n$ and $n < p \leq n+1$.

Suppose $0 < p \leq n$. By the projection property (Theorem D.30), there exists a V in $G(n+1, n)$ such that $\dim_{\mathsf{H}} \pi_{\mathsf{V}}[B] = \dim_{\mathsf{H}} B$. Since V is isometric to $\mathbb{R}^n$ there is a universally null set D in V such that $D \subset \pi_{\mathsf{V}}[B]$ and $\dim_{\mathsf{H}} D = \dim_{\mathsf{H}} \pi_{\mathsf{V}}[B]$. Let E

be a subset of B such that $\pi_V[E] = D$ and $\operatorname{card}(E \cap B \cap \pi_V^{-1}[\{x\}]) = 1$ for every x in D. Then E is a universally null set in $\mathbb{R}^{n+1}$ and $\dim_H E \geq \dim_H \pi_V[E] = \dim_H B$. Thereby the case $0 < p \leq n$ is proved.

Suppose $n < p \leq n+1$. As $n \geq 1$, we have $p - 1 > 0$. By the slicing property (Theorem D.31), there exists a V in $G(n+1, 1)$ such that $\mathsf{H}_1\big(\{x \in V \colon \dim_H\big(B \cap (V^\perp + x)\big) = p-1\}\big) > 0$. By Lemma D.33, we have that $\Phi(x) = \dim_H\big(B \cap (V^\perp + x)\big)$, $x \in V$, is a Borel measurable function on the line V. Hence

$$M = \{x \in V \colon \dim_H\big(B \cap (V^\perp + x)\big) = p - 1\}$$

is a Borel set. As $\dim V^\perp = n - 1$ there is a universally null set E_x in $V^\perp + x$ such that $E_x \subset B \cap (V^\perp + x)$ and $\dim_H E_x = \dim_H\big(B \cap (V^\perp + x)\big)$ whenever $x \in M$. As $t = \min\{p - 1, 1\} > 0$, we have

$$M = \{x \in M \colon \dim_H E_x \geq t\} \quad \text{and} \quad \mathsf{H}_t(M) > 0.$$

We infer from the Besicovitch–Davies–Howroyd theorem that there exists a compact subset M' of M with $0 < \mathsf{H}_t(M') < \infty$. It follows that $\dim_H M' = t$. There exists a universally null set D in the line V that is contained in M' with $\dim_H D = t$. Let

$$E = \bigcup\{E_x \colon x \in D\}.$$

Then E is a universally null set in $\mathbb{R}^n$ by Theorem 1.23. Hence, by Corollary D.39, $\dim_H E \geq p$. Since $E \subset B$ we have $\dim_H E = \dim_H B$ and the inductive step is now proved. □

We have now finished the preparations for the proof of geometric measure theoretic theorem.

Proof of Theorem 5.27. Suppose that A is a nonempty analytic set in $\mathbb{R}^n$. The universally null set E is easily found if $\dim_H A = 0$. For $\dim_H A > 0$, the statement of the theorem follows easily from Lemma 5.31. □

We know from Theorem 5.22 that if X is a separable metric space with $\dim X = \infty$, then there is a universally null set E in X such that $\dim_H E = \dim_H X$. Hence it follows that only analytic metric spaces A with $\dim A < \infty$ are of interest if one wanted to generalize Theorem 5.27 to all analytic spaces. This remark leads to

Question. Is it possible to write every finite dimensional, analytic metric space A as a countable union of analytic subspaces A_i, $i = 1, 2, \ldots$, such that there is a bi-Lipschitzian embedding of each A_i into some $\mathbb{R}^{n_i}$?

We know that every analytic set contained in $\mathbb{R}^n$ is an absolute measurable space. Hence we have the obvious question.

Question. In Theorem 5.27, can the condition that the set A be an analytic set be replaced by co-analytic set or, more generally, by absolute measurable space?

5.6. Zindulka's opaque sets

To investigate singular sets, Zindulka introduced in [**163**] the notion of small opaque sets. We shall develop enough of his notion to prove his theorems concerning Hausdorff dimension and the existence of universally null sets in separable metric spaces. The reader is referred to the original paper for applications to other singular sets.

DEFINITION 5.32. *Let X be a separable metric space and let $\mathcal{C}$ be a family of subsets of X. A subset Y of X is called* $\mathcal{C}$-opaque (*or* opaque with respect to $\mathcal{C}$) *if*

$$C \cap Y \neq \emptyset \text{ whenever } C \in \mathcal{C} \text{ and } \dim C > 0.$$

If the family $\mathcal{C}$ satisfies the mild additional condition that $C \cap F \in \mathcal{C}$ whenever $C \in \mathcal{C}$ and F is a closed set, then $\mathcal{C}$ is said to be closed-complete. *By weakening the condition that F be closed sets to the condition that F be Borel sets, the family $\mathcal{C}$ is said to be* Borel-complete.

Clearly the case where the family $\mathcal{C}$ is the empty one is not very interesting. Similarly, if each C in $\mathcal{C}$ satisfies $\dim C \leq 0$ then every subset of X is $\mathcal{C}$-opaque, in particular, every C in $\mathcal{C}$. So, if $\dim X = 0$, then every subset Y of X is $\mathcal{C}$-opaque for any family $\mathcal{C}$ of subsets of X. Consequently the notion of $\mathcal{C}$-opaque sets will become useful only if $\dim C \geq 1$ for some C in $\mathcal{C}$, whence $\dim X \geq 1$. Also observe that any dense set of real numbers is $\mathcal{C}$-opaque in the space $\mathbb{R}$ for every family $\mathcal{C}$ of subsets of $\mathbb{R}$. Indeed, it appears that the notion of opaque sets is not so interesting for one-dimensional manifolds if one requires that the opaque sets be dense sets. Consequently, for n-dimensional manifolds X, the interesting cases are those with $n \geq 2$. The next proposition gives some properties of $\mathcal{C}$-opaque sets.

PROPOSITION 5.33. *Let Y be a $\mathcal{C}$-opaque set of a separable metric space X.*

(1) *If $\mathcal{C}$ is closed-complete, then $\dim(C \cap Y) \geq \dim C - 1$ for each C in $\mathcal{C}$, in particular, $\dim(C \cap Y) = \infty$ whenever $\dim C = \infty$. Moreover, if $X \in \mathcal{C}$ and $\dim X > 1$, then $\dim Y > 0$.*
(2) *If $\mathcal{C}$ is Borel-complete, then $C \cap Y$ is strongly infinite dimensional for each C in $\mathcal{C}$ that is strongly infinite dimensional.*[5]

PROOF. To prove statement (1) let us assume that it fails and derive a contradiction. So assume that there is a C in $\mathcal{C}$ such that $\dim(C \cap Y) < \dim C - 1$. As $\dim(C \cap Y)$ is finite, there is a G_δ hull G of $C \cap Y$ (that is, $C \cap Y \subset G \subset X$) such that $\dim G = \dim(C \cap Y)$. We have by the addition theorem that

$$\dim C \leq \dim G + \dim(C \setminus G) + 1 < \dim C - 1 + \dim(C \setminus G) + 1.$$

Consequently, $\dim C < \infty$ and $0 < \dim(C \setminus G) < \infty$. Since $X \setminus G$ is an F_σ set, we infer from the sum theorem that there is a closed set F with $C \cap F \subset C \setminus G$ and

[5] A separable metrizable space is strongly infinite dimensional if it cannot be written as a countable union of zero-dimensional sets. See page 245.

$\dim(C \cap F) = \dim(C \setminus G)$. As $\mathcal{C}$ is a closed-complete family, we have that $C \cap F$ satisfies both $C \cap F \in \mathcal{C}$ and $\dim(C \cap F) > 0$. Hence $(C \cap F) \cap Y \neq \emptyset$. But

$$(C \cap F) \cap Y \subset (C \setminus G) \cap Y = (C \cap Y) \setminus G = \emptyset,$$

and the contradiction has been achieved.

The proof of statement (2) is similar to that of statement (1). Assume that there is a C in $\mathcal{C}$ such that C is strongly infinite dimensional and $C \cap Y$ is not. As in the above proof, the set $C \cap Y$ is contained in a $G_{\delta\sigma}$ set G such that G is not strongly infinite dimensional. Clearly, $C \setminus G$ is strongly infinite dimensional. As $\mathcal{C}$ is Borel-complete, we have $C \setminus G \in \mathcal{C}$. Hence $(C \setminus G) \cap Y \neq \emptyset$. But $G \supset C \cap Y$ and the contradiction has occurred. □

5.6.1. Examples. Here are some examples of $\mathcal{C}$-opaque sets that appear in Zindulka [**163**]. Notice that each example is metric independent. The first one motivates the notion of opaqueness.

EXAMPLE 5.34. (*Visibility*) Let $X = \mathbb{R}^n$ with $n \geq 2$, and let $\mathcal{C}$ be the family $\{ C_u \colon u \in \mathbb{R}^n, |u| = 1 \}$, where $C_u = \{ \lambda u \colon \lambda > 0 \}$ is the ray emanating from the origin and passing through the point u. A $\mathcal{C}$-opaque set Y is visible from the origin in every direction u. Conversely, a set Y that is visible from the origin in every direction is $\mathcal{C}$-opaque for this family $\mathcal{C}$. Obviously the whole space $\mathbb{R}^n$ and the usual unit sphere in $\mathbb{R}^n$ are visible from the origin in every direction, rather trivial examples. As $\operatorname{card}(C_u) = \operatorname{card}(\mathbb{R}^n)$ for every C_u, there is a $\mathcal{C}$-opaque set Y such that $Y \cap C_u$ is a singleton set $\{x_u\}$ such that $|x_u| = |x_v|$ if and only if $u = v$. A more complicated example is a totally imperfect subset Y of X such that $X \setminus Y$ is also totally imperfect.

EXAMPLE 5.35. (*Borel opacity*) Let X be a separable metrizable space and let $\mathcal{C}$ be the family of all Borel subsets of X. We shall call a subset Y of X *Borel-opaque* if it is $\mathcal{C}$-opaque. By Proposition 5.33, $\dim Y \geq \dim Z - 1$ for every subset Z of X, whence $\dim Y \geq \dim X - 1$. If X is a compact metrizable space with $\dim X \geq 1$, then each totally imperfect subset Y whose complement is also totally imperfect is Borel-opaque.

EXAMPLE 5.36. (*Arc opacity*) Let X be a separable metrizable space and let $\mathcal{C}$ be the family of all arcs (that is, homeomorphic images of the unit interval $[0, 1]$) contained in X and their Borel subsets. If Y is a subset of X that is $\mathcal{C}$-opaque, then any arc between two distinct points of X meets the set Y. We shall designate these $\mathcal{C}$-opaque sets Y as *arc-opaque* sets. Also $\dim Y \geq \dim K - 1$ for each arc K in X. Clearly, a totally imperfect subset Y of X is arc-opaque whenever $X \setminus Y$ is also totally imperfect.

5.6.2. Construction of opaque sets. As we have seen in the examples listed above, totally imperfect sets often appear as opaque sets in the Zindulka sense. Our interest lies in those totally imperfect sets that are universally null sets in X. So, for example, arc opacity or Borel opacity for the compact spaces $[0, 1]^n$ and $X = [0, 1]^{\mathbb{N}}$ are not good enough to construct universally null sets in these spaces. Some other process must be added to $\mathcal{C}$-opacity. We present here Zindulka's added procedure.

It is known that the real line contains uncountable subsets E that are absolute null spaces. It is also known that the inverse image under continuous bijections onto E are also absolute null spaces. The following definition by Zindulka is designed to take advantage of these facts.

DEFINITION 5.37. *For separable metric spaces X and Y let $\mathcal{I}$ and $\mathcal{J}$ be families of subsets of the respective spaces X and Y. Then $\mathcal{I} \leqslant \mathcal{J}$ is the binary relation determined by the following property:* If A is a subset of X and $f: X \to Y$ is a continuous map such that $f|A: A \to Y$ is an injection and $f[A] \in \mathcal{J}$, then $A \in \mathcal{I}$.

The families $\mathcal{I}$ and $\mathcal{J}$ in the above definition are often σ-ideals. Of particular interest to us is the σ-ideal on a separable metrizable space X formed by the collection of all universally null sets in X. The next theorem provides a sufficient condition for the existence of $\mathcal{C}$-opaque sets of X that are members of the family $\mathcal{I}$.

THEOREM 5.38. *For a separable metrizable space X let $\mathcal{C}$ be a family of subsets of X and $\mathcal{I}$ be a σ-ideal on X; and, for the space $\mathbb{R}$ let E be a uncountable subset of $\mathbb{R}$ and $\mathcal{J}$ be a σ-ideal on $\mathbb{R}$. If $\mathcal{I} \leqslant \mathcal{J}$, $E \in \mathcal{J}$ and* $\operatorname{card}(E) = \operatorname{card}(\mathcal{C})$*, then there exists a subset Y of X such that Y is $\mathcal{C}$-opaque and $Y \in \mathcal{I}$.*

PROOF. We may assume E is a subset of $[0,1]$ and that $\mathcal{C}$ is indexed as $\{ C_r : r \in E \}$. After selecting a metric on X, we have a sequence h_m, $m = 0, 1, 2, \ldots$, of continuous maps and G_δ sets $G(r)$, $r \in (0,1)$, as provided by Theorem 5.20. For each m in ω put

$$E_m = \{ r \in E : C_r \cap {h_m}^{-1}[\{r\}] \neq \emptyset \}.$$

Finally, for each r in E_m, select a point $y(m,r)$ in $C_r \cap {h_m}^{-1}[\{r\}]$ and then define

$$Y_m = \{ y(m,r) : r \in E_m \} \quad \text{and} \quad Y = \textstyle\bigcup_{m \in \omega} Y_m.$$

We assert that Y is in $\mathcal{I}$. Indeed, note that each $h_m|Y_m : Y_m \to [0,1]$ is an injection and $Y_m \subset {h_m}^{-1}[E_m]$. As $E \in \mathcal{J}$ and $\mathcal{I} \leqslant \mathcal{J}$, we have $Y_m \in \mathcal{I}$. Since $\mathcal{I}$ is a σ-ideal we have $Y \in \mathcal{I}$.

It remains to show that Y is $\mathcal{C}$-opaque. To this end let $r \in E$ and assume $C_r \cap Y = \emptyset$. Then, for each m, we have $C_r \cap Y_m = \emptyset$ and hence $C_r \cap {h_m}^{-1}[\{r\}] = \emptyset$. From the definition of $G(r)$ we have $C_r \subset G(r)$; and, from Theorem 5.20 we have $0 \geq \dim G(r) \geq \dim C_r$. Thereby Y is $\mathcal{C}$-opaque. □

Here is an application of Theorem 5.38 to orthogonal projections of $\mathbb{R}^2$. Let L_ϑ be a one-dimensional oriented linear subspace of $\mathbb{R}^2$, where ϑ is the unique unit vector orthogonal to L_ϑ that, together with the orientation of L_ϑ, yields the usual orientation of $\mathbb{R}^2$. Note that ϑ is in the unit sphere $S = \{ x \in \mathbb{R}^2 : |x| = 1 \}$, which carries the usual H_1 measure. Denote by Π_ϑ the orthogonal projection of $\mathbb{R}^2$ onto L_ϑ. Given a subset Y of $\mathbb{R}^2$ there corresponds a set of normal vectors ϑ defined as follows:

$$\{ \vartheta \in S : \lambda_*(L_\vartheta \setminus \Pi_\vartheta[Y]) = 0 \},$$

where λ_* is the Lebesgue inner measure on the line L_ϑ. (See page 46 for Lebesgue inner measure.)

THEOREM 5.39. *There exists a universally null set Y in $\mathbb{R}^2$ such that* $\operatorname{card}(Y) \geq \mathsf{non}\text{-}\mathbb{L}$ *and*

$$\mathsf{H}_{1*}(S \setminus \{\, \vartheta \in S \colon \lambda_*(L_\vartheta \setminus \Pi_\vartheta[Y]) = 0 \,\}) = 0,$$

where H_{1*} is the inner H_1 measure on $\mathbb{R}^2$.

PROOF. Let E be a universally null set in $(0, 1)$ with $\operatorname{card}(E) = \mathsf{non}\text{-}\mathbb{L}$. By Proposition 2.42 there exist full measure subsets A of S and B of $\mathbb{R}$, with respect to their respective measures, such that $\operatorname{card}(A) = \operatorname{card}(B) = \mathsf{non}\text{-}\mathbb{L}$. Provide each line L_ϑ with an isometric copy B_ϑ of B. Let $\mathcal{C}$ be the family $\{\, {\Pi_\vartheta}^{-1}[\{y\}] \colon y \in L_\vartheta, \vartheta \in A \,\}$. Then $\operatorname{card}(\mathcal{C}) = \mathsf{non}\text{-}\mathbb{L}$. By Theorem 5.38 there is a $\mathcal{C}$-opaque subset Y of $\mathbb{R}^2$ such that Y is also a universally null set in $\mathbb{R}^2$. As each element of $\mathcal{C}$ has dimension equal to 1 it also meets the set Y. Hence $\Pi_\vartheta[Y] \supset B_\vartheta$ for each ϑ in A. Clearly $\operatorname{card}(Y) \geq \mathsf{non}\text{-}\mathbb{L}$. □

COROLLARY 5.40. *There is a universally null set Y in $\mathbb{R}^2$ such that*

$$\dim_{\mathsf{H}}(\{\, \vartheta \in S \colon \dim_{\mathsf{H}}(\Pi_\vartheta Y) = 1 \,\}) = 1.$$

The proof of the corollary is straightforward.

5.7. Comments

The cardinalities of the universally null sets E in Theorems 5.27 and 5.30 can be made to satisfy $\operatorname{card}(E) \geq \mathsf{non}\text{-}\mathbb{L}$ whenever the analytic set A is uncountable.

As was already pointed out, the inequality $\dim_{\mathsf{H}} X \geq \dim X$ was known to hold for every separable metric space X and the equality held for some metric for every nonempty separable metrizable space. It was through the recent work of Zindulka [**161, 162**] that the existence of universally null sets E in nonempty separable metric spaces X with $\dim_{\mathsf{H}} E \geq \dim X$ was established. As $\dim_{\mathsf{H}} X \geq \dim_{\mathsf{H}} E$, a completely new proof of the classical result $\dim_{\mathsf{H}} X \geq \dim X$ (Theorem 5.1) follows from Zindulka's existence theorem. The proof relies on an earlier discovered topological dimension theorem (Theorem D.28) for metric spaces (see Zindulka [**163**, Lemma 5.1]) and a careful analysis of the Cantor cubes and the Euclidean space $\mathbb{R}^n$. A straightforward consequence of Theorem 5.2 is that, for each nonempty separable metrizable space X, there is a metric for which $\dim_{\mathsf{H}} X = \dim_{\mathsf{H}} E = \dim X$ for some universally null set E in X (Theorem 5.24).

The important Theorem 5.21 is a consequence of Theorem D.27, a theorem about metric spaces that was proved by Zindulka in [**163**]. The seeds for the latter theorem are found in Zindulka's 1999 paper [**160**, Theorem 2.1] where the functions there were only continuous and not necessarily Lipschitzian. Clearly, for the purpose of Hausdorff dimension, one needs Lipschitzian maps since they preserve p-dimensional Hausdorff measure 0 whereas continuous maps do not, witness the famous Peano curve which is a homeomorphic image of $[0, 1]$ contained in $[0, 1]^2$ with positive Lebesgue measure. In the next chapter we shall turn to the analogue of Theorem 5.21

that results from replacing the Hausdorff dimension with topological dimension. This theorem will require some set theory assumptions.

Zindulka's notion of $\mathcal{C}$-opaque sets also has several applications to various other singular sets. We repeat again that the reader is referred to [**163**] for these applications. In [**163**] Zindulka applied his notion of $\mathcal{C}$-opaque sets to establish the following proposition on the existence of universally null sets in a metric space X. Although this proposition has been superseded by Theorem 5.22, its proof is presented so as to illustrate the application of $\mathcal{C}$-opacity to the σ-ideal $\mathcal{I}$ of universally null sets in X and the σ-ideal $\mathcal{J}$ of universally null sets in $[0, 1]$.

PROPOSITION 5.41. *Let X be a separable metric space and let $n \in \omega$. If $\dim X > n$, then there is a universally null set E in X such that $\mathsf{H}_n(E) = \infty$ and $\mathrm{card}(E) \geq \mathsf{non}\text{-}\mathbb{L}$.*

PROOF. In order to apply Theorem 5.38, two σ-ideals and a family $\mathcal{C}$ must be identified. The σ-ideals $\mathcal{I}$ and $\mathcal{J}$ are $\mathsf{univ}\,\mathfrak{N}(X)$ and $\mathsf{univ}\,\mathfrak{N}(Y)$, respectively, where $Y = [0, 1]$. That $\mathcal{I} \leqslant \mathcal{J}$ follows from Theorem 1.23. The family $\mathcal{C}$ will consist of closed subsets F of X with the property that $\dim F \geq 0$. We infer from the definition of the condition $\dim X > n$ that there are $\mathfrak{c}$ many such sets F. The next computations will result in the family $\mathcal{C}$.

Let $\mathcal{F}$ be a countable family of Lipschitzian maps f of X into $[0, 1]^n$ that is provided by Theorem 5.21. As $\mathsf{H}_n\,|[0, 1]^n$ is a positive, continuous, complete finite Borel measure we infer from the definition of $\mathsf{non}\text{-}\mathbb{L}$ that there is a subset B of $[0, 1]^n$ with $\mathrm{card}(B) = \mathsf{non}\text{-}\mathbb{L}$ and $\mathsf{H}_n(B) > 0$. For each f in $\mathcal{F}$ let

$$B(f) = \{\, r \in B\colon\ \dim f^{-1}[\{r\}] > 0 \,\}.$$

Then $\bigcup_{f\in\mathcal{F}} B(f) = B$. There is a g in $\mathcal{F}$ such that $\mathsf{H}_n(B(g)) > 0$. Clearly, by the properties of the cardinal number $\mathsf{non}\text{-}\mathbb{L}$ and the inclusion $B(g) \subset B$, we have $\mathrm{card}(B(g)) = \mathsf{non}\text{-}\mathbb{L}$. We infer from Theorem 2.41 the existence of a absolute null subspace N of Y with $\mathrm{card}(N) = \mathsf{non}\text{-}\mathbb{L}$. Now let the family $\mathcal{C}$ be the collection $\{\, g^{-1}[\{r\}]\colon r \in B(g) \,\}$. Theorem 5.38 yields a absolute null subspace E of X that is $\mathcal{C}$-opaque.

Let us show that $\mathsf{H}_n(E) = \infty$. We first observe that each C in $\mathcal{C}$ has positive dimension and hence $C \cap E \neq \emptyset$ because E is $\mathcal{C}$-opaque. So g maps E onto $B(g)$ and consequently $\mathsf{H}_n(g[E]) \geq \mathsf{H}_n(B(g)) > 0$. Let L be a Lipschitz constant for g. Then $L^n\,\mathsf{H}_n(E) \geq \mathsf{H}_n(g[E]) > 0$. In order to derive a contradiction assume that $\mathsf{H}_n(E)$ is finite. Then $\mathsf{H}_n\,|E$ induces a nontrivial, continuous, complete finite Borel measure on the absolute null space E, which is not possible. Thereby $\mathsf{H}_n(E) = \infty$. $\square$

It is remarked in [**163, 162**] that this result was known already to D. H. Fremlin for the case of $X = \mathbb{R}^2$.

The development of the material on Hausdorff measure and Hausdorff dimension used in this chapter is very much self contained; it is taken from Zindulka [**162, 161**]. An expanded discussion with some proofs is presented in Appendix D. Concerning the specifics of Zindulka's development that have been presented in this chapter, the key is the choice of metric on the topological space k^ω. The metric $\mathrm{d}_{(k,\alpha)}$ is designed to make the computation of the Hausdorff dimension of $\mathbb{C}(k,\alpha)$ an easy task because the resulting Hausdorff s-measure turns out to be the usual uniform product probability measure on k^{ω}, where $s = \dim_{\mathsf{H}} \mathbb{C}(k,\alpha)$. In order to use the product techniques due to Grzegorek, Zindulka proved the Propositions C.8 and C.9 found in Appendix C on products of metric spaces $\mathbb{C}(k,\alpha)$, where the key is bi-Lipschitzian equivalence rather than metric equality. These remarkably simple but important propositions stand in marked contrast to very technical results on Hausdorff measures of products of sets that were studied in earlier years in the pursuit of geometric properties (see C. A. Rogers [**131**, page 130]). But we do not want to make light of the power of geometric measure theory. Indeed, Theorem 5.27 that establishes the existence of universally null sets E in analytic subsets of $\mathbb{R}^n$ relies very much on geometric measure theory.

Both the classical proof and Zindulka's second proof of Theorem 5.1 are presented in Appendix D; these two proofs are substantially different.

For those who wish to learn more general and more technical aspects of Hausdorff measures, the books by Rogers [**131**], K. J. Falconer [**52**], and Mattila [**99**] are often cited as sources. Another source that deals mainly with $\mathbb{R}^n$ is H. Federer [**55**].

Exercises

5.1. Prove Theorem 5.5 on page 138.

5.2. Prove Theorem 5.7 on page 138.

5.3. Verify Theorem 5.10 on page 139.

5.4. Prove Proposition 5.14 on page 140.

5.5. Show that there is a nonempty, perfect subset of each uncountable compact metric space X such that $\mathsf{H}_s(F) = 0$ whenever $s > 0$. Hint: Let $s_j, j = 1, 2, \ldots$, and a_m, $m = 1, 2, \ldots$, be strictly decreasing sequences in $(0, 1]$ such that $\sum_{m=1}^{\infty} 2^m a_m$ is finite and $\lim_{j\to\infty} s_j = 0$. For each m let $d_m = (a_m)^{1/s_m}$. Verify that $\sum_{m=1}^{\infty} 2^m (d_m)^{s_j}$ is finite for every j. In the usual manner, construct nonempty closed sets $F(m,k)$, $1 \le k \le 2^m$, such that

(a) $\operatorname{diam}(F(m,k)) \le d_m$ for each k,

(b) $F(m,k) \cap F(m,k') = \emptyset$ whenever $k \neq k'$,

(c) $F(m,k) \supset F(m+1, 2k-1) \cup F(m+1, 2k)$ whenever $1 \le k \le 2^m$.

Consider the compact sets $F_m = \bigcup_{k=1}^{2^k} F(m,k)$.

5.6. State and prove the generalization of Theorem 5.39 and its corollary for the space $\mathbb{R}^n$.

6

Martin axiom

Except for two statements in the earlier chapters that used the continuum hypothesis (abbreviated as CH), all the others used only what is now called the usual axioms of set theory – namely, the Zermelo–Frankel axioms plus the axiom of choice, ZFC for short. In this final chapter a look at the use of the continuum hypothesis and the Martin axiom in the context of absolute null space will be made. The discussion is not a thorough coverage of their use – the coverage is only part of the material that is found in the many references cited in the bibliography.

It has been mentioned many times that absolute null space is an example of the so-called singular sets[1]. This example is a topological notion in the sense that it does not depend on the choice of a metric:[2] Two other metric independent singular sets will be included also. They are the Lusin set and the Sierpiński set in a given ambient space X. With regards to ambient spaces, it is known that "absolute null subspace of an ambient space X" is equivalent to "universally null set in X."

The chapter is divided into four sections. The first is a rough historical perspective of the use of the continuum hypothesis in the context of universally null sets in a given space X. The second concerns cardinal numbers of absolute null spaces. The third is a brief discussion of the Martin axiom and its application to the above mentioned singular sets. The fourth is devoted to the dimension theoretic results of Zindulka.

6.1. CH and universally null sets: a historical tour

Let us follow Brown and Cox [**18**] and begin our tour around the time of the "Annexe" article [**15**] in the new series of the *Fundamenta Mathematicae* which appeared in 1937. This article is a commentary on several problems that were proposed in Volume 1 of the old series of the journal. Among these problems is Problem 5 on page 224 which is essentially the start of the notion of universally null sets. Historically, the problems predate 1937. Problem 5 is the following one which was proposed by Sierpiński:

[1] Several of the many other classes of singular sets have been introduced to the reader in Section 4.7 of Chapter 4. A more extensive list of classes of singular sets can be found in the Brown–Cox article [**18**].

[2] In Chapter 5 there was a need to emphasize the metric in the discussion of Hausdorff dimension of universally null sets in a space X because Hausdorff dimension is a metric dependent notion. In this context, topological equivalence was replaced by bi-Lipschitzian equivalence.

PROBLEM 5, *FUNDAMENTA MATHEMATICAE*, VOLUME 1: Does there exist an uncountable subset E of $\mathbb{R}$ such that every homeomorphic image of E contained in $\mathbb{R}$ has Lebesgue measure equal to 0? Can one show its existence on admitting the continuum hypothesis? (Sierpiński)

Of course, the difficulty of the problem comes from the fact that a topological embedding of a subset E of $\mathbb{R}$ into $\mathbb{R}$ need not result from a homeomorphism of $\mathbb{R}$ onto $\mathbb{R}$. Indeed, an arbitrary topological embedding of E need not preserve the induced order on the set E if E is not connected. By 1937 it was known that such subsets of $\mathbb{R}$ were null sets for every continuous, complete, finite Borel measure on $\mathbb{R}$ – we have called them universally null sets in $\mathbb{R}$. Of course, it was not known at that time that the existence of such sets was in fact a set theoretic question.

The second question was positively resolved in 1924 by Lavrentieff [**89**]. He used the existence of Lusin sets in $\mathbb{R}$, which was shown in 1914 to exist by assuming the continuum hypothesis. (See Lusin [**92**]; its existence was also shown by Mahlo [**95**] in 1913.) A construction of Lusin sets using the continuum hypothesis was given in Section 1.3.5 of Chapter 1. That construction used the σ-ideal of all first category subsets of uncountable separable metrizable spaces and was taken from Oxtoby [**120**]. Let us give the definition of Lusin sets. (Note that the notion of 'nowhere dense' is not metric dependent.)

DEFINITION 6.1. *Let X be a separable metrizable space. A subset E of X is called a* Lusin set *in X if* $\operatorname{card}(E) \geq \aleph_1$ *and if* $\operatorname{card}(H \cap E) \leq \aleph_0$ *whenever H is a first category set in X.*[3]

It is not difficult to show, without assuming the continuum hypothesis, that the Lebesgue measure of each topological copy in $\mathbb{R}$ of a Lusin set in $\mathbb{R}$ has Lebesgue measure equal to 0. Let us give two proofs.

The first proof is Lavrentieff's from 1924. First observe that each uncountable G_δ set B in $\mathbb{R}$ can be written as the union $N \cup \bigcup_{n=1}^{\infty} Q_n$, where N is a Lebesgue null set, and each Q_n is a topological copy of the Cantor set. (Observe that $\bigcup_{n=1}^{\infty} Q_n$ is a Zahorski space, see page 112.) Let E be a Lusin set in $\mathbb{R}$ and let $h\colon E \to F$ be a homeomorphism with $F \subset \mathbb{R}$. By his theorem (Theorem A.2 of Appendix A, whose proof does not assume the continuum hypothesis) we have a homeomorphism $H\colon A \to B$ that extends h, where A and B are G_δ sets of $\mathbb{R}$. Then $F = F \cap B = (F \cap N) \cup \bigcup_{n=1}^{\infty} (F \cap Q_n)$. Now $F \cap Q_n = H\big[E \cap H^{-1}[Q_n]\big]$, and $H^{-1}[Q_n]$ is a nowhere dense subset of $\mathbb{R}$, hence $\operatorname{card}(F \cap Q_n) \leq \aleph_0$. Consequently, F is a Lebesgue null set.

The second proof will use the group $\mathsf{HOMEO}(\mathbb{R})$. Let μ be a positive, continuous, complete, finite Borel measure on $\mathbb{R}$ such that $\mathfrak{M}(\mathbb{R}, \mu) = \mathfrak{M}(\mathbb{R}, \lambda)$ and $\mathfrak{N}(\mathbb{R}, \mu) = \mathfrak{N}(\mathbb{R}, \lambda)$, where λ is the Lebesgue measure. We know that μ and $\mathsf{HOMEO}(\mathbb{R})$ generate $\mathrm{MEAS}^{\mathrm{pos,fin}}(\mathbb{R})$, the collection of all positive, continuous, complete finite Borel measures on $\mathbb{R}$. Now observe that first category sets of $\mathbb{R}$ are invariant under the action of $\mathsf{HOMEO}(\mathbb{R})$, whence Lusin sets are invariant under the action of $\mathsf{HOMEO}(\mathbb{R})$.

[3] On page 78 of [**120**] a Lusin set is defined by the property: *An uncountable subset E of X with the property that every uncountable subset of E is a second category set.* The reader is asked to prove the equivalence of these two properties.

Also observe that $\mathbb{R} = N \cup Z$ where $\lambda(N) = 0$ and Z is a Zahorski space. Clearly, Zahorski sets in $\mathbb{R}$ are first category sets. From these observations we see that every Lusin set has μ measure equal to 0. Consequently, Lusin sets in $\mathbb{R}$ are absolute null spaces.

The notions involved in Sierpiński's problem were introduced at the end of the nineteenth and the start of the twentieth centuries. Indeed, E. Borel introduced his sets in 1888, R. Baire introduced sets of first category in 1889, and H. Lebesgue introduced his measure and integral in the years just before the appearance of his 1904 book [**90**].[4] Also, ideas of G. Cantor were very important. Sierpiński's problem is made meaningful by the fact that homeomorphisms do not preserve the Lebesgue measurability of sets in $\mathbb{R}$; this is in contrast to the fact that Borel sets are invariant under homeomorphisms of $\mathbb{R}$.

There is another σ-ideal of subsets of $\mathbb{R}$, namely the collection of Lebesgue null sets. Sierpiński, assuming the continuum hypothesis, proved the following [**140**] in 1924.

THEOREM 6.2. *Assume* CH. *There exists a subset E of $\mathbb{R}$ such that* $\operatorname{card}(E) \geq \aleph_1$ *and such that* $\operatorname{card}(H \cap E) \leq \aleph_0$ *whenever H is a Lebesgue null set.*

A proof is given on page 42. The subspace E is an uncountable non-absolute measurable space for which each universally measurable set in E is a symmetric difference of a Borel set and a universally null set in E.

The theorem leads to the following definition of Sierpiński sets.[5]

DEFINITION 6.3. *A subset E of $\mathbb{R}$ is called a* Sierpiński set *in $\mathbb{R}$ if* $\operatorname{card}(E) \geq \aleph_1$ *and if* $\operatorname{card}(H \cap E) \leq \aleph_0$ *for every Lebesgue null set H.*

Clearly the above constructed Lusin set and the Sierpiński set have cardinality $\mathfrak{c}$ since the continuum hypothesis was assumed. Of interest in this regard is the following 1938 theorem of F. Rothberger [**132**] whose proof does not assume the continuum hypothesis.

THEOREM 6.4. (Rothberger) *If a Lusin set X in $\mathbb{R}$ exists, then* $\operatorname{card}(Y) = \aleph_1$ *whenever Y is a Sierpiński set in $\mathbb{R}$. Also, if a Sierpiński set Y in $\mathbb{R}$ exists, then* $\operatorname{card}(X) = \aleph_1$ *whenever X is a Lusin set in $\mathbb{R}$. Consequently, if a Lusin set X in $\mathbb{R}$ and a Sierpiński set Y in $\mathbb{R}$ simultaneously exist, then* $\operatorname{card}(X) = \operatorname{card}(Y) = \aleph_1$.

The proof relies on the next lemma.

LEMMA 6.5. *If X is not a first category set in $\mathbb{R}$, then $\mathbb{R}$ is the union of κ many Lebesgue null sets, where $\kappa =$* $\operatorname{card}(X)$. *Also, if Y is not a Lebesgue null set, then $\mathbb{R}$ is the union of κ many first category sets, where $\kappa =$* $\operatorname{card}(X)$.

[4] These years were taken from the classic book by Kuratowski [**85**] and also from the recent article by D. Paunić [**124**]. An interesting discussion of the roots of descriptive set theory can be found in the Introduction of the book by Y. N. Moschovakis [**112**, pages 1–9].

[5] Note that the notion of "Lebesgue measure on $\mathbb{R}$" is *not* metric dependent. More generally, the definition can be couched in the context of Borel measure spaces $\mathrm{M}(X, \mu)$, where X is a separable metrizable space. Recall that a Borel measure space is a triple $(X, \mu, \mathfrak{M}(X, \mu))$, where $\mathfrak{M}(X, \mu)$ is the σ-algebra of μ-measurable sets.

PROOF. To prove the first statement, let G be a dense G_δ set with $\lambda(G) = 0$, where λ is the Lebesgue measure on $\mathbb{R}$. Then the collection $\mathcal{C}_X = \{x+G : x \in X\}$ of Lebesgue null sets has $\operatorname{card}(\mathcal{C}_X) \le \kappa$. Suppose $z \notin \bigcup\{x+G : x \in X\}$. Then $(z-G)\cap X = \emptyset$. Clearly $z-G$ is a dense G_δ set. So X is contained in the first category set $\mathbb{R}\setminus(z-G)$, whence X is a first category set in $\mathbb{R}$. But X is not a first category set. Thereby the first statement is proved.

For the proof of the second statement, let G be a first category set in $\mathbb{R}$ such that $\lambda(\mathbb{R}\setminus G) = 0$. This time $\mathcal{C}_Y = \{y+G : y \in Y\}$ is a collection of first category sets in $\mathbb{R}$. Again $\operatorname{card}(\mathcal{C}_Y) \le \kappa$. Suppose $z \notin \bigcup\{y+G : y \in Y\}$. Then $\lambda(\mathbb{R}\setminus(z-G)) = 0$ and $(z-G)\cap Y = \emptyset$. Since the outer Lebesgue measure $\lambda^*(Y)$ is positive, we have $0 < \lambda^*(Y) \le \lambda(\mathbb{R}\setminus(z-G)) = 0$ and a contradiction has appeared. Thereby the second statement is proved. $\square$

PROOF OF THEOREM 6.4. Let X be a Lusin set in $\mathbb{R}$. By definition, $\operatorname{card}(X) \ge \aleph_1$. As every uncountable subset of a Lusin set is also a Lusin set, we may assume $\operatorname{card}(X) = \aleph_1$. The definition of Lusin sets in $\mathbb{R}$ implies that Lusin sets are not sets of the first category in $\mathbb{R}$. It follows that $\mathbb{R} = \bigcup_{\alpha<\omega_1} Z_\alpha$, where Z_α is a Lebesgue null set for each α. Then $\operatorname{card}(Z_\alpha \cap Y) \le \aleph_0$ for each α, whence $\operatorname{card}(Y) \le \aleph_1$ and the first statement follows.

For the second statement, let Y be a Sierpiński set. Again we may assume $\operatorname{card}(Y) = \aleph_1$. Hence Y is not a Lebesgue null set. It follows that $\mathbb{R} = \bigcup_{\alpha<\omega_1} Z_\alpha$, where Z_α is a first category set for each α. Then $\operatorname{card}(Z_\alpha \cap X) \le \aleph_0$ for each α, whence $\operatorname{card}(X) \le \aleph_1$ and the second statement follows.

The remaining statement is now obvious.

The 1938 Rothberger theorem raises the question of the existence of Lusin sets and Sierpiński sets in $\mathbb{R}$ in the absence of the continuum hypothesis. This question will be discussed in the section on consequences of the Martin axiom.

In 1936, Hausdorff proved the existence of an uncountable universally null set in $\mathbb{R}$ without the use of the continuum hypothesis. He introduced the Ω-Ω^* gap in the Cantor ternary set and defined his notion of m-convergence. Using his Ω-Ω^* gap and m-convergence, he produced a universally null set in $\mathbb{R}$ that is contained in the Cantor ternary set with cardinality $\aleph_1$. As the Cantor set is a first category set in $\mathbb{R}$, the universally null set produced by Hausdorff is not a Lusin set in $\mathbb{R}$.

Soon after Hausdorff produced his example, Sierpiński and Szpilrajn produced in 1936 another example without using the continuum hypothesis. They used Hausdorff's m-convergence to show that constituent decompositions of nonanalytic co-analytic spaces X will lead to the existence of an uncountable universally null set in X. Again this example has cardinality $\aleph_1$ because m-convergence will not produce universally null sets in a separable metrizable spaces of cardinality higher than $\aleph_1$. As there are such co-analytic sets contained in the Cantor ternary set, the sets produced by Sierpiński and Szpilrajn also may be non Lusin sets in $\mathbb{R}$.

In 1978, Grzegorek [**68**] produced another example of an uncountable universally null set in the Cantor space $\{0,1\}^{\mathbb{N}}$ without the use of the continuum hypothesis. Actually Grzegorek had a different goal in mind, it was the construction of certain kinds of σ-fields. The purpose of the example was to give a positive resolution of a question of Banach [**7**, P 21] (see footnote 10 on page 19 for a statement of this problem). The cardinal number non-$\mathbb{L}$ appears in the paper, though under a different symbol. It is shown that there is a universally null set X in $\mathbb{R}$ such that $\mathrm{card}(X) = \text{non-}\mathbb{L}$. Grzegorek acknowledges using a method of K. Prikry [**128**]. It is interesting to note that Haydon's space given in [**58**, Example 5.7, page 975] (see also R. Haydon [**74**]) has calculations that resemble those found in the Grzegorek paper. Grzegorek's example has been instrumental in substantial advances in the study of universally null sets in ambient spaces.

Grzegorek's example leads to the following equivalent theorem. That is, the statement "there exists a universally null set X in $\mathbb{R}$ such that $\mathrm{card}(X) = \text{non-}\mathbb{L}$," is equivalent to

THEOREM 6.6. *There exists a cardinal number κ with the property: there are subsets X_1 and X_2 of $\mathbb{R}$ such that* $\mathrm{card}(X_1) = \mathrm{card}(X_2) = \kappa$, $\lambda^*(X_1) > 0$, *and X_2 is an absolute null space.*

(The proof of the equivalence is left as an exercise.) This theorem is stated (in a different form) in [**68**]. The cardinal numbers that satisfy the requirements of the theorem are called *Banach cardinals*. Clearly non-$\mathbb{L}$ is a Banach cardinal. It is also clear that any cardinal number κ smaller than non-$\mathbb{L}$ is not a Banach cardinal. Indeed, if $\mathrm{card}(X_1) = \mathrm{card}(X_2) = \kappa < \text{non-}\mathbb{L}$ with $\lambda^*(X_1) > 0$ and X_2 is an absolute null space, then the contradiction $\text{non-}\mathbb{L} \leq \mathrm{card}(X_1) = \kappa < \text{non-}\mathbb{L}$ will result. The question of the uniqueness of the Banach cardinal was posed in the above cited paper with the announcement that J. Cichoń [**30**] had shown that Banach cardinals are not unique.[6] We shall return to Banach cardinals shortly.

It is known that $\aleph_1 \leq \text{non-}\mathbb{L} \leq \mathfrak{c}$. Regarding $\text{non-}\mathbb{L} = \mathfrak{c}$, we mention a 1976 result of R. Laver [**88**, page 152] that asserts the existence of a model for ZFC + $\neg$ CH in which no universally null set E in $\mathbb{R}$ can exist with $\mathrm{card}(E) > \aleph_1$. Hence $\text{non-}\mathbb{L} = \mathfrak{c}$ cannot be proved in ZFC + $\neg$ CH. This discussion will be expanded in the next section.

Hausdorff's notion of m-convergence uses the cardinal number ω_1 as an indexing set. Also, in Grzegorek's existence proof, the cardinal number non-$\mathbb{L}$ is used as an indexing set. By replacing cardinal numbers with ordinal numbers as indexing sets to well order absolute null spaces, Recław in 2001 [**130**] and Plewik in 1993 [**127**] used relations on absolute measurable spaces and absolute Borel spaces, respectively, as sufficient conditions for the existence of absolute null spaces. For an uncountable absolute Borel space, Plewik devised a schema which resulted in many applications. Let us turn to the applications that were mentioned in Section 1.5. The first application of Plewik's schema is the Hausdorff Ω-Ω^* gap. Following Plewik, we shall work on the compact space $P(\omega)$ of all subsets of the cardinal number

[6] The reader will find in [**68**] an interesting discussion of the theorem which resolves Banach's problem. There are four theorems that relate to Banach's problem, all of them equivalent to each other. The first one is the resolution of Banach's problem, the next two concerns the existence of cardinal numbers, and the fourth concerns the cardinal number non-$\mathbb{L}$. In his discussion he shows that the cardinal number of the third theorem, which is equivalent to Theorem 6.6, cannot be simply chosen as $\aleph_1$ or $\mathfrak{c}$. Grzegorek calls these cardinal numbers *Banach cardinals*. The fourth theorem is the one of interest in our book. In [**67**] Grzegorek gave an earlier proof of the third theorem under the assumption of the Martin axiom.

$\omega = [0, \omega) = \{\alpha : \alpha < \omega\}$.[7] Define the *quasi-order* $\subseteq^*$ *on* $P(\omega)$ by $F \subseteq^* G$ if $F \setminus G$ is finite. Define $F \subset^* G$ to mean $F \subseteq^* G$ holds and $G \subseteq^* F$ fails. With the notations of Section 1.5 on page 24, define the relation $\leq_n$ on $P([0, n))$ as follows: $\leq_0 = \emptyset$; $X \leq_n Y$ if and only if $n - 1 \in Y$ for $n \neq 0$. Then the relation $\prec$, as defined in that section, is the relation $\leq^*$ as given by the Hausdorff construction. We can now prove Hausdorff's assertion that each Ω-Ω^* gap $[f_\alpha, g_\alpha]$, $\alpha < \omega_1$, has the property that the set $X = \{x_\alpha : \alpha < \omega_1\}$, where $x_\alpha \in [f_\alpha, g_\alpha]$, for each α, is an absolute null space (see page 12 for the m-convergence version). This is obvious since X can be written as the union of two sets X_0 and X_1, where the first one is well ordered by the relation $\leq^*$ and the second is well ordered by the reverse of $\leq^*$.

The second application concerns the cardinal number $\mathfrak{b}$. In order to define this cardinal number we will need to introduce some notation. Denote by ${}^\omega\omega$ the collection of all functions from ω into ω and define the *quasi-order* $\leqslant^*$ *on* ${}^\omega\omega$ by

$$f \leqslant^* g \text{ if } f(n) \leq g(n) \text{ for all but finitely many } n.$$

Clearly $\langle {}^\omega\omega, \leqslant^* \rangle$ has no maximal elements; indeed, if f is in ${}^\omega\omega$, then $g(n) = \sum_{i \leq n}(f(i) + 1)$, $n \in \omega$, is a strictly increasing function in ${}^\omega\omega$ such that $f \leqslant^* g$, even more, $f(n) < g(n)$ for every n. There is a natural topology on ${}^\omega\omega$ that is homeomorphic to ω^ω (which is the same as $\mathbb{N}^\mathbb{N}$). A subset B of ${}^\omega\omega$ is said be *unbounded* if it is unbounded with respect to the quasi-order $\leqslant^*$. The cardinal number $\mathfrak{b}$ is defined to be

$$\mathfrak{b} = \min\{\operatorname{card}(B) : B \text{ is an unbounded subset of } {}^\omega\omega\}$$

It is easy to construct an unbounded subset B of ${}^\omega\omega$ with $\operatorname{card}(B) = \mathfrak{b}$ that consists only of strictly increasing functions. Let us turn to the members X of $P(\omega)$ with $\operatorname{card}(X) = \omega$. Observe that the natural indexing $X = \{n_i : i < \omega\}$ will define a unique function $i \mapsto n_i$ in ${}^\omega\omega$ and that the subset

$$[\omega]^\omega = \{X \in P(\omega) : \operatorname{card}(X) = \omega\}$$

is a Borel subset of $P(\omega)$. Hence there is a natural $\mathfrak{B}$-homeomorphism φ of $[\omega]^\omega$ onto the collection of all strictly increasing functions in ${}^\omega\omega$. It follows that φ induces a natural quasi-order $\leqslant_*$ on $[\omega]^\omega$ as follows:

$$\text{For } (X, Y) \in [\omega]^\omega \times [\omega]^\omega, \quad X \leqslant_* Y \text{ if and only if } \varphi(X) \leqslant^* \varphi(Y).$$

Hence, if $X = \{x_k : k \in \omega\}$ and $Y = \{y_k : k \in \omega\}$ are two infinite subsets of ω with the natural well ordering, then $X \leqslant_* Y$ if and only if $x_k \leq y_k$ for all k larger than some m. This quasi-order can be extended to all of $P(\omega)$ by declaring $X \leqslant_* Y$ if X is a finite subset of ω and Y is any subset of ω. It is known (see P. L. Dordal **[46]**) that there is a subset $\{f_\alpha : \alpha < \lambda\}$ of $[\omega]^\omega$ which is well ordered by the quasi-order $\leqslant^*$

[7] The notation developed by E. K. van Douwen **[154]** will be used. Using characteristic functions of subsets of ω, one can identify $P(\omega)$ with ${}^\omega\{0, 1\}$ which can be identified with $\{0, 1\}^\mathbb{N}$. Hence $P(\omega)$ has a compact topology.

of ${}^{\omega}\omega$ and satisfies $\mathfrak{b} \leq \lambda$, and that the inequality $\mathfrak{b} < \lambda$ can hold in some model for ZFC. As we have seen above, each function f_α may be assumed to be strictly increasing. Hence there is a collection of subsets $\{X_\alpha : \alpha < \lambda\}$ that is well ordered by the relation $\leqslant_*$. It remains to show that the relation $\leqslant_*$ can be produced by the schema – that is, we must define appropriate relations $\leq_n$ on $P([0, n))$ for each n in ω in order to define a relation $\prec$ that is the same as $\leqslant_*$. We do this for even and for odd cases separately. Let us begin with the odd n's (these relations $\leq_n$ are used to guarantee that $X \prec Y$ whenever $\operatorname{card}(X) < \omega$ and to guarantee that $\operatorname{card}(Y) = \omega$ whenever $\operatorname{card}(X) = \omega$ and $X \prec Y$). If n is odd, then the relation $\leq_n$ is defined by "$X \leq_n Y$ if $n - 1 \in Y$ whenever $n - 1 \in X$." Consider $n = 2m$, where m is in ω (these relations $\leq_n$ are used to guarantee that $X \prec Y$ if and only if $X \leqslant_* Y$ whenever $\operatorname{card}(X) = \omega$). Let X and Y be in $P([0, 2m))$ with $m > 0$. Define $X \leq_{2m} Y$ to fail if $\operatorname{card}(X) < m$ or $\operatorname{card}(Y) < m$. If $m \leq \operatorname{card}(X)$ and $m \leq \operatorname{card}(Y)$, then $X = \{x_i : 0 \leq i \leq s\}$ and $Y = \{y_i : 0 \leq i \leq t\}$. Then define $X \leq_{2m} Y$ to hold if and only if "$s = t = m$ and $x_i \leq y_i$ all i." It is left to the reader to verify that $\prec$ and $\leqslant_*$ are the same.

It now follows that there is an absolute null space whose cardinality is $\mathfrak{b}$. It is noted by D. H. Fremlin [**56**] that one cannot prove $\mathfrak{b} \leq \mathsf{non}\text{-}\mathbb{L}$ using ZFC axioms alone, hence $\mathsf{non}\text{-}\mathbb{L} \leq \mathfrak{b} \leq \mathfrak{c}$. It now follows that the existence of an absolute null space with cardinality $\mathsf{non}\text{-}\mathbb{L}$ has a second proof in ZFC. From $\mathsf{non}\text{-}\mathbb{L} \leq \mathfrak{b}$ we infer the existence of a subset M of $[0, 1]$ with $\operatorname{card}(M) = \mathfrak{b}$ and with positive outer Lebesgue measure (see Exercise 6.2). Consequently $\mathfrak{b}$ is a Banach cardinal. Hence in any model for ZFC for which $\mathsf{non}\text{-}\mathbb{L} < \mathfrak{b}$ there are more than one Banach cardinal. This gives a second proof of Cichoń's 1981 result [**30**], namely,

Theorem 6.7 (Cichoń). *It is consistent with* ZFC *that there exists more than one Banach cardinal.*

The original proof by Cichoń did not use the cardinal number $\mathfrak{b}$. In his article Cichoń proposed the following question.

Question. Is the supremum of the collection of all Banach cardinals a Banach cardinal?

Let us return to Recław's theorem. This theorem uses relations that are absolute measurable spaces whereas Plewik's theorem used the more restrictive class of absolute Borel spaces. Let us give Recław's example of a relation that is a co-analytic space. Recall the following: *If B is a Borel subset of $\mathbb{R} \times \mathbb{R}$, then the relation $R = \{(x, y) : B_x \subset B_y\}$ is a co-analytic space*, where the set B_t is defined to be $\{s \in \mathbb{R} : (t, s) \in B\}$ for t in $\mathbb{R}$. He used this example to derive several interesting set theoretic consequences which are tangent to the goal of our book and hence discussions of these consequences will be omitted.

It is shown in Chapter 2 that Hausdorff's m-convergence, which was defined in 1936, is not only a sufficient condition for the existence of uncountable absolute null spaces but is also a necessary condition–indeed, with a judicious use of absolute measurable spaces, a necessary and sufficient condition has been proved for the existence of absolute null spaces contained in arbitrary separable metrizable spaces by means of m-convergence (Theorem 1.35). Also Recław's theorem has been turned into a necessary and sufficient condition for the existence of absolute null spaces contained in arbitrary separable metrizable spaces (Theorem 1.54). Each of these results is achieved in ZFC.

Let us turn to the results of Darst that were covered in the earlier chapters. All but one of the theorems attributed to him during the years 1968–1971 were originally proved in ZFC + CH, the exception being Theorem 4.2 in which only ZFC is assumed. In each of the original proofs, Darst used the following proposition.

PROPOSITION 6.8. *Let K be an uncountable Borel subset of $\mathbb{R}$ and let C be the Cantor ternary set. If A is a universally null set in C and E is a subset of $C \times K$ such that $\pi_C|E$ is a bijection of E onto A and $\pi_K[E] = K$, where π_C and π_K are the natural projections, then E is a universally null set in $\mathbb{R}^2$ and K is not a universally null set in $\mathbb{R}$.*

The proposition is correct; but, Darst could not prove that the hypothesis was not empty unless the continuum hypothesis was used. He essentially observed that a certain Lusin set A contained in C would satisfy the hypothesis. In 1981, Grzegorek and Ryll-Nardzewski [**71**] made the simple observation that the replacement of the condition "$\pi_K[E]$ equals K" by the condition "$\pi_K[E]$ is a non Lebesgue measurable subset of K" would result in a correct proposition in ZFC which would work equally well in the proof of Darst's main theorem. Earlier, Mauldin [**101**, Theorem 5.5] showed in 1978 that Darst's main theorem was also true if the continuum hypothesis was replaced with the Martin axiom.

For the final theorem in ZFC + CH we turn to a simple proof of the following theorem given by Zindulka in 1999 [**160**, Proposition 3.9]. The earlier 1937 proof by S. Mazurkiewicz and E. Szpilrajn [**106**] will be discussed in the comment section.

THEOREM 6.9 (CH-dimension theorem). *Assume* CH. *For each nonempty separable metrizable space X there is a universally null set E in X such that* $\dim E = \dim X - 1$.

PROOF. We may assume $\dim X > 1$ as the contrary case is correct in ZFC. Clearly $\operatorname{card}(X) = \mathfrak{c}$ follows from the definition of small inductive dimension (see page 245). Indeed, there are $\mathfrak{c}$ many closed sets H such that $\dim H > 0$. Let $\mathfrak{H}$ denote the collection of all such closed sets.

Let μ be a continuous, complete, finite Borel measure on X such that $\mu(X) > 0$. By Proposition D.7, there is a zero-dimensional F_σ set F such that $\mu(X \setminus F) = 0$. Denote by $\mathfrak{F}$ the collection of all such sets F corresponding to all such measures.

Let $H_\alpha, \alpha < \mathfrak{c}$, be a well ordering of $\mathfrak{H}$, and let $F_\alpha, \alpha < \mathfrak{c}$ be a map of $\{\alpha : \alpha < \mathfrak{c}\}$ onto $\mathfrak{F}$. Since the continuum hypothesis has been assumed, we have, by the sum theorem of dimension theory, that the set $M_\alpha = H_\alpha \setminus \bigcup_{\beta \leq \alpha} F_\beta$ is not empty for each α. For each α select an x_α in M_α. Define $E_0 = \{x_\alpha : \alpha < \mathfrak{c}\}$.

Let us show that E_0 is a universally null set in X. To this end, let μ be a continuous, complete, finite Borel measure on X with $\mu(X) > 0$. There exists an α such that $\mu(X \setminus F_\alpha) = 0$. As

$$E_0 = (E_0 \cap (X \setminus F_\alpha)) \cup (E_0 \cap F_\alpha) \subset (X \setminus F_\alpha) \cup \{x_\beta : \beta < \alpha\}$$

we have $\mu(E_0) = 0$ and thereby E_0 is a universally null set in X.

It remains to show that there is a subset E of E_0 such that $\dim E = \dim X - 1$. This follows from the lemma below. □

The next lemma is used in the above proof and in the proof of Theorem 6.26 (Zindulka's Martin axiom version of the CH-dimension theorem) which will be presented later. The proof of the lemma does not use the continuum hypothesis.

LEMMA 6.10. *Let X be a nonempty separable metrizable space. If E_0 is a subset of X such that $E_0 \cap H$ is not empty for every closed subset H of X with* $\dim H > 0$, *then there is a subset E of E_0 such that* $\dim E = \dim X - 1$.

PROOF. The statement is obviously true if $\dim X \leq 1$. So let n be an integer such that $1 \leq n < \dim X$ and suppose $\dim E_0 < n$. By the G_δ-hull property of topological dimension (see page 245) there is a G_δ set Y such that $E_0 \subset Y$ and $\dim Y = \dim E_0$. We assert $\dim(X \setminus Y) \geq 1$. Indeed, assume the contrary. Then the addition theorem yields $n < \dim X \leq \dim Y + \dim(X \setminus Y) + 1 \leq (n-1) + 0 + 1 = n$, a contradiction. As $X \setminus Y$ is an F_σ set with $\dim(X \setminus Y) \geq 1$, the sum theorem yields a closed set H such that $\dim H > 0$ and $H \cap E_0 = \emptyset$. There is an α such that $H_\alpha = H$. Then $\emptyset = E_0 \cap H = E_0 \cap H_\alpha \neq \emptyset$ and a contradiction has appeared. Hence $\dim E_0 < n$ is not possible. That is, $\dim E_0 \geq n$. If $\dim X = \infty$, then let $E = E_0$. So suppose $\dim X < \infty$. Then $0 \leq \dim X - 1 \leq \dim E_0 \leq \dim X$. From the definition of the small inductive dimension we infer the existence of the subset E of E_0 with $\dim E = \dim X - 1$. The lemma is proved. □

This brings us to the end of our historical tour of CH and universally null sets. We turn our attention in the remainder of the chapter to the last half of the twentieth century.

6.2. Absolute null space and cardinal numbers

The results in this section are set theoretic implications in the study of singular sets; their statements with a brief discussion seem appropriate at this juncture. Set theoretic results impact in a very strong way on the study of singular sets and they cannot be ignored. It is not the purpose of the book to delve deeply into these set theoretic results. We shall take a rather casual approach to the task of describing those results that seem to be important for the book.

The results that concern us in this section are not topological, they are results on the cardinality of certain singular sets. The four important cardinal numbers for us are $\aleph_0$, $\aleph_1$, $\aleph_2$ and $\mathfrak{c}$. The singular sets that have appeared in the book are absolute null spaces, Lusin sets, Sierpiński sets, and strong measure zero sets. We have made many references to the 1976 *Acta Mathematica* article [**88**] by Laver titled "On the consistency of Borel's conjecture." The Borel conjecture of the title concerns strong measure zero sets. Borel conjectured in [**13**, page 123] that "all strong measure zero sets are countable." The main result of the paper is

THEOREM 6.11 (Laver). *If* ZFC *is consistent, so is* ZFC + Borel's conjecture.

As the title of our book indicates, the singular sets of prime interest are the absolute null spaces. Laver's article has in its introduction a very important paragraph which we shall quote in its entirety. (The cited papers in the quote are [17]=[**133**] and [24]=[**144**].)

Regarding universal measure zero sets, Hausdorff's theorem is best possible in the sense that there is a model of ZFC + $2^{\aleph_0} > \aleph_1$ in which there are no universal measure zero sets of power $\aleph_2$ (add Sacks reals ([17]) or Solovay reals ([24]) to a model of $2^{\aleph_0} = \aleph_1$). This fact (not proved here) is an extension of an unpublished result of Baumgartner which states that adding Sacks real to a model of $2^{\aleph_0} = \aleph_1$ gives a model in which there are no strong measure zero sets of power $\aleph_2$.

A. W. Miller states in his 1984 article [**110**, page 214] that "... it is a theorem of Baumgartner and Laver that in the random real model every universal measure zero set has cardinality less and or equal to ω_1." (Here, universal measure zero set = universally null set in $\mathbb{R}$.) A proof of this assertion is provided by Miller in [**109**, pages 576–578]. Let us extract the relevant parts of his proof. He first proves (which he says is known but has not been published)

THEOREM 6.12. *If ω_2 random reals are added to a model of* CH, *then in the extension every set of reals of cardinality ω_2 contains a subset of cardinality ω_2 which is the continuous image of a Sierpiński set.*

From this theorem follows the above quoted Baumgartner–Laver assertion

THEOREM 6.13 (Baumgartner–Laver). *In the random real model every universally null set in $\mathbb{R}$ has cardinality less than or equal to $\aleph_1$.*

PROOF. Recall that an uncountable subset of a Sierpiński set is not a universally null set in $\mathbb{R}$. Suppose that there is a universally null set N in $\mathbb{R}$ with $\operatorname{card}(N) \geq \aleph_2$. Then there is a subset N' of N with cardinality $\aleph_2$ and a continuous surjection $f : S \to N'$ of some Sierpiński set S. Let S' be a subset of S such that the restriction $f|S'$ is a continuous bijection of S' onto N'. As N' is a universally null set in $\mathbb{R}$ we have that S' is also a universally null set as well as a Sierpiński set, which is not possible, and a contradiction has appeared. □

Miller's main theorem of the paper [**109**] is

THEOREM 6.14. *In the iterated perfect set model every set of reals of cardinality ω_2 can be mapped continuously onto the closed unit interval.*

We have the corollary.

COROLLARY 6.15. *In the iterated perfect set model, the inequalities $\aleph_1 \leq \mathfrak{c} \leq \aleph_2$ hold. Hence $\mathfrak{c} = \aleph_2$ if $\mathfrak{c} \neq \aleph_1$.*

PROOF. Suppose $\aleph_2 < \mathfrak{c}$. Then there exists a subset X of $\mathbb{R}$ with $\operatorname{card}(X) = \aleph_2$. Let $f : X \to [0,1]$ be a continuous surjection provided by the theorem. Then $\mathfrak{c} = \operatorname{card}([0,1]) \leq \operatorname{card}(X) = \aleph_2 < \mathfrak{c}$, which is a contradiction. □

To amplify the above Laver quote, we quote again from the above 1984 Miller paper. He states there that in the iterated perfect set model "… $\mathfrak{c} = 2^{\omega_1}$ so there are only continuum many universal measure zero sets … ." Let us state and prove the last part of the quote as a proposition.

PROPOSITION 6.16. *In the iterated perfect set model, the cardinality of the collection* $\mathsf{univ}\,\mathfrak{N}(\mathbb{R})$ *of all universally null sets in* $\mathbb{R}$ *is* $\mathfrak{c}$.

PROOF. Since the cardinality of each universally null set in $\mathbb{R}$ is at most $\aleph_1$, we have $\mathfrak{c} \le \mathrm{card}\big(\mathsf{univ}\,\mathfrak{N}(\mathbb{R})\big) \le \mathfrak{c}^{\aleph_1} = (2^{\aleph_0})^{\aleph_1} = 2^{\aleph_1} = \mathfrak{c}$. □

The often quoted Laver assertion about the cardinality of a universally null set in $\mathbb{R}$, no doubt true, is not explicitly demonstrated in the literature. Hence no explict citation of the proof due to Laver has been provided. The recent tract **[31]** by K. Ciesielski and L. Pawlikowski does present an explicit proof of the Laver assertion. They define a covering property axiom called CPA and prove that the iterated perfect set model satisfies the covering property axiom. Among the many consequences of this axiom that are established in the tract is Laver's cardinality assertion [**31**, Section 1.1].

In a private conversation, Mauldin asked if the cardinality of a universally null set in $\mathbb{R}$ was known. Clearly, in ZFC + CH, the answer is $\mathfrak{c}$ because $\mathsf{non}\text{-}\mathbb{L} = \mathfrak{c}$. For ZFC + $\neg$ CH, by the iterated perfect set model, the answer is $\aleph_1$ which is not $\mathfrak{c}$. To sharpen the question, define the cardinal number $\mathfrak{n}$ to be the least cardinal number that is greater than or equal to $\mathrm{card}(X)$ of every absolute null space X. As every absolute null space can be embedded in the Hilbert cube, we have $\aleph_1 \le \mathsf{non}\text{-}\mathbb{L} \le \mathfrak{n} \le \mathfrak{c}$. Mauldin's question becomes:

QUESTION. In ZFC + $\neg$ CH, is there an absolute null space X such that $\mathrm{card}(X) = \mathfrak{n}$?

Observe that this question is analogous to Cichoń's question about Banach cardinals.

Zindulka, in a private conversation, provided the author with the following analysis of the above weak inequalities. It will be shown in the next section that there is a model of ZFC + $\neg$ CH such that $\mathsf{non}\text{-}\mathbb{L} = \mathfrak{c}$; hence $\aleph_1 < \mathsf{non}\text{-}\mathbb{L}$ happens for this model. Cichoń has shown in **[30]** that there is a model of ZFC such that there are at least two Banach numbers; hence $\mathsf{non}\text{-}\mathbb{L} < \mathfrak{n}$ happens for this model.

Related to this is another question attributed to Mauldin which asks if there always are more than $\mathfrak{c}$ universally measurable sets in $\mathbb{R}$? (See page 214 of **[110]** and page 422 of **[111]**.) Clearly, $\mathrm{card}\big(\mathsf{univ}\,\mathfrak{M}(\mathbb{R})\big) \ge \mathfrak{c}$. Proposition 6.16 reveals the difficulties encountered in this question, which is

QUESTION (Mauldin). Is it consistent to have $\mathrm{card}\big(\mathsf{univ}\,\mathfrak{M}(\mathbb{R})\big) = \mathfrak{c}$?

In Section 2.4 of Chapter 2, the symmetric difference property of universally measurable set in a space was discussed. Recall that a Lebesgue measurable set in $\mathbb{R}$ is the symmetric difference of a Borel set and a Lebesgue null set. For universally measurable sets in $\mathbb{R}$ this property holds for every continuous, complete, finite Borel measure on $\mathbb{R}$. It is tempting to claim that each universally measurable set in $\mathbb{R}$ is the symmetric difference of a Borel set and a universally null set in $\mathbb{R}$. Then, in the iterated perfect set model, the cardinality of the collection of universally measurable sets in $\mathbb{R}$ would be $\mathfrak{c}$. But this analogue of the symmetric difference property fails as Theorem 2.26 shows. If the consistency question of Mauldin is to have a positive resolution, another approach will be needed. Of course, the property in the question could possibly fail or be independent of ZFC.

6.3. Consequences of the Martin axiom

We begin with the definition of the Martin axiom, actually an equivalent topological form. The definition uses the notion of the countable chain condition which will be defined first, again in the topological context.

DEFINITION 6.17. *A topological space X has the* countable chain condition (c.c.c.) *if and only if there is no uncountable family of pairwise disjoint open sets of X.* (See K. Kunen [**83**, page 50].)

DEFINITION 6.18 (**Martin axiom: topological form**). *No compact Hausdorff space with* c.c.c. *is the union of less than $\mathfrak{c}$ closed nowhere dense sets (equivalently, $\bigcap\{\, U_\alpha : \alpha < \kappa \,\} \neq \emptyset$ whenever $\{\, U_\alpha : \alpha < \kappa \,\}$ is a family of dense open sets and $\kappa < \mathfrak{c}$.)* (See Kunen [**83** pages 52 and 65].)

Often we will designate the Martin axiom as MA. It is known that the continuum hypothesis implies the Martin axiom (see Kunen [**83**, page 55]).

6.3.1. Properties of separable metrizable spaces. Two theorems that are consequences of the Martin axiom will be singled out. We shall take the statements from Kunen [**83**, pages 58 and 59] without providing proofs. The first is

THEOREM 6.19. *Assume* MA. *If $\{\, M_\alpha : \alpha < \kappa \,\}$, where $\kappa < \mathfrak{c}$, is a family of first category sets of $\mathbb{R}$, then $\bigcup\{\, M_\alpha : \alpha < \kappa \,\}$ is a first category set in $\mathbb{R}$.*

The proof uses only the second countability of $\mathbb{R}$, that is, there is a countable base for the open sets of $\mathbb{R}$. It is remarked in Kunen [**83**, Exercise 13, page 87] that $\mathbb{R}$ in the above theorem can be replaced by any separable metrizable space.

The next lemma was provided to the author by K. P. Hart.

LEMMA 6.20. *Assume* MA. *If $M \subset \mathbb{R}$ and* $\operatorname{card}(M) < \mathfrak{c}$, *then M is a Lebesgue null set.*

PROOF. Let G be a dense G_δ set of Lebesgue measure zero and denote its complement by D. Obviously, D is a first category set in $\mathbb{R}$. Let us show $\mathbb{R} = \bigcup_{x\in M}(x + D)$ whenever $\lambda^*(M) > 0$. To this end, let $y \in \mathbb{R}$ and $\lambda^*(M) > 0$. As $\lambda^*(y - M) > 0$ we have $D \cap (y - M) \neq \emptyset$. Let z be a point in the intersection and t be a point in M such that $z = y - t$. Then $y \in t + D \subset \bigcup_{x\in M}(x + D)$. By contraposition we have shown that $\mathbb{R} \neq \bigcup_{x\in M}(x + D)$ implies $\lambda^*(M) = 0$. By the preceding theorem we have that $\bigcup_{x\in M}(x + D)$ is a first category set; the completeness of $\mathbb{R}$ provides the final step of the proof. □

The lemma yields the following theorem. See also Fremlin [**57**, Corollary 22H(d), page 49].

THEOREM 6.21. *Assume* MA. *A separable metrizable space X is an absolute null space whenever* $\operatorname{card}(X) < \mathfrak{c}$.

PROOF. We may assume that X is contained in $[0,1]^{\mathbb{N}}$. As $[0,1]^{\mathbb{N}}$ and $\mathbb{R}$ are $\mathfrak{B}$-homeomorphic, only those subsets X of $\mathbb{R}$ with $\operatorname{card}(X) < \mathfrak{c}$ must be considered. For each h in $\mathsf{HOMEO}(\mathbb{R})$ we have $h^{-1}[X]$ is Lebesgue measurable. Since λ and $\mathsf{HOMEO}(\mathbb{R})$ generate $\mathsf{univ}\,\mathfrak{M}(\mathbb{R})$, we have $X \in \mathsf{univ}\,\mathfrak{M}(\mathbb{R})$. Hence X is an absolute measurable space. As $\operatorname{card}(X) < \mathfrak{c}$ we have X is totally imperfect. Consequently, X is an absolute null space by Theorem 1.20. □

Observe that the theorem implies the lemma.

There is a stronger result than the above Lemma 6.20 that appears in Kunen [**83**, Theorem 2.21, page 59]. This is the second of the two theorems that we said would be taken from Kunen's book without proof.

THEOREM 6.22. *Assume* MA. *If* $\{ M_\alpha : \alpha < \kappa \}$, *where* $\kappa < \mathfrak{c}$, *is a family of Lebesgue null sets of* $\mathbb{R}$, *then* $\bigcup\{ M_\alpha : \alpha < \kappa \}$ *is a Lebesgue null set in* $\mathbb{R}$.

There is a nice corollary of this theorem. The case where X is $\mathbb{R}$ was observed by Miller [**110**, page 219].

COROLLARY 6.23. *Assume* MA. *Let X be a separable mertizable space such that* $X = \bigcup_{\alpha<\kappa} X_\alpha$, *where X_α is an absolute null space for each α and where $\kappa < \mathfrak{c}$. Then X is an absolute null space.*

PROOF. As in the proof of Theorem 6.21, we may assume that X is a subset of $\mathbb{R}$. Let μ be a positive, complete, continuous, finite Borel measure on $\mathbb{R}$ such that $\mathfrak{M}(\mathbb{R},\lambda) = \mathfrak{M}(\mathbb{R},\mu)$ and $\mathfrak{N}(\mathbb{R},\lambda) = \mathfrak{N}(\mathbb{R},\mu)$. Suppose ν is a positive, continuous, complete, finite Borel measure on $\mathbb{R}$ and let h be in $\mathsf{HOMEO}(\mathbb{R})$ such that $\nu = c \cdot h_\# \mu$ for an appropriate positive number c. Now $h^{-1}[X_\alpha]$ is an absolute null space and hence a Lebesgue null set in $\mathbb{R}$. By the above theorem, $h^{-1}[X]$ is a Lebesgue null set. So, $\nu(X) = h_\#\mu(X) = \mu(h^{-1}[X]) = 0$. Consequently, $X \in \mathsf{univ}\,\mathfrak{N}(\mathbb{R})$ and therefore X is an absolute null space. □

6.3.2. Universally null sets in $\mathbb{R}$. Grzegorek has shown, with only ZFC assumed, the existence of a universally null set E in $\mathbb{R}$ with $\operatorname{card}(E) = \mathsf{non}\text{-}\mathbb{L} \geq \aleph_1$. It was mentioned in Section 6.1 that there are models of ZFC + ¬ CH in which no universally null sets E in $\mathbb{R}$ with $\operatorname{card}(E) > \aleph_1$ exist. Also from Theorem 6.21 we have that every subset of $\mathbb{R}$ with cardinality less than $\mathfrak{c}$ is an absolute null space in ZFC + MA. We now have the following assertion from Grzegorek and Ryll-Nardzewski [**71**].

THEOREM 6.24 (Grzegorek–Ryll-Nardzewski). *Neither the statement "there exists an absolute null space with cardinality* $\mathfrak{c}$*" nor its negation is provable from* ZFC + ¬ CH.

PROOF. As each absolute null space is $\mathfrak{B}$-homeomorphic to an absolute null space contained in $\mathbb{R}$, it is enough to consider only those absolute null spaces contained in $\mathbb{R}$. First observe that the statement "every subset of $\mathbb{R}$ with cardinality less than $\mathfrak{c}$ is Lebesgue measurable" implies the statement "$\mathsf{non}\text{-}\mathbb{L}$ is $\mathfrak{c}$." Indeed, suppose

$\mathsf{non}\text{-}\mathbb{L} < \mathfrak{c}$ and every subset of $\mathbb{R}$ with cardinality less than $\mathfrak{c}$ is Lebesgue measurable. Then there exists a set E such that $\mathrm{card}(E) = \mathsf{non}\text{-}\mathbb{L}$ and $\lambda^*(E) > 0$. From $\mathsf{non}\text{-}\mathbb{L} < \mathfrak{c}$ we have that the set E is Lebesgue measurable and satisfies $\lambda(E) > 0$. As every positive Lebesgue measurable set contains a copy of the Cantor set we have $\mathrm{card}(E) = \mathfrak{c}$. A contradiction has appeared. Hence there is a universally null set X in $\mathbb{R}$ with $\mathrm{card}(X) = \mathfrak{c}$ in ZFC + MA. It is known that ZFC + MA + $\neg$ CH is consistent provided ZFC is consistent. Hence the negation of the statement "there exists an absolute null space X with $\mathrm{card}(X) = \mathfrak{c}$" is not provable from ZFC + $\neg$ CH.

It was mentioned earlier in Section 6.1 that Laver has shown the existence of a model for ZFC + $\neg$ CH for which $\mathrm{card}(X) \leq \aleph_1$ whenever X is a universally null set in $\mathbb{R}$. Hence the statement is also not provable from ZFC + $\neg$ CH. Thereby we have shown that the statement in the theorem is independent of ZFC + $\neg$ CH. □

An immediate consequence of the theorem is the fact that one cannot drop the CH assumption in the CH-dimension theorem. Indeed, suppose it can be shown in ZFC that there is a universally null set E in X with $\dim E = \dim X - 1$ whenever X is a separable metrizable space. Then this would be true also in ZFC + $\neg$ CH. So it can be shown that there would exist an absolute null space E with $\mathrm{card}(E) = \mathfrak{c}$ because $\dim X > 1$ implies $\dim E \geq 1$, which in turn implies $\mathrm{card}(E) = \mathfrak{c}$. This denies the above theorem.

We have the following interesting consequence of Theorem 6.24.

THEOREM 6.25. *The statement "there is an absolute null space that cannot be topologically embedded in the Cantor ternary set" is not provable from* ZFC + $\neg$ CH.

PROOF. Assume ZFC + $\neg$ CH implies the existence of an absolute null space that cannot be topologically embedded in the Cantor ternary set. It is well-known that a separable metrizable space has dimension less than 1 if and only if the space can be topologically embedded in the Cantor ternary set. Hence it would be possible to prove that ZFC + $\neg$ CH implies the existence of an absolute null space with cardinality $\mathfrak{c}$, contradicting Theorem 6.24. □

Another consequence of the theorem is that ZFC + $\neg$ CH implies that there are no Lusin sets E in $\mathbb{R}$ with $\mathrm{card}(E) = \mathfrak{c}$. For suppose it can be shown that ZFC + $\neg$ CH implies the existence of a Lusin set E in $\mathbb{R}$ with $\mathrm{card}(E) = \mathfrak{c}$. The set E is a universally null set in $\mathbb{R}$ with cardinality $\mathfrak{c}$ because, in ZFC, Lusin sets in $\mathbb{R}$ are always universally null sets in $\mathbb{R}$ whenever they exist.

We have seen in the previous subsection that, in ZFC + MA, every subset E of $\mathbb{R}$ with $\mathrm{card}(E) < \mathfrak{c}$ is simultaneously a first category set and a Lebesgue null set. By definition, if E is a Lusin set in $\mathbb{R}$, then $\mathrm{card}(E) \geq \aleph_1$. Suppose that the existence of a Lusin set E in $\mathbb{R}$ can be shown in ZFC + MA+ $\neg$ CH. As uncountable subsets of Lusin sets are also Lusin sets, we may assume $\mathrm{card}(E) = \aleph_1$. Then $\aleph_1 = \mathrm{card}(E) = \mathrm{card}(E \cap E) \leq \aleph_0$ and thereby a contradiction has occurred. We have shown that no uncountable subset of $\mathbb{R}$ can be a Lusin set. A corresponding argument will show that no uncountable subset of $\mathbb{R}$ can be a Sierpiński set. Hence, in ZFC + MA + $\neg$ CH, each uncountable set in $\mathbb{R}$ is not a Lusin set and is not a Sierpiński set. Observe that this proof did not use the preceding paragraph.

More will be said about the existence of Lusin sets and Sierpiński sets in $\mathbb{R}$ in the comment section at the end of the chapter.

6.4. Topological dimension and MA

Let us begin this section with a short discussion of the dimensions of separable metrizable spaces X and universally null sets in X. In Chapter 5 we learned that a nonempty space X has a metric such that $\dim X = \dim_{\mathsf{H}} X$. For this metric the Hausdorff dimension of a universally null set E in X is bounded above by the topological dimension of the ambient space X. We also know that the topological dimension of E is no bigger than the Hausdorff dimension of E. Consequently, it still can happen that $\dim E < \dim_{\mathsf{H}} E$ for every universally null set E in X, indeed, $\dim E = 0$ is still possible for every nonempty universally null set E in X. In fact, this is so for $X = \mathbb{R}$ since every nonempty universally null set E in $\mathbb{R}$ is totally imperfect and thereby is zero-dimensional. Actually, it is known that every n-dimensional subset of $\mathbb{R}^n$ must contain a nonempty open set and hence is not a universally null set in $\mathbb{R}^n$ because universally null sets in any space X are totally imperfect. This shows $\dim E \leq n - 1$ whenever E is a universally null set in $\mathbb{R}^n$. We have learned earlier that the statement "there is a universally null set E in X such that $\dim E = \dim X - 1$" is provable in ZFC + CH and is not provable in ZFC + $\neg$ CH. Hence, in ZFC + $\neg$ CH, some assumption on the space X as well as some further set theoretic assumption must be made in order for the statement to be true. In light of this discussion Zindulka proved the following result [**160**, Theorem 3.5] for analytic spaces in ZFC + MA.

THEOREM 6.26. (MA-dimension theorem) *Assume* MA. *If X is an analytic space, then there exists a universally null set E in X such that* $\dim E = \dim X - 1$.

The proof will depend on a lemma. It is a consequence of a theorem due to J. Stern [**146**, Theorem 3] which we state and prove [8] next.

THEOREM 6.27 (Stern). *Assume* MA. *Every cover of* $\mathbb{N}^{\mathbb{N}}$ *by fewer than* $\mathfrak{c}$ *closed sets has a countable subcover.*

PROOF. Since $\mathbb{N}^{\mathbb{N}}$ is homeomorphic to $\mathcal{N}$ we shall use $\mathcal{N}$ in place of $\mathbb{N}^{\mathbb{N}}$. As any set F that is of the first category in $\mathcal{N}$ is also a first category set in the open interval $(0, 1)$, we may replace the space $\mathbb{R}$ in Theorem 6.19 by $\mathcal{N}$ to get a corollary that is the analogue for the space $\mathcal{N}$. Let $\mathfrak{F}$ be a cover of $\mathcal{N}$ by fewer than $\mathfrak{c}$ closed sets. Define $\mathfrak{O}$ to be the family of all open sets of $\mathcal{N}$ that are covered by countably many members of $\mathfrak{F}$. By the above mentioned corollary, the collection $\mathfrak{O}$ is not empty. Indeed, each nonempty, simultaneously open and closed subset U of $\mathcal{N}$ has the property that some F in $\mathfrak{F}$ contains a nonempty open subset of U. Since $\bigcup \mathfrak{O}$ is the union of a countable subfamily of $\mathfrak{O}$, we have $V = \bigcup \mathfrak{O}$ is a member of $\mathfrak{O}$ and V is dense in $\mathcal{N}$. Let us show that $Y = \mathcal{N} \setminus V$ is empty. Assume the contrary case. Then we must have

[8] The proof given here was provided to the author by K. P. Hart.

card$(Y) > \aleph_0$ since countable subsets of $\mathcal{N}$ are covered by countable subfamilies of $\mathfrak{F}$. So the uncountable absolute Borel space Y will contain a topological copy Y' of $\mathcal{N}$. Hence the family $\{ F \cap Y' : F \in \mathfrak{F} \}$ will have a member $F' = F \cap Y'$ such that its interior relative to Y' is not empty. That is, the interior of F relative to $\mathcal{N}$ will contain a point y of Y. Consequently, $y \in V$ and a contradiction has arrived. Thereby we have shown $\mathcal{N} = V$ and the theorem is proved. □

The lemma needed to prove Zindulka's theorem is an immediate consequence of the above theorem because analytic spaces are continuous images of the space $\mathcal{N}$.

LEMMA 6.28. *Assume* MA. *Every cover of every analytic space by fewer than* $\mathfrak{c}$ *closed sets has a countable subcover.*

PROOF OF THEOREM 6.26. The proof will have two parts, just as the proof of the CH-dimension theorem had. The construction of the required set E is similar to the CH version. The proof that E is a universally null set in X is a little more involved – it uses the condition that X is an analytic space. The computation of $\dim E$ is exactly the same for the MA version as for the CH version – it does not require that X be an analytic space.

Let X be an analytic space. The statement of the theorem is clearly obvious if $\dim X \leq 1$. So assume $1 < \dim X$. Then card$(X) = \mathfrak{c}$, whence the collection $\mathfrak{F}$ of all zero-dimensional F_σ subsets of X has cardinality $\mathfrak{c}$ because X contains a topological copy of the the Cantor space. The collection $\mathfrak{H}$ of all closed subsets H of X with $\dim H > 0$ also has cardinality $\mathfrak{c}$. The last assertion follows from the definition of the small inductive dimension. Let F_α, $\alpha < \mathfrak{c}$, and let H_α, $\alpha < \mathfrak{c}$, be well orderings of $\mathfrak{F}$ and $\mathfrak{H}$, respectively. Inductively select points x_α, $\alpha < \mathfrak{c}$, such that $x_\alpha \in H_\alpha \setminus \bigcup_{\beta \leq \alpha} F_\beta$. This is possible because H_α is an analytic space and the assumption of $H_\alpha \setminus \bigcup_{\beta \leq \alpha}(H_\alpha \cap F_\beta) = \emptyset$ together with the above lemma and the sum theorem of dimension theory imply $0 < \dim H_\alpha \leq \sup \{ \dim(H_\alpha \cap F_\beta) : \beta \leq \alpha \} \leq 0$, a contradiction. Define E_0 to be the set $\{ x_\alpha : \alpha < \mathfrak{c} \}$. By Lemma 6.10 there is a subset E of E_0 such that $\dim E = \dim X - 1$.

Let us verify that E is a universally null set in X. To this end, let μ be a continuous, complete, finite Borel measure on X. By Corollary 4.31 there is a Zahorski space Z such that $Z \subset X$ and $\mu(X \setminus Z) = 0$. As Z is a zero-dimensional absolute F_σ space, there is an α such that $F_\alpha = Z$. So

$$E_0 = (E_0 \cap (X \setminus Z)) \cup (E_0 \cap F_\alpha) \subset (X \setminus Z) \cup \{ x_\beta : \beta \leq \alpha \}.$$

As $\{ x_\beta : \beta \leq \alpha \}$ is an absolute null space we have $\mu(E_0) = 0$. Consequently E is a universally null set in X. The theorem is proved. □

We have the following consequence of the construction of the set E_0 in the proof of the above theorem.

THEOREM 6.29. *Assume* MA. *If* X *is analytic space with* $\dim X \geq 1$, *then there is a universally null set* E_0 *in* X *such that* $E_0 \cap H \neq \emptyset$ *whenever* H *is a closed set in* X *with* $\dim H \geq 1$.

Let us give three corollaries. Recall that a projection $\pi : X \to X$ of a Banach space X is a linear map such that composition $\pi\,\pi$ is π.

COROLLARY 6.30. *Assume* MA. *Let X be a separable Banach space. There exists a universally null set E in X such that $\pi[E] = \pi[X]$ for all nontrivial projections $\pi : X \to X$.*

PROOF. If X is a one-dimensional Banach space, there are no nontrivial projections π. So consider the contrary case. Let E be the set E_0 from the theorem. As π is a nontrivial projection, $H = \pi^{-1}[\{y\}]$ is a closed set with $\dim H > 0$ whenever $y \in \pi[X]$. Then $H \cap E \neq \emptyset$ and the corollary follows. □

The next corollary has an equally easy proof.

COROLLARY 6.31. *Assume* MA. *There exists a universally null set E in $\mathbb{R}^2$ such that E projects onto each line.*

Recall that a curve is the continuous image of $[0, 1]$. Hence a nontrivial curve C has $\dim C > 0$. As curves are compact the next corollary is obvious.

COROLLARY 6.32. *Assume* MA. *Each analytic space X contains a universally null set E such that E meets each nontrivial curve in X.*

REMARK 6.33. The MA assumption in Zindulka's MA-dimension theorem cannot be dropped. Indeed, if the statement was provable in ZFC alone, then it would be provable in ZFC + ¬ CH, and hence it would be possible to prove that ZFC + ¬ CH implies the existence of an absolute null space E whose cardinality is $\mathfrak{c}$. But, just as we have already seen for the CH-dimension theorem, this will lead to a contradiction of Theorem 6.24. Obviously, the same remark also holds for the above corollaries.

6.5. Comments

As has been mentioned in the introduction to the chapter, many of the citations on the use of the continuum hypothesis that are interspersed in the development appear in the article by Brown and Cox [**18**]. Also, large parts of the development are derived from articles in the *Handbook of Set-Theoretic Topology* [**84**]. We have used Kunen's book [**83**] as a primary source for the Martin axiom so that a newcomer to the subject will not be overwhelmed by the need to search too many references. Our commentary will be ordered by the sections in the chapter. The last two subsections are musings of the author which set out material that have been left out of the book – that is, tasks for further study.

6.5.1. Historical comment. Two proofs have been given of the fact that Lusin sets in $\mathbb{R}$ are universally null in $\mathbb{R}$. The first one was Lavrentieff's original proof in which he used his famous G_δ extension theorem of homeomorphisms. This proof does not use the continuum hypothesis, which was used only to prove the existence of Lusin sets in $\mathbb{R}$. The second proof, which also does not use the continuum hypothesis, uses the group $\mathsf{HOMEO}(\mathbb{R})$. This second proof seems to avoid the use of Lavrentieff's theorem; but, this is not the case since absolute Borel spaces are used in the study of absolute null spaces and Lavrentieff's theorem is used to characterize the collection of absolute Borel spaces. A third proof was given in Chapter 2 that uses the continuum hypothesis and Hausdorff's notion of m-convergence, see Section 1.3.5. Zahorski spaces appear in the above proofs, this is a consequence of the fact that F_σ sets X of the first category in $\mathbb{R}$ contain Zahorski spaces Z such that $\operatorname{card}(X \setminus Z) \leq \aleph_0$. We have already observed that the collection of Zahorski spaces is topologically invariant, a very nice property.

The proof of the CH-dimension theorem (Theorem 6.9) that was given in the chapter is due to Zindulka. With regards to priority, it was mentioned that another proof had appeared earlier. We now comment on this earlier proof by Mazurkiewicz and Szpilrajn [**106**]. In the cited paper the authors provide a general method of constructing examples of certain singular spaces X with $\dim X = n$ for $n = 1, 2, \ldots,$ and ∞ whenever such singular spaces with cardinality $\mathfrak{c}$ existed as subsets of $\mathbb{R}$. Before we give the statement and proof of their theorem let us state, without proof, the dimension theoretic theorem of A. Hilgers [**75**] which is key to their method (see also [**85**, page 302]).

THEOREM 6.34 (Hilgers). *For $n \neq -1$, every separable metrizable space Y with* $\operatorname{card}(Y) = \mathfrak{c}$ *is a continuous bijective image of some separable metrizable space X with* $\dim X = n$.

The Hilgers theorem does not use the continuum hypothesis. Consequently, the Mazurkiewicz–Szpilrajn theorem does not use the continuum hypothesis. We give only the absolute null space part of their theorem.

THEOREM 6.35. *If Y is a universally null set in $\mathbb{R}$ with* $\operatorname{card}(Y) = \mathfrak{c}$, *then for each n there exists an absolute null space X with* $\dim X = n$.

PROOF. Let X be such that $\dim X = n$ and let $f : X \to Y$ be a continuous bijection as provided by Hilgers' theorem. Theorem 1.23 completes the proof since Y is an absolute null space. □

Theorem 6.9 is now a corollary since the continuum hypothesis implies the existence of an absolute null space Y with $\operatorname{card}(Y) = \mathfrak{c}$, which has many proofs. In their proof, Mazurkiewicz and Szpilrajn observed that the Sierpiński–Szpilrajn example (see Section 1.3.3) provided an absolute null space Y with $\operatorname{card}(Y) = \aleph_1$. Hence they concluded that the continuum hypothesis yielded an example that satisfied the hypothesis of their theorem, thereby proving the existence of higher dimensional absolute null spaces in ZFC + CH.

It should be mentioned that the proof due to Zindulka that was given earlier is very different from the above proof.

6.5.2. Cardinal numbers. Although the general relationship of the cardinal numbers $\aleph_0$, $\aleph_1$ and $\mathsf{non}\text{-}\mathbb{L}$ to absolute null spaces require only the usual set axioms ZFC, the literature has shown that cardinal numbers of some classes of absolute null spaces are very much influenced by set theoretic assumptions. Since reference to these assumptions were not needed for our discussion of these general relationships, the more subtle aspects of set theoretic methods such as the powerful "forcing" were not included in the earlier chapters. Hence discussions of the Banach numbers were delayed to the last chapter. But our need to use special classes of absolute null spaces such as Lusin sets and strong measure zero sets require that the reader be aware of the set theoretic implications about these sets. Along this line of thought, the unresolved Mauldin's cardinal number question is certainly an interesting one.

In Chapter 1 we referred to the survey article [**134**] by Scheepers on Ω-Ω^* gaps. This article deals not only with Ω-Ω^* gaps using just ZFC but also using forcing and using special axioms. One sees from the article that there is an extensive literature on Ω-Ω^* gaps. Scheepers also discusses several applications of Ω-Ω^* gaps.

6.5.3. Martin axiom. In addition to Kunen's book the reader might find other sources for the Martin axiom helpful or useful. We mention some that have been used: A. W. Miller [**110**], J. R. Shoenfield [**137**], R. J. Gardner and W. F. Pfeffer [**58**], and the first few pages of W. Weiss [**156**].

Theorem 6.24 is just a comment in [**71**], with a proof included, by Grzegorek and Ryll-Nardzewski. We have formalized it as a theorem, using their proof, since it appears in many other papers in comment form. An application of this theorem showed that the existence of a Lusin set in $\mathbb{R}$ with cardinality $\mathfrak{c}$ is not provable from ZFC + $\neg$ CH. One might say that Rothberger's theorem already denied the possible existence of a Lusin set in $\mathbb{R}$ with cardinality $\mathfrak{c}$ in ZFC + $\neg$ CH. But this is a misreading of Rothberger's theorem which denies the simultaneous existence of a Lusin set and a Sierpiński set in $\mathbb{R}$ for which one of their cardinalities is greater than $\aleph_1$. It is still possible that ZFC + $\neg$ CH implies no Lusin set and no Sierpiński set in $\mathbb{R}$ exist. But, as reported in Miller's article [**110**, page 205], it is consistent with ZFC + $\neg$ CH that a Lusin set and a Sierpiński set in $\mathbb{R}$ exist. That is, there is a model for ZFC + $\neg$ CH in which a Lusin set and a Sierpiński set both exist. Hence, by Rothberger's theorem, no Lusin set and no Sierpiński set can have cardinality $\mathfrak{c}$ in this model for ZFC + $\neg$ CH. Better yet, every Lusin set and every Sierpiński set in $\mathbb{R}$ has cardinality $\aleph_1$ in this model for ZFC + $\neg$ CH. Also, it was shown earlier that ZFC + MA + $\neg$ CH implies that no uncountable subset of $\mathbb{R}$ can be a Lusin set or a Sierpiński set. Consider now the statement "there exists a Lusin set X in $\mathbb{R}$." It was already mentioned that ZFC + MA + $\neg$ CH is consistent provided ZFC is consistent. Hence the statement is not true for some model of ZFC + $\neg$ CH. Moreover, from the above mentioned Miller article, the statement is true for some other model for ZFC + $\neg$ CH. Hence the statement is independent of ZFC + $\neg$ CH. The corresponding assertion may be made for Sierpiński sets in $\mathbb{R}$. Let us summarize the above discussion as a proposition.

PROPOSITION 6.36. *Each of the statements "there exists a Lusin set X in $\mathbb{R}$" and "there exists a Sierpiński set X in $\mathbb{R}$" is independent of* ZFC + $\neg$ CH.

By no means is the discussion of set theoretic results given in the book claimed to be exhaustive or up to date. The methods used in the subject are not easily mastered and we have made no attempt to describe many of them. But one thing is certain, the study of absolute measurable spaces cannot ignore set theoretic results that have been proved in the last half century. More generally, the study of singular sets cannot ignore them, also.

6.5.4. Topological dimension. The main theorem on topological dimension is Zindulka's MA-dimension theorem. The proof given in this chapter is slightly shorter than the original one due to the introduction of Zahorski spaces. The original proof was based on Zindulka's theorem on proper, σ-additive, saturated ideals on a space X (see [**160**, Theorem 2.8]).

Zindulka proved that the CH-dimension theorem and the MA-dimension theorem are not theorems in ZFC. To achieve this he proved the following theorem [**160**, theorem 3.10] whose proof is based on a suggestion by S. Todorčević.

THEOREM 6.37. *It is relatively consistent with* ZFC *that* $\mathfrak{c} = \aleph_2$ *and that each absolute null space is zero-dimensional.*

Of course, this theorem is related to Remark 6.33 and to Theorem 6.24. As the reader can see, we have used Theorem 6.24, proved earlier by Grzegorek and Ryll-Nardzewski in 1981, to show that the CH-dimension theorem and the MA-dimension theorem are not theorems in ZFC.

As every analytic space is an absolute measurable space, we have the question

QUESTION. In Theorem 6.26, can the requirement that X be an analytic space be replaced by the requirement that X be an absolute measurable space?

Topological dimension is a very nice and intuitive way to classify topological spaces. Dimension theory begins with the empty space being assigned the dimension -1. Extensions of this process has been successfully investigated in topology (see J. M. Aarts and T. Nishiura [**1**]), where the empty space has been replaced by various classes $\mathcal{P}$ of topologically invariant spaces. The class of absolute measurable spaces and the class of absolute null spaces are certainly good possibilities for extensions of dimension theory, especially transfinite dimension theory. Perhaps an interesting interplay between the set theoretic implications that have been discovered so far and transfinite dimension and its extensions can be found.

6.5.5. Nonmetrizable spaces. The main setting of the book has been separable metrizable spaces. Our purpose was to develop a theory of absolute measurable spaces. We were able to carry out this development in ZFC. The very early history of the subject seemed to have the need to apply the continuum hypothesis to establish the existence of an uncountable absolute null space. But this was shown not to be necessary for its existence in ZFC; the continuum hypothesis yielded the existence of absolute null spaces whose cardinalities were as large as possible, namely $\mathfrak{c}$. The cardinality of an absolute null space is intimately tied to the assumptions made about set axioms in addition to ZFC. We have used the continuum hypothesis, its negation and the Martin axiom as applied to the real line. Of course the continuum hypothesis and the Martin axiom apply to other topological spaces. Indeed, the definition of the Martin axiom uses compact Hausdorff spaces. Many advances in the topology of general spaces have been collected together in the late twentieth century under the name of set theoretic topology.

It is not obvious to the author that a reasonable theory of absolute measurable spaces and absolute null spaces can be developed for general topological spaces. Clearly, it is the lack of examples in the non-metrizable setting that hampers the project of extending the theory of absolute measurable spaces and absolute null spaces beyond separable metrizable spaces. Perhaps one should begin by considering topologies that are finer than the Euclidean topology on $\mathbb{R}^n$, especially those that are connected to the Lebesgue measure. One such topology is the density topology.[9] The density topology satisfies the countable chain condition, also. Notice that homeomorphisms of $\mathbb{R}^n$ with respect to the usual topology need not be a homeomorphism with respect to the density topology. A simple example is easily found. But a bi-Lipschitzian map of $\mathbb{R}^n$ onto $\mathbb{R}^n$ is a homeomorphism with respect to the density topology. Indeed, there is the following stronger theorem due to Z. Buczolich [**25**].

THEOREM 6.38. *If $f : S \to T$ is a bi-Lipschitzian map of a Lebesgue measurable set S onto a Lebesgue measurable set T of $\mathbb{R}^n$, then f maps the set of density points of S onto the set of density points of T and maps the set of dispersion points of S onto the set of dispersion points of T.*[10]

6.5.6. Other singular sets. There are many other singular sets that were not considered in the book. The main reason for excluding them was the metric dependence of their definitions. But another reason is that the notions of absolute measurable spaces and of universally measurable sets in a space lead naturally to classes of functions which can be investigated from the point of view of analysis without resorting to additional set theoretic axioms beyond ZFC as illustrated by Chapter 4. Their exclusion does not mean that they are not worthy of a systematic treatment from the basics of descriptive set theory, axiomatic set theory, metric topology, and analysis. Such a book has been left for others to write; good beginnings can be found in the literature–waiting to be organized into book form.

6.5.7. Challenge revisited. The preface begans with the challenge to investigate the role of the notions of absolute measurable space and of universally measurable set in analysis, topology and geometry in the context of separable metrizable space. The book has presented several ways in which this challenge has been met. There must be more – where does one find potential sources of possibilities? Perhaps they are hidden is the many mathematical handbooks[11] or mathematical history books or mathematical textbooks that have recently appeared – many such books have been cited in the bibliography.

[9] There is an extensive literature on the density topology: C. Goffman, C. J. Neugebauer and T. Nishiura [**61**], Oxtoby [**120**], F. D. Tall [**153**], S. Scheinberg [**135**], and J. Lukeš, J. Malý and L. Zajíček [**91**], to list a few. The last book contains many other citations.

[10] A point x of $\mathbb{R}^n$ is a density (dispersion) point of a Lebesgue measurable set S if the Lebesgue density of S at x is 1 (respectively, 0).

[11] An interesting one is the recent [**123**] by E. Pap.

To find new conjectures, examples, counterexamples, positive results, negative results – all are part of the challenge.

Exercises

6.1. Prove that the statement "there exists an absolute null space X such that $\operatorname{card}(X) = \mathsf{non}\text{-}\mathbb{L}$" is equivalent to Theorem 6.6 on page 161.

6.2. Find necessary and sufficient conditions on a cardinal number κ to be such that there is a subset X of $\mathbb{R}$ with the property that $\lambda^*(X) > 0$ and $\operatorname{card}(X) = \kappa$.

6.3. Verify, as requested of the reader on page 163, that the relations $\prec$ and $\leqslant_*$ are the same.

6.4. Prove Corollary 6.31 on page 173.

Appendix A

Preliminary material

Only separable metrizable spaces will be considered. Appendix A will be used to gather various notions and facts that are found in well-known reference books (for example, K. Kuratowski [**85**]) with the goal of setting consistent notation and easing the citing of facts. Also, the final Section A.7 will be used to present a proof, which includes a strengthening due to R. B. Darst [**37**], of a theorem of R. Purves [**129**].

A.1. Complete metric spaces

A metric that yields the topology for a metrizable space need not be complete. But this metric space can be densely metrically embedded into another metric space that is complete. A space that possesses a complete metric will be called *completely metrizable*. It is well-known that a G_δ subspace of a completely metrizable space is also completely metrizable (see J. M. Aarts and T. Nishiura [**1**, page 29]). Consequently,

THEOREM A.1. *A separable metrizable space is completely metrizable if and only if it is homeomorphic to a G_δ subset of a separable completely metrizable space.*

The collection of all separable metrizable spaces will be denoted by MET and the collection of all completely metrizable spaces in MET will be denoted by $\mathsf{MET}_{\mathrm{comp}}$.

A.1.1. Extension of a homeomorphism. In the development of absolute notions it is often useful to extend a homeomorphism between two subsets of completely metrizable spaces to a homeomorphism between some pair of G_δ subsets containing the original subsets. This is accomplished by means of the M. Lavrentieff theorem [**89**].

THEOREM A.2 (Lavrentieff). *Suppose that X and Y are complete metric spaces. Then every homeomorphism between subspaces A and B of X and Y, respectively, can be extended to a homeomorphism between G_δ sets of X and Y.*

In passing, we remark that Lavrentieff's theorem applies to arbitrary metric spaces (see Aarts and Nishiura [**1**, page 31]).

A.1.2. Base for the topology. We shall assume that the reader is familiar with the notions of base and subbase for the open sets of a topology. Useful properties for separable metrizable spaces are the existence of a countable base for the open sets, and the Lindeloff property, namely, every open cover of a subset has a countable subcover.

A.1.3. Borel set. Let X be a separable metrizable space, not necessarily completely metrizable. A member of the smallest σ-algebra that contains all open subsets of X (equivalently, all closed sets) is called a *Borel set* of X and this σ-algebra will be denoted by $\mathfrak{B}(X)$. The collection of all open subsets of X is denoted by G_0 and the collection of all closed subsets of X is denoted by F_0. The collection G_0 is closed under the operation of taking countable unions and the collection F_0 is closed under the operation of taking countable intersections. By transfinitely repeating the operations of taking countable intersections or countable unions, one defines the collections G_α and F_α for each ordinal number α less than the first uncountable ordinal number ω_1.

A.1.4. Analytic space. We shall use the following definition of analytic sets in a separable metrizable space X. We begin with the product space $\mathbb{N}^{\mathbb{N}}$, where $\mathbb{N}$ is the set of natural numbers. It is well-known that this product space is homeomorphic to the space $\mathcal{N}$ of all irrational numbers between 0 and 1. A subset of X is called *analytic* if it is the continuous image of the space $\mathcal{N}$. (See H. Federer [**55**, page 65] or [**85**, page 478].) Motivated by this definition, we define a separable metrizable space to be an *analytic space* if it is the continuous image of the space $\mathcal{N}$ (for convenience, the empty space also will be considered an analytic space).

THEOREM A.3. *In a separable completely metrizable space, a nonempty subset is a Borel set if and only if it is an injective, continuous image of some G_δ subset of $\mathcal{N}$. Hence a Borel subspace of a separable completely metrizable space is an analytic space.*

The reader is referred to [**85**, Remark 1, page 488] for this theorem. A useful property of analytic sets is the following ([**85**, Theorem 0, page 479]).

THEOREM A.4. *Every uncountable analytic set in a separable metrizable space contains a topological copy of the classical Cantor ternary set.*

The collection of all analytic spaces will be denoted by ANALYTIC. A word of warning is appropriate at this juncture. Analytic sets are associated with sieves of the Suslin operations $(\mathcal{A})$ which can be applied also to noncomplete separable metric spaces. The class ANALYTIC may possibly not contain such subspaces of non completely metrizable spaces.

A.1.5. Co-analytic space. Associated with the collection of analytic spaces is the collection of spaces that are the complements, relative to completely metrizable spaces, of the analytic spaces. That is, a space X is *co-analytic* if it is homeomorphic to $Y \setminus A$ for some separable completely metrizable space Y and some analytic subspace A

of Y. It is well-known that not every co-analytic space is analytic. The collection of all co-analytic spaces will be denoted by CO-ANALYTIC.

A straightforward application of Lavrentieff's theorem will show that X *is in* CO-ANALYTIC *if and only if* $Y \setminus X$ *is in* ANALYTIC *for every separable completely metrizable space* Y *that contains* X. Consequently, the class CO-ANALYTIC is invariant under homeomorphisms.

Of importance to us are two theorems concerning co-analytic spaces. Let us first introduce some terminology concerning sieves for analytic spaces A. Using the above definition for a co-analytic space X, we let $X = Y \setminus A$, where Y is a separable completely metrizable space and A is an analytic subset of Y. When A is sieved by a sieve W composed of closed sets (or, more generally, of Borel sets) there is a collection $\{A_\alpha : \alpha < \omega_1\}$ of mutually disjoint subsets of Y such that

$$X = Y \setminus A = \bigcup_{\alpha < \omega_1} A_\alpha, \tag{A.1}$$

where the sets A_α are called *constituents* of the set X determined by the sieve W.

The following theorems can be found in [**85**, pages 500–502].

THEOREM A.5. *The constituents* A_α *are Borel sets of* Y.

It is asserted and proved in [**85**, page 502] that, in uncountable, separable, completely metrizable spaces, there are co-analytic subspaces such that the Borel classes of their constituents A_α, $\alpha < \omega_1$, are unbounded.

The next theorem is called the covering theorem and is due to N. Lusin.

THEOREM A.6 (Lusin). *In equation* (A.1) *for a co-analytic space* X, *let* E *be an analytic subspace of* Y *such that*

$$E \subset X = Y \setminus A = \bigcup_{\alpha < \omega_1} A_\alpha. \tag{A.2}$$

Then there exists an index α_0 *less than* ω_1 *such that*

$$E \subset \bigcup_{\alpha < \alpha_0} A_\alpha. \tag{A.3}$$

An immediate corollary of the last theorem is

COROLLARY A.7. *If a separable metrizable space is both analytic and co-analytic, then it is a Borel set in every completely metrizable extension of it.*

The next theorem is often useful (see [**85**, page 485] for a proof that does not rely on the above Lusin theorem).

THEOREM A.8. *If* A *and* B *are disjoint analytic sets contained in a separable metrizable space* X, *then there is a Borel set* E *in* X *such that* $A \subset E$ *and* $B \subset X \setminus E$.

A.1.6. Absolute Borel space. Some subspace properties of Borel sets are easily shown. For example, if X is a separable metrizable space and Y is a subspace of X, then $B \cap Y$ is a Borel subset of Y whenever B is a Borel subset of X. Conversely, a subset A of Y is a Borel set of Y if there is a Borel subset B of X such that $A = B \cap X$. Hence, if Y is a Borel set of X, then every Borel subset of the subspace Y is also a Borel subset of X.

The modifier "absolute" in the title of this section concerns the process of topological embedding.[1] A space X is said to be an *absolute Borel space* if, for every separable metrizable space Y, all topological copies of X in Y are members of $\mathfrak{B}(Y)$. The class of all absolute Borel spaces will be denoted by abBOR. There is the following theorem.

THEOREM A.9. *A separable metrizable space X is an absolute Borel space if and only if X is a Borel set in a completely metrizable extension of it, whence X is homeomorphic to a Borel subset of some completely metrizable space.*

The theorem is a consequence of Lavrentieff's theorem. Observe that our definition of analytic spaces has built into it the absolute property.

A very useful theorem (see [**85**, Corollary 1c, page 450]) concerning uncountable absolute Borel spaces is

THEOREM A.10. *A separable metrizable space X is an uncountable absolute Borel space if and only if there exists a continuous injection $f : \mathcal{N} \to X$ such that $X \setminus f[\mathcal{N}]$ is countable. Moreover, the inverse bijection from $f[\mathcal{N}]$ onto $\mathcal{N}$ is Borel measurable.*

Absolute G_0 (that is, absolute open) spaces can be defined in a similar manner. It is quite clear that the empty space is the only such space. Absolute F_0 (or, absolute closed) spaces are precisely the compact spaces; absolute G_1 (or, absolute G_δ) spaces are the separable completely metrizable spaces; and, the absolute F_1 (or absolute F_σ) spaces are precisely the σ-compact spaces. For a discussion of the above in arbitrary metrizable spaces see Aarts and Nishiura [**1**, page 114].

The collection of all spaces that are homeomorphic to some G_δ subset of some completely metrizable space will be denoted by $\mathsf{ab}\,\mathsf{G}_\delta$. From Theorem A.1 we have that $\mathsf{ab}\,\mathsf{G}_\delta$ is precisely the collection $\mathsf{MET}_{\mathrm{comp}}$ of all separable completely metrizable spaces.

A.2. Borel measurable maps

Let X and Y be separable metrizable spaces. A map $f : X \to Y$ is said to be *Borel measurable* (often abbreviated as *B-measurable*) if $f^{-1}[U] \in \mathfrak{B}(X)$ for every open set U of Y. The above condition is equivalent to the condition $f^{-1}[B] \in \mathfrak{B}(X)$ if $B \in \mathfrak{B}(Y)$. We have the following useful theorem [**85**, Theorem 1, page 384].

THEOREM A.11. *Let X and Y be separable metrizable spaces. If $f : X \to Y$ is a Borel measurable map, then* $\operatorname{graph}(f) \in \mathfrak{B}(X \times Y)$.

COROLLARY A.12. *If $f : X \to Y$ is a Borel measurable map from an absolute Borel space X into a separable metrizable space Y, then* $\operatorname{graph}(f)$ *is an absolute Borel space.*

[1] A *topological embedding* of a space X into a space Y is a mapping $e : X \to Y$ such that $e : X \to e[X]$ is a homeomorphism. The image $e[X]$ is called a *topological copy* of X.

A.2.1. Invariance of analytic spaces. The following invariance theorem is stated in [**85**, page 478] and proved in Federer [**55**, Section 2.2.14, page 70].

THEOREM A.13. *Let $f : X \to Y$ be a Borel measurable map from an analytic space X into a separable metrizable space Y. Then the image $f[X]$ is an analytic space.*

The proof given in [**55**] applies since the analytic space X is the continuous image of the completely metrizable space $\mathcal{N}$ and compositions of Borel measurable maps are again Borel measurable. Concerning the graph of mappings, we infer the following from [**85**, Theorem 2, page 489].

THEOREM A.14. *Let $f : X \to Y$ be a map, where X and Y are separable metrizable spaces. If* $\operatorname{graph}(f)$ *is an analytic space, then X is an analytic space and f is Borel measurable.*

A.2.2. Invariance of absolute Borel spaces. As absolute Borel spaces are analytic spaces, Theorem A.13 yields that $f[X]$ is analytic whenever f is a Borel measurable map and X is an absolute Borel space. A sharper result holds when the map is also injective.

THEOREM A.15. *Let $f : X \to Y$ be a Borel measurable injection of an absolute Borel space X into a separable metrizable space Y. Then the image $f[X]$ is an absolute Borel space.*

For a proof see [**85**, Theorem 1, page 489]. A consequence of this theorem is

COROLLARY A.16. *Let $f : X \to Y$ be a map, where X and Y be separable metrizable spaces. If* $\operatorname{graph}(f)$ *is an absolute Borel space, then X is an absolute Borel space and f is Borel measurable.*

Let X and Y be separable metrizable spaces. A bijection $f : X \to Y$ is said to be a $\mathfrak{B}$-*homeomorphism*[2] if both f and f^{-1} are Borel measurable. We infer from [**85**, Theorem 2, page 450] the following.

THEOREM A.17. *If X and Y are uncountable absolute Borel spaces, then there exists a $\mathfrak{B}$-homeomorphism of X onto Y.*

It is known in topology that the natural projection of an open set of a product space into a factor space is an open set of that factor space. That is, the projection map is an open-map – a map that sends open sets to open sets. Also, the projection map of a product of a space with a compact space onto the first factor space is a closed-map – a map that sends closed sets to closed sets.[3] Analogous to the definitions of open-maps and closed-maps, we define $\mathfrak{B}$-maps as follows.

[2] The maps defined here also have been called *generalized homeomorphisms*. See, for example, the book [**85**] by Kuratowski.

[3] The definitions of open-map and closed-map in topology also include the requirement that the maps be continuous. In [**85**], open continuous maps are called *bicontinuous*.

DEFINITION A.18. *Let X and Y be separable metrizable spaces. A Borel measurable map $f : X \to Y$ is a $\mathfrak{B}$-map*[4] *iff $f[B] \in \mathfrak{B}(Y)$ whenever $B \in \mathfrak{B}(X)$.*

Obviously the composition of two $\mathfrak{B}$-maps is again a $\mathfrak{B}$-map. With this definition Theorem A.15 can be restated as

THEOREM A.19. *If $f : X \to Y$ is a Borel measurable injective map into a separable metrizable space, then it is a $\mathfrak{B}$-map whenever X is an absolute Borel space. Consequently, if $f : X \to Y$ is a Borel measurable bijection, then f is a $\mathfrak{B}$-homeomorphism whenever X is an absolute Borel space.*

Concerning the notion of $\mathfrak{B}$-maps, there are very nice theorems [**85**, Theorem 2, page 496 and Corollaries 2 and 5, page 498] that deal with countable-to-one maps. One of the theorems will be given here. The theorem uses the following notation.

NOTATION A.20. *Let $f : X \to Y$ be any map. The set of all points y of Y for which* $\mathrm{card}(f^{-1}[\{y\}]) > \aleph_0$ *is called the* set of uncountable order of f *and is denoted by $U(f)$. The set of all points y of Y for which* $0 < \mathrm{card}(f^{-1}[\{y\}]) \le \aleph_0$ *is called the* set of countable order of f *and is denoted by $D(f)$.*

The sets $U(f)$ and $D(f)$ have the following properties.

THEOREM A.21. *Let Y be a separable metrizable space.*

(1) *If $f : X \to Y$ is a Borel measurable map from an analytic space X, then the set of uncountable order $U(f)$ is an analytic space.*
(2) *If $f : X \to Y$ is a Borel measurable map from an absolute Borel space X, then the set of countable order $D(f)$ is a co-analytic space.*
(3) *If $f : X \to Y$ is a continuous surjection of a separable completely metrizable space X such that $Y = D(f)$, then Y is an absolute Borel space. Moreover, X is the union of a sequence B_n, $n = 1, 2, \dots,$ of Borel subsets of X such that $f|B_n$ is a homeomorphism for each n.*

R. Purves in [**129**] observed the following sufficient condition for a Borel measurable map to be a $\mathfrak{B}$-map.

THEOREM A.22. *Let $f : X \to Y$ be a Borel measurable map from an absolute Borel space into a separable metrizable space. If the set of uncountable order satisfies* $\mathrm{card}\big(U(f)\big) \le \aleph_0$, *then $f[A]$ is an absolute Borel space whenever $A \in \mathfrak{B}(X)$, whence f is a $\mathfrak{B}$-map.*

PROOF. If $\mathrm{card}(U(f)) \le \aleph_0$, then $f^{-1}\big[U(f)\big]$ is a Borel subset of X, whence $B = X \setminus f^{-1}\big[U(f)\big]$ is an absolute Borel space. Suppose that A is a Borel subset of X. If $A \cap B$ is a countable set, then $f[A]$ is a countable set; so assume that $C = A \cap B$ is uncountable. The graph of $f|C$ is an uncountable absolute Borel space. Hence there is a continuous injection $g : \mathcal{N} \to \mathrm{graph}(f|C)$ such that $\mathrm{graph}(f|C) \setminus g[\mathcal{N}]$ is

[4] In [**129**] Purves calls such maps *bimeasurable* with the extra requirement that X and $f[X]$ be absolute Borel spaces. It appears that this terminology has been used to preserve somewhat its analogy with the definition of bicontinuous found in [**85**].

countable. The composition $h = \pi g$, where π is the natural projection of graph$(f|C)$ onto $f[C]$, is a continuous map on $\mathcal{N}$ into $f[C]$ such that $U(h) = \emptyset$. By statement (3) of the previous theorem, we have that $\mathcal{N}$ is the countable union of absolute Borel spaces B_n such that $h|B_n$ is a homeomorphism for each n. Now, as each $h[B_n]$ is an absolute Borel space contained in $f[C] = \pi\big[\text{graph}(f|C) \setminus g[\mathcal{N}]\big] \cup \bigcup_{n=1}^{\infty} h[B_n]$, we have shown that $f[A] = f[C] \cup \big(f[A] \cap U(f)\big)$ is an absolute Borel space. □

Purves's converse and Darst's extension of this theorem will be proved in Section A.7. Darst's extension plays a critical role in Section 2.1.2 of Chapter 2.

A.2.3. Baire property. Let X be a separable metrizable space. A very nice property possessed by Borel subsets B of X is that there exists an open set G such that $B \setminus G$ and $G \setminus B$ are sets of the first category of X. Subsets of X with this property are said to have the *Baire property*. The collection of all subsets of X having the Baire property forms a σ-algebra of X. See [**85**, pages 87–88].

Let X and Y be separable metrizable spaces. Observe that the characteristic function χ_B of a Borel subset of X has the property that $\chi_B{}^{-1}[U]$ has the Baire property for every open set U of $\mathbb{R}$. A Borel measurable map $f : X \to Y$ also has the very nice property that $f^{-1}[U]$ has the Baire property whenever U is an open set of Y. Any map with this property is said to have the *Baire property*. The collection of all such maps from X into Y is closed under pointwise convergence. For a discussion of maps having the Baire property, see [**85**, pages 399–403]. There is the following important theorem [**85**, page 400].

THEOREM A.23. *Let X and Y be separable metrizable spaces. A map $f : X \to Y$ has the Baire property if and only if there is a set F of the first category in X such that $f|(X \setminus F)$ is continuous.*

A.3. Totally imperfect spaces

Totally imperfect spaces occur in a natural way in the study of singular sets (see J. B. Brown and G. V. Cox [**18**]).

DEFINITION A.24. *A separable metrizable space is said to be* totally imperfect *if it contains no nonempty, compact, perfect subsets (hence, contains no topological copies of the classical Cantor ternary set).*

The existence of such spaces is easily established by a transfinite construction. Let us state the well-known Bernstein theorem. (See, for example, [**85**, Theorem 1, page 514].)

THEOREM A.25 (Bernstein). *Each uncountable completely metrizable space is the union of two disjoint totally imperfect subsets each with the power of the continuum.*

A simple consequence of the theorem is that each of two disjoint totally imperfect sets whose union is the closed interval $[0, 1]$ is not Lebesgue measurable since they cannot contain topological copies of the classical Cantor set.

Let us close with a simple proposition whose proof will be left to the reader.

PROPOSITION A.26. *Subspaces of totally imperfect spaces are totally imperfect; finite products of totally imperfect spaces are totally imperfect; and, a separable metrizable space that is a countable union of totally imperfect, closed subspaces is totally imperfect.*

A.3.1. Inclusion properties. Observe the following inclusions for separable metrizable spaces.

$$\begin{aligned}\mathsf{MET} &\supset \mathsf{ANALYTIC} \cup \mathsf{CO\text{-}ANALYTIC} \\ &\supset \mathsf{ANALYTIC} \cap \mathsf{CO\text{-}ANALYTIC} = \mathsf{abBOR} \\ &\supset \mathsf{MET}_{\mathrm{comp}} = \mathsf{ab}\,\mathsf{G}_\delta .\end{aligned}$$

The second inclusion is Theorem A.3. It is well-known that the second and third inclusions are proper ones. Totally imperfect spaces show the first inclusion is proper by Theorem A.4.

A.4. Complete Borel measure spaces

We shall assume that the reader knows the definition of a countably additive measure μ on an abstract set X, the definition of a complete measure, and the definition of a σ-finite measure μ (where μ is a nonnegative, extended-real-valued function defined on a σ-algebra $\mathfrak{M}(X,\mu)$ of subsets of X). A subset M of X is called *μ-measurable* if $M \in \mathfrak{M}(X,\mu)$. We denote a measure space by $\mathrm{M}(X,\mu)$ and the domain of the measure function μ by $\mathfrak{M}(X,\mu)$. As $\mathfrak{M}(X,\mu)$ is important to our discussions, we shall refer to the *measure space* $\mathrm{M}(X,\mu)$ as the triple $\big(X,\mu,\mathfrak{M}(X,\mu)\big)$.

Again we emphasize that all topological spaces are to be separable and metrizable. On such a space X, a measure space $\mathrm{M}(X,\mu)$ is called a *Borel measure space* (often called *regular Borel measure space*)

(1) if the σ-algebra $\mathfrak{M}(X,\mu)$ contains the collection $\mathfrak{B}(X)$ of all Borel sets, and
(2) if, for each M in $\mathfrak{M}(X,\mu)$, there exists a B in $\mathfrak{B}(X)$ such that $M \subset B$ and $\mu(B) = \mu(M)$.

Observe that if $\mathrm{M}(X,\mu)$ is a Borel measure space and M is a μ-measurable set with $\mu(M) < \infty$, then there are Borel sets A and B such that $A \subset M \subset B$ and $\mu(A) = \mu(M) = \mu(B)$ (equivalently, $\mu(B \setminus A) = 0$). Hence, if $\mathrm{M}(X,\mu)$ is a σ-finite measure space and M is a μ-measurable set, then there are Borel sets A and B such that $A \subset M \subset B$ and $\mu(B \setminus A) = 0$.

The *null collection* $\{Z \in \mathfrak{M}(X,\mu) : \mu(Z) = 0\}$ of a measure space $\mathrm{M}(X,\mu)$ will be denoted by $\mathfrak{N}(X,\mu)$. Clearly, $\mathfrak{N}(X,\mu)$ is a σ-ideal whenever the measure space is complete. The collection of all complete Borel measure spaces will be denoted by $\mathrm{MEAS}_{\mathrm{comp}}$.

A Borel measure space $\mathrm{M}(X,\mu)$ is *continuous*[5] if $\mu(\{x\})=0$ whenever $x\in X$. In general, the set $\mathcal{D}(X,\mu)=\{x\in X:\ \mu(\{x\})>0\}$ is not empty. Of particular interest will be those Borel measure spaces for which $\mathrm{card}(\mathcal{D}(X,\mu))\le\aleph_0$. Clearly σ-finite Borel measure spaces $\mathrm{M}(X,\mu)$ satisfy $\mathrm{card}(\mathcal{D}(X,\mu))\le\aleph_0$. The collection of all σ-finite Borel measure spaces will be denoted by $\mathrm{MEAS}_{\mathrm{sigma}}$, and the subcollection of all such measure spaces with $\mu(X)<\infty$ will be denoted by $\mathrm{MEAS}_{\mathrm{finite}}$. Observe that all Hausdorff measures[6] H_k on $\mathbb{R}^n$ with $0\le k<n$ are not σ-finite, and $\mathcal{D}(\mathbb{R}^n,\mathsf{H}_0)=\mathbb{R}^n$. As the book deals mainly with complete Borel measure spaces that are either σ-finite or finite we denote the collections of all such measure spaces by

$$\mathrm{MEAS}=\mathrm{MEAS}_{\mathrm{sigma}}\cap\mathrm{MEAS}_{\mathrm{comp}}, \tag{A.4}$$

and

$$\mathrm{MEAS}^{\mathrm{finite}}=\mathrm{MEAS}_{\mathrm{finite}}\cap\mathrm{MEAS}_{\mathrm{comp}}. \tag{A.5}$$

The *topological support* of a Borel measure space $\mathrm{M}(X,\mu)$ is the smallest closed set F, denoted by $\mathrm{support}(\mu)$, for which $\mu(X\setminus F)=0$.

A.4.1. G_δ hull and F_σ kernel. If a Borel measure space $\mathrm{M}(X,\mu)$ is complete and σ-finite, then the condition (2) above implies (see Exercise A.1)

(3) $M\in\mathfrak{M}(X,\mu)$ if and only if there exist members A and B of $\mathfrak{B}(X)$ such that $A\subset M\subset B$ and $\mu(B\setminus A)=0$.

In the light of this, it will be convenient to introduce the following terminology: a Borel set H of a space X such that $W\subset H$ is called a *Borel hull* of W in the space X; and, a Borel set K of a space X such that $K\subset W$ is called a *Borel kernel* of W in the space X. Consequently, if a Borel measure space $\mathrm{M}(X,\mu)$ is in MEAS, then a set M is μ-measurable if and only if there is a Borel kernel A and a Borel hull B of M such that $\mu(B\setminus A)=0$. The last requirement can be strengthened to use G_δ hulls and F_σ kernels whenever $\mathrm{M}(X,\mu)$ is in $\mathrm{MEAS}^{\mathrm{finite}}$.

THEOREM A.27. *Let X be a separable metrizable space and $\mathrm{M}(X,\mu)$ be a complete, finite Borel measure space. If M is a Borel subset of X, then there exists an F_σ kernel A and there exists a G_δ hull B of M such that $\mu(B\setminus A)=0$. Consequently, each μ-measurable set M also has an F_σ kernel A and a G_δ hull B such that $\mu(B\setminus A)=0$.*

PROOF. Consider the collection $\mathfrak{M}$ of all subsets M of X such that there exists an F_σ kernel A and a G_δ hull B of M such that $\mu(B\setminus A)=0$. Clearly, all closed sets and all open sets belong to $\mathfrak{M}$. It is obvious that the complement of every member of $\mathfrak{M}$ is also a member of $\mathfrak{M}$. Let us prove that the collection is closed under countable union. Let $M=\bigcup_{i=1}^{\infty}M_i$, where M_i, $i=1,2,\ldots$, is a sequence of members of $\mathfrak{M}$. For each M_i, there is a G_δ hull B_i and an F_σ kernel A_i such that $\mu(B_i\setminus A_i)=0$. Let $\varepsilon>0$. As

[5] In some books a point x for which $\mu(\{x\})>0$ is called an *atom*, whence a continuous measure is sometimes referred to as a *nonatomic measure*. Another name in the literature for these measures is *diffused measure*.

[6] For Hausdorff measures see Chapter 5 and Appendix D.

$\mu(X) < \infty$, there is an open set U_i that contains B_i with $\mu(U_i \setminus B_i) < \varepsilon\, 2^{-i}$ for each i. Hence $U = \bigcup_{i=1}^{\infty} U_i$ satisfies $\mu\big(U \setminus \bigcup_{i=1}^{\infty} B_i\big) < \varepsilon$ and $U \supset \bigcup_{i=1}^{\infty} B_i \supset M$. Since ε is an arbitrary positive number, there is a G_δ hull B of $\bigcup_{i=1}^{\infty} B_i$ such that $\mu\big(B \setminus \bigcup_{i=1}^{\infty} B_i\big) = 0$. Obviously $A = \bigcup_{i=1}^{\infty} A_i$ is an F_σ kernel of M. Hence B is a G_δ hull of M such that $\mu(B \setminus A) \leq \mu\big(B \setminus \bigcup_{i=1}^{\infty} B_i\big) + \mu\big(\bigcup_{i=1}^{\infty} B_i \setminus A\big) \leq \sum_{i=1}^{\infty} \mu(B_i \setminus A_i) = 0$. We have $\mu(B \setminus A) = 0$, whence $M \in \mathfrak{M}$. It now follows that $\mathfrak{B}(X) \subset \mathfrak{M}$ and that $\mathfrak{M}$ is the completion of $\mathrm{M}(X, \mu)$. Since $\mathrm{M}(X, \mu)$ is complete, the theorem is proved. □

Remark A.28. A simple trick will permit us to extend the last theorem to the larger collection MEAS of all complete, σ-finite Borel measure spaces. Indeed, for a σ-finite $\mathrm{M}(X, \nu)$, let X_n, $n = 1, 2, \ldots$, be a countable collection of ν-measurable subsets of X such that $\nu(X_n) < \infty$ for each n and $X = \bigcup_{n=1}^{\infty} X_n$. For each M in $\mathfrak{M}(X, \nu)$ define

$$\mu(M) = \sum_{n=1}^{\infty} \frac{1}{2^n(1+\nu(X_n))}\, \nu(X_n \cap M).$$

It is easily seen that μ is a finite measure on the σ-algebra $\mathfrak{M}(X, \nu)$ and that $\nu(N) = 0$ if and only if $\mu(N) = 0$. By the previous theorem, for each M in $\mathfrak{M}(X, \nu)$, there is an F_σ kernel A and a G_δ hull B of M such that $\mu(B \setminus A) = 0$, whence $\nu(B \setminus A) = 0$.

It is well-known that topological copies M of analytic spaces are μ-measurable in every complete, σ-finite Borel measure space $\mathrm{M}(X, \mu)$, [**55**, page 69]. Hence there is an F_σ kernel A of M such that $\mu(A) = \mu(M)$. The next lemma is a stronger statement. It is a consequence of a more general theorem by M. Sion [**143**, Theorem 4.2, page 774]. We shall give the proof by Sion adapted to the context of our book – it does not assume that analytic spaces satisfy the property stated at the start of the paragraph. This lemma will permit a second proof of that property in the next section. Recall that $\mathcal{N}$ is the collection of all irrational numbers in $I = [0, 1]$.

Lemma A.29. *Let $\mathrm{M}(X, \mu)$ be a complete, σ-finite Borel measure space. If $g \colon \mathcal{N} \to X$ is a continuous surjection, then there exists a σ-compact kernel A of X such that $\mu(X \setminus A) = 0$.*

Proof. By the previous remark there is no loss in assuming the measure space is complete and finite. Let $\eta < \mu(X)$. Let $F_{i,j}$ be compact subsets of I such that $\mathcal{N} = \bigcap_{i=1}^{\infty} \bigcup_{j=1}^{\infty} F_{i,j}$. Obviously, for each i, we have $\mathcal{N} = \bigcup_{j=1}^{\infty} \big(\mathcal{N} \cap F_{i,j}\big)$. Hence, for $i = 1$, there is an integer j_1 such that $\mu\big(g[\mathcal{N} \cap \bigcup_{j=1}^{j_1} F_{1,j}]\big) > \eta$. Let n be a positive integer and suppose that there are integers j_i such that $\mu\big(g[\mathcal{N} \cap \bigcap_{i=1}^{n} \bigcup_{j=1}^{j_i} F_{i,j}]\big) > \eta$. Then there is an integer j_{n+1} such that

$$\mu\big(g[(\mathcal{N} \cap \bigcap_{i=1}^{n} \bigcup_{j=1}^{j_i} F_{i,j}) \cap \bigcup_{j=1}^{j_{n+1}} F_{n+1,j}]\big) > \eta.$$

For each n, the set $C_n = \bigcap_{i=1}^{n} \bigcup_{j=1}^{j_i} F_{i,j}$ is a compact subset of I and $C_n \supset C_{n+1}$. Let $C = \bigcap_{n=1}^{\infty} C_n$. Clearly C is a compact subset of $\mathcal{N}$, whence $K = g[C]$ is a compact subset of X. Let us show that $\mu(K) \geq \eta$. To this end, let U be a neighborhood of K. Then $V = g^{-1}[U]$ is a neighborhood of C in the topological space $\mathcal{N}$. Let W be an open set in I such that $W \cap \mathcal{N} = V$. As $C = \bigcap_{n=1}^{\infty} C_n$ and each C_n is compact and $C_{n+1} \subset C_n$, there is an n_0 such that $C_{n_0} \subset W$. Hence $C \subset \mathcal{N} \cap C_{n_0} \subset V$, whence $g[C] \subset g[\mathcal{N} \cap C_{n_0}] \subset U$. We now have $g[C] = \bigcap_{n=1}^{\infty} g[\mathcal{N} \cap C_n]$ and $\mu\big(g[\mathcal{N} \cap C_n]\big) > \eta$ for each n. Hence $\mu(K) \geq \eta$. The lemma follows easily. □

A.4.2. Subspaces. Corresponding to a μ-measurable set Y of a Borel measure space $\mathrm{M}(X, \mu)$ are two natural Borel measure spaces. The first one is the *restriction* measure space $\mathrm{M}(Y, \mu|Y)$, where the σ-algebra $\mathfrak{M}(Y, \mu|Y)$ is defined to be $\{E \cap Y : E \in \mathfrak{M}(X, \mu)\}$ and the measure $\mu|Y$ on Y is given by $\big(\mu|Y\big)(M) = \mu(M)$ whenever $M \in \mathfrak{M}(Y, \mu|Y)$. The second is the *limited* Borel measure space $\mathrm{M}(X, \mu \llcorner Y)$ where the measure $\mu \llcorner Y$ on the space X is given by

$$\big(\mu \llcorner Y\big)(M) = \mu(M \cap Y) \text{ whenever } M \in \mathfrak{M}(X, \mu).$$

Observe that the limited Borel measure $\mu \llcorner Y$ on X need not be complete even when μ is complete; an additional step is needed to make the limited Borel measure space complete. For convenience, we shall assume that the completion step has been taken; that is, for a complete, σ-finite Borel measure space $\mathrm{M}(X, \mu)$,

$$\mathfrak{M}(X, \mu \llcorner Y) = \{M : \mu\big((B \setminus A) \cap Y\big) = 0 \text{ for some } A \text{ and } B \text{ in } \mathfrak{B}(X) \text{ with } A \subset M \subset B\}.$$

A.4.3. Outer measures. Often it is convenient to define the restriction of a measure μ to arbitrary subsets Y of the space X. To accomplish this we define the outer measure of arbitrary subsets of X.

Definition A.30. *Let $\mathrm{M}(X, \mu)$ be a complete, σ-finite Borel measure space and let W be a subset of X. The* outer measure $\mu^*(W)$ *of W is the extended-real number*

$$\mu^*(W) = \min\{\mu(B) : W \subset B \text{ and } B \in \mathfrak{B}(X)\}.$$

Lemma A.31. *Let $\mathrm{M}(X, \mu)$ be a finite Borel measure space and let Y be a subset of X. There is a complete Borel measure space $\mathrm{M}(Y, \nu)$ such that $\nu(B) = \mu^*(B)$ for every B in $\mathfrak{B}(Y)$, whence $\nu(M) = \mu^*(M)$ whenever $M \in \mathfrak{M}(Y, \nu)$.*

Proof. let $\mathfrak{M}$ be the collection of all subsets M of Y such that there is a Borel kernel A and a Borel hull B in the space Y such that $\mu^*(B \setminus A) = 0$. Obviously, $\mathfrak{B}(Y) \subset \mathfrak{M}$. It is clear that if $M \in \mathfrak{M}$, then $Y \setminus M \in \mathfrak{M}$. We need to show that $\mathfrak{M}$ is a σ-algebra for Y. Suppose that M_i, $i = 1, 2, \ldots$, is a sequence in $\mathfrak{M}$. Let A_i and B_i be a Borel kernel and a Borel hull of M_i in Y with $\mu^*(B_i \setminus A_i) = 0$ for each i. Then $A = \bigcup_{i=1}^{\infty} A_i$ and $B = \bigcup_{i=1}^{\infty} B_1$ are a Borel kernel and a Borel hull of $M = \bigcup_{i=1}^{\infty} M_i$ in Y. As $B \setminus A \subset \bigcup_{i=1}^{\infty}(B_i \setminus A_i)$, we have $\mu^*(B \setminus A) = 0$. Thereby $\mathfrak{M}$ is a σ-algebra for Y. For each M in $\mathfrak{M}$ let $\nu(M) = \mu^*(M)$. It remains to show that ν satisfies the conditions for a complete Borel measure on Y. It is sufficient to show that ν is a measure on $\mathfrak{B}(Y)$. Let B_i, $i = 1, 2, \ldots$, be a countable collection of mutually disjoint members of $\mathfrak{B}(Y)$. For each i let H_i be a member of $\mathfrak{B}(X)$ such that $\nu(B_i) = \mu(H_i)$ and $B_i = H_i \cap Y$. As the collection of B_i is mutually disjointed, we may further suppose that the collection of H_i is also mutually disjointed. Finally, let H be a Borel hull of $B = \bigcup_{i=1}^{\infty} B_i$ in the space X such that $\nu(B) = \mu(H)$. With very little effort it can be seen that $H = \bigcup_{i=1}^{\infty} H_i$ may be assumed. Consequently, $\nu(B) = \mu(H) = \sum_{i=1}^{\infty} \mu(H_i) = \sum_{i=1}^{\infty} \nu(B_i)$. It now follows that ν is a measure on $\mathfrak{B}(Y)$, and thereby on the σ-algebra $\mathfrak{M}$. That is, $\mathfrak{M}(X, \nu) = \mathfrak{M}$. □

The resulting measure whose existence is assured by the lemma will be denoted by $\mu|Y$ also.

We can now combine Lemma A.29 with the above lemma to prove that analytic spaces Y are μ-measurable for each complete, σ-finite Borel measure space $\mathrm{M}(X, \mu)$ with $Y \subset X$.

THEOREM A.32. *Let* $\mathrm{M}(X, \mu)$ *be a complete, σ-finite Borel measure space. If* $Y \in$ ANALYTIC *and* $Y \subset X$, *then* $Y \in \mathfrak{M}(X, \mu)$.

PROOF. Consider the restriction $\mathrm{M}(Y, \mu|Y)$. There exists a σ-compact subset A of Y such that $\big(\mu|Y\big)(Y \setminus A) = 0$. As $\mu^*(Y \setminus A) = \big(\mu|Y\big)(Y \setminus A) = 0$ and $\mathrm{M}(X, \mu)$ is complete, we have $Y \in \mathfrak{M}(X, \mu)$. □

A.4.4. Measures induced by measurable maps. For separable metrizable spaces X and Y, let $\mathrm{M}(X, \mu)$ be a Borel measure space and let $f : X \to Y$ be a Borel measurable map. The map f induces a measure $f_\# \mu$ on the space Y and the σ-algebra $\mathfrak{B}(Y)$ by the formula

$$f_\# \mu(B) = \mu(f^{-1}[B]) \quad \text{for} \quad B \in \mathfrak{B}(Y)$$

which has a completion $(Y, f_\# \mu, \mathfrak{M}(Y, f_\# \mu))$. That is, $M \in \mathfrak{M}(Y, f_\# \mu)$ if and only if there exist A and B in $\mathfrak{B}(Y)$ such that

$$A \subset M \subset B \quad \text{and} \quad \mu(f^{-1}[B \setminus A]) = 0.$$

We shall use the notation $f_\# \mathfrak{M}(X, \mu)$ to denote the induced σ-algebra $\mathfrak{M}(Y, f_\# \mu)$. Additionally, if $g : Y \to Z$ is Borel measurable, where Z is also a separable metrizable space, then the composition $gf : X \to Z$ satisfies

$$(gf)_\# \mu = g_\# f_\# \mu,$$

$$(gf)_\# \mathfrak{M}(X, \mu) = g_\# f_\# \mathfrak{M}(X, \mu).$$

If $\mathrm{M}(X, \mu)$ is a complete, finite Borel measure space, then $\mathrm{M}(Y, f_\# \mu)$ is also; clearly, the requirement of finite cannot be replaced with σ-finite as the constant map will show.

Due to the "if and only if" requirement in the definition of the σ-algebra $f_\# \mathfrak{M}(X, \mu)$ the Borel measure space $\mathrm{M}(X, \mathrm{id}_\# \mu)$ induced by the identity map $\mathrm{id} : X \to X$ is the measure completion of the Borel measure space $\mathrm{M}(X, \mu)$.

For subspaces X of separable metrizable spaces Y, the identity map $\mathrm{id} : X \to Y$ will provide a means of extending Borel measure spaces. Indeed, we have the following proposition.

Proposition A.33. *If X is a subspace of a separable metrizable space Y and* $\mathrm{M}(X, \mu)$ *is a complete, σ-finite Borel measure space, then there exists a complete, σ-finite Borel measure space* $\mathrm{M}(Y, \nu)$ *such that* $\big(\nu|X\big)(B) = \mu(B)$ *whenever* $B \in \mathfrak{B}(X)$.

Observe that X need not be a member of $\mathfrak{M}(Y, \nu)$ in the above proposition.

A simple application of measures induced by measurable maps is the following product lemma.

Lemma A.34. *Let* $\mathrm{M}(Y_1 \times Y_2, \mu)$ *be a complete, finite Borel measure space for a separable metrizable product space* $Y_1 \times Y_2$ *and let* P_1 *be the natural projection of* $Y_1 \times Y_2$ *onto* Y_1. *If* X_1 *is a subset of* Y_1 *such that* $X_1 \in \mathfrak{M}(Y_1, P_{1\#}\mu)$, *then* $X_1 \times Y_2 \in \mathfrak{M}(Y_1 \times Y_2, \mu)$.

Proof. Let A' and B' be Borel sets in Y_1 such that $A' \subset X_1 \subset B'$ and $P_{1\#}\mu(B' \backslash A') = 0$. Then $A = {P_1}^{-1}[A']$ and $B = {P_1}^{-1}[B']$ are Borel sets in $Y_1 \times Y_2$ such that $A \subset X_1 \times Y_2 \subset B$ and $\mu(B \setminus A) = 0$. As μ is a complete Borel measure, we have $X_1 \times Y_2 \in \mathfrak{M}(Y_1 \times Y_2, \mu)$ and the lemma is proved. □

Remark A.35. In the definition of the induced measure $f_{\#}\mu$ we have assumed that the map $f : X \to Y$ was Borel measurable. The Borel measurability restriction was made so that compositions of maps could be considered. If one were not concerned with compositions, then it is possible to consider μ-measurable maps, for $f^{-1}[B]$ will be a μ-measurable subset of X whenever $B \in \mathfrak{B}(Y)$. More precisely, let $\mathrm{M}(X, \mu)$ be a complete Borel measure space and let $f : X \to Y$ be a μ-measurable map of X into a separable metrizable space Y. Then the formula that defines $f_{\#}\mu$ on $\mathfrak{B}(Y)$ will result in a complete Borel measure space $\mathrm{M}(Y, f_{\#}\mu)$. This approach was used in Definition 1.3 on page 2 for complete, σ-finite Borel measure spaces $\mathrm{M}(X, \mu)$.

A.4.5. Embedding spaces. In Proposition A.33 a special embedding, namely the inclusion map, was used. We also have the following proposition for an arbitrary topological embedding whose proof is the same.

Proposition A.36. *Let* $f : X \to Y$ *be a topological embedding of a space X into a separable metrizable space Y and let* $\mathrm{M}(X, \mu)$ *be a Borel measure space. Then* $\nu = f_{\#}\mu$ *is a Borel measure on Y.*

As an application, let C denote the usual Cantor ternary set in the interval $[0, 1]$. The well-known Cantor function on C defines a continuous measure μ on C. It is also well-known that a nonempty, totally disconnected, perfect, compact metrizable space X is homeomorphic to the Cantor ternary set C. Suppose that Y is a separable metrizable space that contains a topological copy X of C and let $h : C \to X$ be a homeomorphism. Then one can easily see that there is a complete Borel measure ν on Y such that ν is the completion of $h_{\#}\mu$ and such that $\operatorname{support}(\nu) = X$. Clearly, the measure ν is continuous and $\nu(Y) = 1$. Let us summarize this discussion as a lemma.

LEMMA A.37. *Let X be nonempty, totally disconnected, perfect, compact subset of a separable metrizable space Y. Then there exists a continuous, complete Borel measure space* $\mathrm{M}(Y,\nu)$ *such that* $\nu(Y)=1$ *and* $\operatorname{support}(\nu)=X$.

Another application of the inclusion map is the proof that co-analytic spaces are μ-measurable for every complete, σ-finite Borel measure space $\mathrm{M}(X,\mu)$. The proof of the second statement of the theorem uses an argument given in [**142**] by Sierpiński and Szpilrajn.

THEOREM A.38. *Let* $\mathrm{M}(Y,\mu)$ *be a complete, σ-finite Borel measure space and let X be a co-analytic space. If $X \subset Y$, then $X \in \mathfrak{M}(Y,\mu)$. Moreover, X has a σ-compact kernel K such that $\mu(X \setminus K)=0$.*

PROOF. Let Y' be a separable completely metrizable space such that $Y \subset Y'$. From equation (A.1) of page 181, we have

$$X = Y' \setminus A = \bigcup\nolimits_{\alpha<\omega_1} A_\alpha,$$

where A is an analytic space. From Theorem A.5, Corollary A.7 and Theorem A.9, we have that the constituents A_α are absolute Borel spaces. Let $\mathrm{M}(Y,\mu) \in \mathrm{MEAS}$ and $f\colon Y \to Y'$ be the inclusion map. Then $\mathrm{M}(Y', f_\# \mu)$ is also in MEAS. As $Y' \setminus X$ is an analytic space we have that X is $f_\#\mu$-measurable, whence $\big(f_\#\mu\big)|X = \mu|X$. There is an M in $\mathfrak{B}(Y')$ such that $M \subset X$ and $f_\#\mu(X\setminus M)=0$. Since f is the inclusion map, we have $\mu(X \setminus M)=0$. By equations (A.2) and (A.3) of Theorem A.6, there is a ordinal number β such that $\beta < \omega_1$ and $M \subset \bigcup_{\alpha<\beta} A_\alpha$, whence $\mu\big(\bigcup_{\beta\le\alpha<\omega_1} A_\alpha\big)=0$. Each set A_α contains a σ-compact kernel K_α such that $\mu(A_\alpha \setminus K_\alpha)=0$. Hence $K = \bigcup_{\alpha<\beta} K_\alpha$ is a σ-compact kernel of X such that $\mu(X \setminus K)=0$. □

A.5. The sum of Borel measures

For a fixed separable metrizable space X let $\mathrm{M}(X,\mu)$ and $\mathrm{M}(X,\nu)$ be finite Borel measure spaces. Let us explain what is meant by the sum of the Borel measures μ and ν. For each Borel subset B of X the sum $\mu(B)+\nu(B)$ is well defined. Hence this function $\mu+\nu$ defined on the σ-algebra $\mathfrak{B}(X)$ is a Borel measure on X. The σ-algebra $\mathfrak{M}(X,\mu+\nu)$ of $(\mu+\nu)$-measurable sets consists of those subsets M for which there are Borel sets A and B such that $A \subset M \subset B$ and $\big(\mu+\nu\big)(B \setminus A)=0$. Clearly, the measure so defined is complete and is continuous whenever μ and ν are continuous. The same procedure will work for an infinite sequence μ_n, $n=1,2,\ldots$, of Borel measures on X. Clearly, $\mathfrak{N}(X, \sum_{n=1}^\infty \mu_n) = \bigcap_{n=1}^\infty \mathfrak{N}(X,\mu_n)$.

As we have mentioned earlier in Remark A.28, σ-finite Borel measure spaces are associated in a natural way to finite Borel measure spaces. Indeed, for a σ-finite Borel measure space $\mathrm{M}(X,\mu)$, let X_i, $i=1,2,\ldots$, be a countable collection of μ-measurable sets with $\mu(X_i)<\infty$ for all i such that $X=\bigcup_{i=1}^\infty X_i$. The finite-valued measure is defined as

$$\nu = \sum\nolimits_{i=1}^\infty \frac{1}{2^i(1+\mu(X_i))}\, \mu \llcorner X_i .$$

Clearly, the σ-algebras satisfy the equality $\mathfrak{M}(X,\mu) = \mathfrak{M}(X,\nu)$ and the collections of null sets satisfy $\mathfrak{N}(X,\mu) = \mathfrak{N}(X,\nu)$.

Continuous measures will play an important role in our investigations of sets of measure zero. Since we consider only complete, σ-finite Borel measure spaces, the set $\mathcal{D}(X,\mu)$ consisting of all singletons with positive μ-measure is countable. Consequently, the measure μ can be written uniquely as

$$\mu = \mu_0 + \sum_{x\in\mathcal{D}(X,\mu)} \mu \llcorner \{x\},$$

where μ_0 is a continuous measure.

A.5.1. Approximation by Borel measurable maps. We state the well-known theorem

THEOREM A.39. *Let X be a separable metrizable space. If μ is a complete, finite Borel measure on X and $f\colon X \to [0,1]$ is μ-measurable function on X, then there is a Borel class 2 function $g\colon X \to [0,1]$ such that $f = g$ μ-almost everywhere.*

Suppose that Y is a subspace of the Hilbert cube $[0,1]^{\mathbb{N}}$. Denote by Π_n the coordinate projection of the Hilbert cube onto the n-th coordinate space. If $f\colon X \to Y$ is a μ-measurable map, then $f_n = \Pi_n f$ is a μ-measurable map for each n. From this observation we infer that there is a Borel class 2 map $g\colon X \to Y$ such that $f = g$ μ-almost everywhere. Indeed, for each i, there is a Borel class 2 map $g_i\colon X \to [0,1]$ such that $\mu\big(\{x\colon f_i(x) \neq g_i(x)\}\big) = 0$. Clearly $g = (g_1, g_2, \dots)$ is a Borel class 2 map and $\{x\colon f(x) \neq g(x)\} \subset \bigcup_{i=1}^{\infty}\{x\colon f_i(x) \neq g_i(x)\}$.

A.6. Zahorski spaces

A subset of a separable metrizable space X is called a *Zahorski set in X* if it is the empty set or it is the union of a countable collection of topological copies of the Cantor set. A *Zahorski space* is a space such that it is a Zahorski set in itself.[7] Zahorski spaces appear in a very prominent way in many proofs.

As Zahorski spaces are absolute F_σ spaces, we have

PROPOSITION A.40. *Every Zahorski set E in X is a Borel set and hence $E \in \mathfrak{M}(X,\mu)$ for every Borel measure μ on X.*

There are some easily shown facts about Zahorski sets.

PROPOSITION A.41. *Let X be a separable metrizable space.*

(1) *If E_i, $i = 1,2,\dots$, is a sequence of Zahorski sets in X, then $\bigcup_{i=1}^{\infty} E_i$ is a Zahorski set in X.*
(2) *Every nonempty Zahorski set in X can be written as a countable union of disjoint topological copies of the Cantor set.*

[7] In his study **[158, 159]** of the derivative function, Z. Zahorski used a special class of subsets of the real line called sets of the type M_1. These sets are those that densely contain the union of countably many topological copies of the Cantor set.

(3) *If E is a Zahorski set in X and U is an open subset of X, then $E \cap U$ is a Zahorski set in X. If E is a nonempty Zahorski set in X, then there exists a closed subset F of X such that $E \cap F$ is not a Zahorski set in X.*
(4) *If E is a nonempty Zahorski set in X, then there is a Zahorski set E' in X such that E' is densely contained in E, and E' is a set of the first Baire category of X.*
(5) *If E is a nonempty Zahorski set in X, then there exists a continuous, complete, finite Borel measure μ on X such that $\mu(X \setminus E) = 0$ and such that $\mu(E \cap U) > 0$ whenever U is an open set in X with $E \cap U \neq \emptyset$.*

Proof. The proof relies on two well-known facts about the Cantor space $\{0, 1\}^{\mathbb{N}}$. First, every nonempty simultaneously closed and open subset of $\{0, 1\}^{\mathbb{N}}$ is homeomorphic to $\{0, 1\}^{\mathbb{N}}$. Hence, if A and B are simultaneously closed and open subsets of $\{0, 1\}^{\mathbb{N}}$ with $A \setminus B \neq \emptyset$, then $A \setminus B$ is homeomorphic to $\{0, 1\}^{\mathbb{N}}$. Second, the collection of all simultaneously closed and open subsets of $\{0, 1\}^{\mathbb{N}}$ is a base for the open sets of $\{0, 1\}^{\mathbb{N}}$. Hence every nonempty open subset of $\{0, 1\}^{\mathbb{N}}$ is the countable union of mutually disjoint simultaneously closed and open subsets of $\{0, 1\}^{\mathbb{N}}$.

The first statement is obvious. The second statement follows from the above two well-known facts.

To prove the third statement let U be an open subset of X and E be a Zahorski set in X. Then $E = \bigcup_{n=1}^{\infty} E_n$ where E_n is a topological copy of $\{0, 1\}^{\mathbb{N}}$ for every n. As $U \cap E_n$ is a Zahorski set for each n it follows that $E \cap U$ is a Zahorski set.

The fourth statement is a consequence of the fact that the set of non-end-points of the Cantor ternary set contains a dense Zahorski subset of the Cantor ternary set.

To prove the last statement let $E = \bigcup_{n=1}^{\infty} E_n$, where E_n is a topological copy of the Cantor ternary set. By Lemma A.37 there is a continuous, complete, finite Borel measure μ_n on X such that $\mu_n(X) = 1$ and support$(\mu_n) = E_n$. It is easy to see that the measure $\mu = \sum_{n=1}^{\infty} \frac{1}{2^n} \mu_n$ fulfills the requirements of statement (5). □

The above statement (5) leads to

Definition A.42. *Let E be a nonempty Zahorski set contained in a separable metrizable space X. A* Zahorski measure *determined by E is a continuous, complete, finite Borel measure on X such that $\mu(X \setminus E) = 0$ and $\mu(E \cap U) > 0$ whenever U is an open set in X with $E \cap U \neq \emptyset$.*

A.7. Purves's theorem

Let us turn our attention to Borel measurable maps that preserve Borel measurability, that is, $\mathfrak{B}$-maps (see Definition A.18). The images of absolute Borel spaces under Borel measurable maps are always analytic spaces. A sufficient condition was given in Theorem A.22 for the invariance of the class of absolute Borel spaces under Borel measurable maps. The R. Purves theorem [**129**] asserts the necessity of this condition. We state the theorem.

Theorem A.43 (Purves). *Let X be an absolute Borel space and let $f : X \to Y$ be a Borel measurable map of X into a separable metrizable space Y. Then a necessary* *and sufficient condition for f to be a $\mathfrak{B}$-map is* $\mathrm{card}(U(f)) \leq \aleph_0$, *where $U(f)$ is the set of uncountable order of f. Moreover, Y may be assumed to be the Hilbert cube.*

The proof of the necessity of the condition will be split into four subsections, namely, preliminaries and three reductions to special cases. The next definition will facilitate the reductions.

DEFINITION A.44. *For absolute Borel spaces X and X^* let f and f^* be mappings of X and X^* into separable metrizable spaces Y and Y^*, respectively. Then f^* is said to be a B-*successor *of f, denoted $f \succ_B f^*$, if there are injective $\mathfrak{B}$-maps $\Theta : X^* \to X$ and $\vartheta : f^*[X^*] \to f[X]$ such that*

$$\begin{array}{ccccc} X & \xrightarrow{f} & f[X] & \xrightarrow{\subset} & Y \\ \Theta\uparrow & & \uparrow\vartheta & & \\ X^* & \xrightarrow{f^*} & f^*[X^*] & \xrightarrow{\subset} & Y^* \end{array}$$

is a commutative diagram and $\vartheta^{-1} : \vartheta f^*[X^*] \to Y^*$ is a $\mathfrak{B}$-map.

Observe that if $f : X \to Y$ is a $\mathfrak{B}$-map then so is $f|B : B \to f[X]$ whenever B is a Borel subset of X. Hence the following transitivity statement is easily proved.

PROPOSITION A.45. *If $f \succ_B f^*$ and $f^* \succ_B f^{**}$, then $f \succ_B f^{**}$.*

Let us show that the B-successor relation preserves $\mathfrak{B}$-maps.

LEMMA A.46. *Let $f \succ_B f^*$, whence $f^*[X^*]$ is a Borel subset of Y^*. If f is a $\mathfrak{B}$-map, then f^* is a $\mathfrak{B}$-map.*

PROOF. Observe that the above commutative diagram yields $f\Theta[X^*]=\vartheta f^*[X^*]$. As $\vartheta^{-1} : \vartheta f^*[X^*] \to Y^*$ is a $\mathfrak{B}$-map, we have f^* is a $\mathfrak{B}$-map. □

In our applications of $f \succ_B f^*$, the maps f^* are surjections, that is $f^*[X^*] = Y^*$.

A.7.1. Preparatory lemmas.

A.7.1. Preparatory lemmas. We begin with some lemmas that will be used in the proof. The first lemma is due to Purves [**129**, Lemma 6, page 155].

LEMMA A.47. *For an analytic space S let $\pi : S \to Y$ be a continuous map into a separable metrizable space. If $\pi[S]$ is uncountable, then there is a compact subset K of S such that $\pi[K]$ is uncountable.*

PROOF. As S is an analytic space, there is a continuous surjection $f : \mathcal{N} \to S$. Hence the composition $g = \pi f : \mathcal{N} \to \pi[S]$ is continuous surjection. Let μ be a non-trivial, continuous, complete, finite Borel measure on the analytic space $\pi[S]$. By Lemma A.29 there is a compact set K_0 contained in $\mathcal{N}$ such that $\mu(g[K_0]) > 0$. Then $K = f[K_0]$ is compact and $\pi[K]$ is uncountable. □

We turn next to a selection lemma. The usual selection theorems result in a measurable selection. As a continuous selection is needed, we shall give a proof of a measurable selection lemma from which a continuous one will result.

LEMMA A.48. *Let V and W be compact metrizable spaces and let K be a closed subset of $V \times W$ such that $\pi_V[K] = V$ and $\pi_W[K] = W$, where π_V and π_W are the natural projection maps. Then there is a Borel measurable map $s\colon V \to W$ such that $s(v) \in \pi_W\big[(\pi_V|K)^{-1}[\{v\}]\big]$ for every v in V. Moreover, if V is uncountable, then there is a nonempty, perfect, compact subset V_0 of V such that $s|V_0$ is continuous.*

PROOF. The proof follows the one in L. Cesari [**28**, 8.2.i, page 275]. It is well-known that there exists a continuous surjection $\varphi\colon \mathcal{C} \to W$, where $\mathcal{C}$ is the Cantor ternary set. Clearly $h = \mathrm{id} \times \varphi\colon V \times \mathcal{C} \to V \times W$ is also a continuous surjection. Denote by $g\colon h^{-1}[K] \to V$ the natural projection $\pi_1\colon V \times \mathcal{C} \to V$ restricted to $h^{-1}[K]$. Then, for v in V, the set $F_v = \pi_2\big[g^{-1}[\{v\}]\big]$ is closed in $\mathcal{C}$, where π_2 is the natural projection of $V \times \mathcal{C}$ onto $\mathcal{C}$. For each v in V denote by $\sigma(v)$ the smallest real number in F_v. The function $\sigma\colon V \to \mathcal{C}$ is a lower semi-continuous real-valued function, whence Borel measurable. The composition $s = \varphi\sigma$ is the desired function in the lemma.

Suppose that V is uncountable. As graph(s) is an uncountable Borel set, there is a topological copy M of the Cantor ternary set $\mathcal{C}$ contained in graph(s). Let $V_0 = \pi_V[M]$. Clearly, graph$(s|V_0) = M$ and $\pi_V|M$ is the inverse function of $s|V_0$. Hence $s|V_0$ is continuous. □

The next lemma will use the hyperspace 2^W of nonempty compact subsets of a compact metrizable space W. The members of the hyperspace 2^W are precisely those subsets of W that are nonempty and closed. There is a compact metrizable topology on 2^W associated with the topology of W. Indeed, let ρ be a metric for W and H_ρ be the Hausdorff metric on 2^W. That is, if F_1 and F_2 are in 2^W, then

$$H_\rho(F_1, F_2) = \min\{r \in \mathbb{R} : r \geq \rho(x, F_1) \text{ and } r \geq \rho(x, F_2) \text{ whenever } x \in F_1 \cup F_2\}.$$

As W is compact, the topology of 2^W is independent of the choice of the metric ρ on W. It is easy to show that the set $\bigcup_{F \in \mathcal{F}} F$ is a closed subset of the space W whenever $\mathcal{F}$ is a closed subset of the hyperspace 2^W. The following lemma is proved in Kuratowski [**86**, Theorem 3, page 50].

LEMMA A.49. *Let $\mathcal{P}$ be the collection of all nonempty, perfect subsets of a compact metrizable space W. Then $\mathcal{P}$ is a G_δ subset of the hyperspace 2^W.*

We close this preparatory section with Purves's construction of a canonical homeomorphism of $\{0,1\}^{\mathbb{N}}$ onto each nonempty perfect subset P of $\{0,1\}^{\mathbb{N}}$. It will be convenient to choose a particular metric on $\{0,1\}^{\mathbb{N}}$, namely,

$$\rho(x, x') = \textstyle\sum_{n=1}^{\infty} \frac{|x_n - x'_n|}{n} 2^{-n}, \quad x, x' \in \{0,1\}^{\mathbb{N}}.$$

A nice feature of this metric is that $\rho(x, x') < \frac{1}{n} 2^{-n}$ if and only if $x_i = x'_i$ for $1 \leq i \leq n$.

Definition A.50. *Let P be a nonempty perfect subset of $\{0,1\}^{\mathbb{N}}$. A positive integer m is said to be a* free coordinate *of P if there are z and z' in P such that $z_m \neq z'_m$. Let $m(P) = \min\{m\colon m$ is a free coordinate of $P\}$.*

Let $\mathcal{I}$ be the set of all finite sequences $c = (c_1, c_2, \dots, c_m)$ of 0's and 1's. (The length m of c will be denoted by $l(c)$.) For each c in $\mathcal{I}$, the set

$$N(c) = \{x \in \{0,1\}^{\mathbb{N}} \colon x_i = c_i \quad \text{for} \quad 1 \leq i \leq l(c)\}$$

is a closed and open set in the space $\{0,1\}^{\mathbb{N}}$, and $N(d) \subset N(c)$ if and only if c is an initial finite sequence of d. Clearly, $\operatorname{diam} N(c) \leq \frac{2}{l(c)} 2^{-l(c)}$.

Observe that if P a nonempty perfect subset of $\{0,1\}^{\mathbb{N}}$ and z and z' are in P, then $z_i = z'_i$ for $1 \leq i < m(P)$. Hence there exist $c(P)$ and $c'(P)$ in $\mathcal{I}$ with $l(c(P)) = l(c'(P)) = m(P)$ such that their coordinates satisfy $c(P)_i = c'(P)_i$ for $1 \leq i < m(P)$, $c(P)_{m(P)} = 0$, $c'(P)_{m(P)} = 1$, and such that

(*a*) $P \cap N(c(P)) \neq \emptyset$ and $P \cap N(c'(P)) \neq \emptyset$,
(*b*) $P = \big(P \cap N(c(P))\big) \cup \big(P \cap N(c'(P))\big)$.

Moreover, $c(P)$ and $c'(P)$ are unique in the sense that if d and d' in $\mathcal{I}$ are such that $l(d) = l(d') = m$, $d_i = d'_i$ for $1 \leq i < m$, $d_m = 0$, $d'_m = 1$, and

(*a'*) $P \cap N(d) \neq \emptyset$ and $P \cap N(d') \neq \emptyset$,
(*b'*) $P = \big(P \cap N(d)\big) \cup \big(P \cap N(d')\big)$,

then $m = m(P)$, $d = c(P)$ and $d' = c'(P)$. To see the uniqueness, suppose that v is in $P \cap N(d)$ and v' is in $P \cap N(d')$. Then $v_i = d_i = d'_i = v'_i$ for $1 \leq i < m$ and $v_m = d_m = 0 \neq 1 = d'_m = v'_m$. Hence $m(P) \leq m$ and $c(P)_i = d_i$ for $1 \leq i < m(P)$. By condition (*b*), each of v and v' are in either $P \cap N(c(P))$ or $P \cap N(c'(P))$. In either case we have $v_j = v'_j = c(P)_j = c'(P)_j$ for $1 \leq j < m(P)$. So, $m \leq m(P)$ and thereby $m = m(P)$. It is now evident that $d = c(P)$ and $d' = c'(P)$. Hence we have a map $\varphi\colon P \mapsto (c(P), c'(P), m(P))$ of $\mathcal{P}$ into the subset $\overline{\mathcal{I}}$ of $\mathcal{I} \times \mathcal{I} \times \mathbb{N}$ such that (c, c', m) satisfies $l(c) = l(c') = m$, $c_i = c'_i$ for $1 \leq i < m$, $c_m = 0$ and $c'_m = 1$. Finally, with $\overline{\mathcal{I}}$ supplied with the discrete topology and $\mathcal{P}$ supplied with the Hausdorff metric, let us show that $\varphi\colon \mathcal{P} \to \overline{\mathcal{I}}$ is continuous at P. From the definition of the Hausdorff metric, the inequality $H(P', P) < \frac{1}{m(P)} 2^{-m(P)}$ implies that $P' \cap N(c(P))$ and $P' \cap N(c'(P))$ are not empty and that P' is a subset of $N(c(P)) \cup N(c'(P))$. The uniqueness of $m(P')$, $c(P')$ and $c'(P')$ gives $m(P') = m(P)$, $c(P') = c(P)$ and $c'(P') = c'(P)$ and thereby the continuity is proved. The following lemma is now easily shown.

Lemma A.51. *Let $P \in \mathcal{P}$, where $\mathcal{P}$ is the collection of all nonempty perfect subsets of $\{0,1\}^{\mathbb{N}}$ supplied with the Hausdorff metric H. With* (0) *and* (1) *in $\mathcal{I}$, there are unique $y[P,(0)]$ and $y[P,(1)]$ in $\mathcal{I}$ of length $m(P)$ such that their i-th coordinates are equal for $1 \leq i < m(P)$ and their last coordinates are respectively equal to* 0 *and* 1, *and such that*

(*a*) $P \cap N(y[P,(0)]) \neq \emptyset$ *and* $P \cap N(y[P,(1)]) \neq \emptyset$,
(*b*) $P = \big(P \cap N(y[P,(0)])\big) \cup \big(P \cap N(y[P,(1)])\big)$.

Moreover, if $H(P',P) < \frac{1}{m(P)} 2^{-m(P)}$ *for* P' *in* $\mathcal{P}$, *then* $m(P') = m(P)$, $y[P',(0)] = y[P,(0)]$ *and* $y[P',(1)] = y[P,(1)]$, *whence*

(c) $P' \cap N(y[P,(0)]) \neq \emptyset$ *and* $P' \cap N(y[P,(1)]) \neq \emptyset$,
(d) $P' = \big(P' \cap N(y[P,(0)])\big) \cup \big(P' \cap N(y[P,(1)])\big)$.

Let us summarize the features of the above lemma that are essential for our inductive procedure. Let d be in $\mathcal{I}$ with $l(d) = 1$ and let P be in $\mathcal{P}$. Then there is a unique $y[P,d]$ in $\mathcal{I}$ that satisfies the following properties.

(1.0) $l(y[P,d]) \geq 1$ whenever $l(d) = 1$.
(1.1) $N(y[P,d]) \cap N(y[P,d']) = \emptyset$ whenever $l(d) = l(d') = 1$ and $d \neq d'$.
(1.2) $N(y[P,d]) \subset N(y[P,d'])$ if and only if d' is an initial finite sequence of d with $l(d) = 1$.
(1.3) $P \cap N(y[P,d]) \neq \emptyset$ whenever $l(d) = 1$.
(1.4) $P \subset \bigcup_{l(d)=1} N(y[P,d])$.
(1.5) There is a positive number η_1 such that if P' is in $\mathcal{P}$ and $H(P',P) < \eta_1$ then $y[P',d] = y[P,d]$ whenever $l(d) = 1$.

Apply the lemma to each nonempty perfect sets $P \cap N\big(y[P,\tilde{d}]\big)$ with $l\big(\tilde{d}\big) = 1$. Then, for each d in $\mathcal{I}$ with $l(d) = 2$, there is a unique $y[P,d]$ in $\mathcal{I}$ that satisfies the following properties.

(2.0) $l(y[P,d]) \geq 2$ whenever $l(d) = 2$.
(2.1) $N(y[P,d]) \cap N(y[P,d']) = \emptyset$ whenever $l(d) = l(d') = 2$ and $d \neq d'$.
(2.2) $N(y[P,d]) \subset N(y[P,d'])$ if and only if d' is an initial finite sequence of d with $l(d) = 2$.
(2.3) $P \cap N(y[P,d]) \neq \emptyset$ whenever $l(d) = 2$.
(2.4) $P \subset \bigcup_{l(d)=2} N(y[P,d])$.
(2.5) There is a positive number η_2 such that if P' is in $\mathcal{P}$ and $H(P',P) < \eta_2$ then $y[P',d] = y[P,d]$ whenever $l(d) = 2$.

The following lemma is now evident.

Lemma A.52. *For each* P *in* $\mathcal{P}$ *and for each* d *in* $\mathcal{I}$ *there is a unique* $y[P,d]$ *in* $\mathcal{I}$ *such that*

(0) $l(y[P,d]) \geq l(d)$;
(1) $N(y[P,d]) \cap N(y[P,d']) = \emptyset$ *whenever* $d' \in \mathcal{I}$, $d \neq d'$ *and* $l(d) = l(d')$;
(2) $N(y[P,d]) \subset N(y[P,d'])$ *if and only if* d' *is an initial finite sequence of* d;
(3) $P \cap N(y[P,d]) \neq \emptyset$;
(4) $P \subset \bigcup_{l(d)=n} N(y[P,d])$ *for* $n \in \mathbb{N}$;
(5) *for each* n *in* $\mathbb{N}$ *there is a positive number* η_n *such that if* P' *is in* $\mathcal{P}$ *and* $H(P',P) < \eta_n$ *then* $y[P',d] = y[P,d]$ *whenever* $l(d) = n$.

For x in $\{0,1\}^{\mathbb{N}}$ let $d(n,x)$ be the initial finite sequence of x whose length is n. Then, for each P in $\mathcal{P}$, the sequence of compact sets

$$N\big(y[P,d(n,x)]\big), \quad n = 1,2,\ldots,$$

is nested and the diameters converge to 0. Hence there is a unique y in $\{0,1\}^{\mathbb{N}}$ contained in each member of the nested sequence. Since $P \cap N(y[P,d(n,x)]) \neq \emptyset$ for each n, we also have $y \in P$. Clearly, $y[P,d(n,x)]$ is an initial finite sequence of y for each n.

Definition A.53. *Define* $\psi\colon \{0,1\}^{\mathbb{N}} \times \mathcal{P} \to \{0,1\}^{\mathbb{N}} \times \mathcal{P}$ *to be the map given by* $(x,P) \mapsto (y,P)$, *where* y *is the above described unique member of* $P \cap \bigcap_{n=1}^{\infty} N(y[P,d(n,x)])$.

Lemma A.54. *The map* $\psi\colon \{0,1\}^{\mathbb{N}} \times \mathcal{P} \to \{0,1\}^{\mathbb{N}} \times \mathcal{P}$ *is a continuous injection. Moreover, the map* $\psi(\cdot,P)$ *is a homeomorphism of* $\{0,1\}^{\mathbb{N}} \times \{P\}$ *onto* $P \times \{P\}$ *for each* P *in* $\mathcal{P}$, *and*

$$\psi\big[\{0,1\}^{\mathbb{N}} \times \mathcal{P}\big] = \{(y,P)\colon y \in P \in \mathcal{P}\}$$

is an absolute Borel space.

Proof. Let (x,P) and $\varepsilon > 0$ be given. By condition (0) of the above lemma there is an n such that $\operatorname{diam} N(y[P,d(n,x)]) < \varepsilon/2$. Note that $d(n,x') = d(n,x)$ whenever $x' \in N(d(n,x))$. By condition (5) of the same lemma there is an η_n such that $y[P',d(n,x')] = y[P,d(n,x')]$ whenever P' is such that $H(P',P) < \eta_n$. Hence we have

$$N(y[P,d(n,x)]) = N(y[P,d(n,x')]) = N(y[P',d(n,x')]),$$

whence $P' \cap N(y[P',d(n,x')]) = P' \cap N(y[P,d(n,x)])$. Consequently, $\psi(x',P') \in N(y[P,d(n,x)]) \times \{P'\}$ whenever $H(P',P) < \eta_n$. Also if $H(P',P) < \min\{\varepsilon/2, \eta_n\}$, then $H(P',P) + \rho(\psi(x',P'), \psi(x,P)) < \varepsilon$ whenever $x' \in N(d(n,x))$. Thereby, the continuity of ψ at (x,P) has been established.

To show that ψ is injective, let $(x,P) \neq (x',P')$. If $P \neq P'$, then $\psi(x,P) \neq \psi(x',P')$ is obvious. So suppose that $P = P'$ and $x \neq x'$. There is an n such that $d(n,x) \neq d(n,x')$. By condition (1) of the above lemma we have $\psi(x,P) \neq \psi(x',P)$.

Let $P \in \mathcal{P}$. Then $\psi(\cdot,P)$ is a continuous injective function into $P \times \{P\}$. Conditions (2) and (4) of the previous lemma yields the needed onto condition to show that $\psi(\cdot,P)$ is a homeomorphism. That the image of ψ is an absolute Borel space follows from the fact that $\mathcal{P}$ is a G_δ subset of $2^{\{0,1\}^{\mathbb{N}}}$. □

Corollary A.55. *For Borel measurable injections* $h\colon \{0,1\}^{\mathbb{N}} \to \mathcal{P}$, *let* H *be the map* $(x,t) \mapsto (x,h(t))$ *and* $M = \bigcup_{t \in \{0,1\}^{\mathbb{N}}} \{(y,t)\colon y \in h(t)\}$. *Then* H *is a Borel measurable injection such that the map* $\Psi = H^{-1}\psi H$ *is a* $\mathfrak{B}$*-homeomorphism of* $\{0,1\}^{\mathbb{N}} \times \{0,1\}^{\mathbb{N}}$ *onto* M. *Moreover, if* h *is continuous, then* Ψ *is a homeomorphism.*

Proof. Clearly H is a $\mathfrak{B}$-homeomorphism of $\{0,1\}^{\mathbb{N}} \times \{0,1\}^{\mathbb{N}}$ onto $M_0 = \bigcup_{t \in \{0,1\}^{\mathbb{N}}} \{(x,h(t))\colon x \in \{0,1\}^{\mathbb{N}}\}$. Since $\psi[M_0] \subset M_0$, we have that the restriction of H^{-1} to $\psi[M_0]$ is also a $\mathfrak{B}$-homeomorphism of $\psi[M_0]$ onto M, thereby the first assertion is proved. The second assertion of the corollary is now obvious. □

A.7.2. First reduction of the proof. The first reduction will be stated by two propositions.

PROPOSITION A.56. *Let X be an absolute Borel space. If $f: X \to Y$ is a Borel measurable map with* $\operatorname{card}(U(f)) > \aleph_0$, *then there exists a continuous surjection $f^*: \mathbb{N}^\mathbb{N} \to \mathbb{N}^\mathbb{N}$ and there exist continuous injections $\Theta: \mathbb{N}^\mathbb{N} \to X$ and $\vartheta: \mathbb{N}^\mathbb{N} \to Y$ such that $f \succ_B f^*$ and $U(f^*) = \mathbb{N}^\mathbb{N}$.*

PROOF. The following diagram will help in the proof, where the indicated maps are defined in the development below. Of course, we want to prove the diagram is commutative.

$$\begin{array}{ccccccc}
\mathbb{N}^\mathbb{N} & \xrightarrow{\ g\ } & M & \xrightarrow{\ \pi_1|M\ } & \pi_1[M] & \xrightarrow{\ \subset\ } & X \\
{\scriptstyle \pi_2 g}\downarrow & & {\scriptstyle \pi_2}\downarrow & & {\scriptstyle f|(\pi_1[M])}\downarrow & & {\scriptstyle f}\downarrow \\
\pi_2[M] & = & \pi_2[M] & \xrightarrow[\text{id}]{} & f\pi_1[M] & \xrightarrow[\subset]{} & Y \\
{\scriptstyle h}\uparrow & & & & & & \\
\mathbb{N}^\mathbb{N} & & & & & &
\end{array}$$

As $U(f)$ is an uncountable analytic space there is a topological copy Y_0 of $\mathbb{N}^\mathbb{N}$ that is contained in $U(f)$. Recall $\operatorname{graph}(f) \subset X \times Y$ and let π_1 and π_2 be the natural projections of $X \times Y$ onto X and Y, respectively. As $\operatorname{graph}(f)$ is an absolute Borel space, the intersection $M_0 = \pi_2{}^{-1}[Y_0] \cap \operatorname{graph}(f)$ is an uncountable absolute Borel space. Note that $\mathcal{N}$ and $\mathbb{N}^\mathbb{N}$ are homeomorphic. Hence there is a continuous bijection $g: \mathbb{N}^\mathbb{N} \to M_0$ such that $D = M_0 \setminus g[\mathbb{N}^\mathbb{N}]$ is a countable set. For convenience of exposition, let $M = g[\mathbb{N}^\mathbb{N}]$, an absolute Borel space. The projection $\pi_2: M \to Y_0$ is a continuous map. Remember that Y_0 is homeomorphic to $\mathbb{N}^\mathbb{N}$ and observe that $Y_0 \setminus \pi_2[D] \subset \pi_2[M] \subset Y_0$. As $\pi_2[D]$ is countable, we have that $\pi_2[M]$ and $\mathbb{N}^\mathbb{N}$ are homeomorphic. Let $h: \mathbb{N}^\mathbb{N} \to \pi_2[M]$ be a homeomorphism. Observe that $\pi_1[M] \subset X$ and $f[\pi_1[M]] = \pi_2[M]$. The commutativity of the diagram is easily seen because $M \subset \operatorname{graph}(f)$. Let $f^* = h^{-1}\pi_2 g$, $\Theta = \pi_1 g$ and $\vartheta = h$. The composition $\Theta = \pi_1 g: \mathbb{N}^\mathbb{N} \to \pi_1[M]$ is a continuous bijection since $M \subset \operatorname{graph}(f)$. Clearly, $U(f^*) = h^{-1}h[U(f^*)] = h^{-1}[U(hf^*)] = h^{-1}[U(f|(\pi_1[M]))] = h^{-1}[\pi_2[M]] = \mathbb{N}^\mathbb{N}$. □

PROPOSITION A.57. *If $f: \mathbb{N}^\mathbb{N} \to \mathbb{N}^\mathbb{N}$ is a continuous surjection with $U(f) = \mathbb{N}^\mathbb{N}$, then there is a continuous surjection $f^*: \{0,1\}^\mathbb{N} \to \{0,1\}^\mathbb{N}$ and there are continuous injective maps Θ and ϑ of $\{0,1\}^\mathbb{N}$ to $\mathbb{N}^\mathbb{N}$ such that $f \succ_B f^*$ and $U(f^*) = \{0,1\}^\mathbb{N}$.*

PROOF. Let us first show that there exists a nonempty, perfect, compact subset F of $\mathbb{N}^\mathbb{N}$ such that

(1) $U(f|F) = f[F]$,
(2) $\operatorname{card}(U(f|F)) > \aleph_0$,
(3) $f[F]$ is a nonempty perfect set.

Observe that a subset K of $\mathbb{N}^\mathbb{N}$ is compact if and only if there is a b in $\mathbb{N}^\mathbb{N}$ such that $1 \leq d_i \leq b_i$ for every i whenever $d \in K$. We shall write $d \leq b$ whenever $1 \leq d_i \leq b_i$ for every i.

Consider the product space

$$\mathrm{domain}(f) \times \mathrm{domain}(f) \times \mathrm{range}(f) = \mathbb{N}^{\mathbb{N}} \times \mathbb{N}^{\mathbb{N}} \times \mathbb{N}^{\mathbb{N}}.$$

The set $H = \{(x,b,y) \in \mathbb{N}^{\mathbb{N}} \times \mathbb{N}^{\mathbb{N}} \times \mathbb{N}^{\mathbb{N}} : x \leq b \text{ and } x \in f^{-1}[\{y\}]\}$ is a closed set. Denote the projection $(x,b,y) \mapsto (b,y)$ by π_0. Then $U(\pi_0|H)$ is the set

$$S = \{(b,y) \in \mathrm{domain}(f) \times \mathrm{range}(f) :$$
$$\mathrm{card}(\{x : x \leq b\} \cap f^{-1}[\{y\}]) > \aleph_0\},$$

an uncountable analytic space. S has the property that if (b,y) is in S and $b \leq b'$ then (b',y) is in S. Let π_1 be the projection $(b,y) \mapsto y$. Clearly, $\pi_1\pi_0$ is the projection $(x,b,y) \mapsto y$. So $\pi_1|S$ maps S onto range(f). Hence $U(\pi_1|S) = \mathbb{N}^{\mathbb{N}}$. By Lemma A.47 there is a compact subset K of S such that $\pi_1[K]$ is uncountable. Let B be the perfect part of $\pi_1[K]$. Let π_2 be the projection $(b,y) \mapsto b$. As $\pi_2[K]$ is a compact subset of $\mathbb{N}^{\mathbb{N}}$, there is a b_0 such that $\pi_2[K] \subset \{x \in \mathbb{N}^{\mathbb{N}} : x \leq b_0\}$. Let A be the compact set $\{x \in \mathbb{N}^{\mathbb{N}} : x \leq b_0\} \cap f^{-1}[B]$. Clearly, $A \times B \subset \mathrm{domain}(f) \times \mathrm{range}(f)$. We assert that the continuous map $f|A$ is a surjective map of A to B. Let $y \in B$. Then $(b,y) \in K$ for some b in $\mathbb{N}^{\mathbb{N}}$. As $b \leq b_0$, we have $\{x : \mathrm{card}(\{x : x \leq b\} \cap f^{-1}[\{y\}]) > \aleph_0\} \subset A$. Hence we have that $f|A$ is surjective map and also $U(f|A) = B$. Let F be the perfect part of the uncountable compact set $A = \mathrm{domain}(f|A)$. As $A \setminus F$ is a countable set, we have that F satisfies the conditions (1) – (3) listed above.

Let us define f^*. Let $h_1 : \{0,1\}^{\mathbb{N}} \to F$ and $h_2 : \{0,1\}^{\mathbb{N}} \to f[F]$ be homeomorphisms. Then $f^* = {h_2}^{-1}(f|F)h_1$ is a B-successor of $f|F$. □

A.7.3. Second reduction of the proof. The first reduction has led us to a continuous surjection $f : \{0,1\}^{\mathbb{N}} \to \{0,1\}^{\mathbb{N}}$ with $U(f) = \{0,1\}^{\mathbb{N}}$. The second reduction will again be stated as a proposition.

PROPOSITION A.58. *If $f : \{0,1\}^{\mathbb{N}} \to \{0,1\}^{\mathbb{N}}$ is a continuous surjection with $U(f) = \{0,1\}^{\mathbb{N}}$, then there exists a continuous surjection $f^* : \{0,1\}^{\mathbb{N}} \to \{0,1\}^{\mathbb{N}}$ and there exist continuous injections Θ and ϑ of $\{0,1\}^{\mathbb{N}}$ to $\{0,1\}^{\mathbb{N}}$ such that $f \succ_B f^*$ and $U(f^*) = \{0,1\}^{\mathbb{N}}$, and such that the map h given by $h(y) = (\vartheta f^*)^{-1}[\{y\}]$ is a continuous map of $\{0,1\}^{\mathbb{N}}$ into $\mathcal{P}$ satisfying $h(y) \subset \Theta^{-1}f^{-1}[\{y\}]$ for each y in $\{0,1\}^{\mathbb{N}}$.*

PROOF. Let L be the subset of $2^{\{0,1\}^{\mathbb{N}}} \times \{0,1\}^{\mathbb{N}}$ consisting of all points (F,y) such that $F \subset f^{-1}[\{y\}]$, where F is a nonempty closed set. As f is continuous, the set L is compact. The set $S = L \cap (\mathcal{P} \times \{0,1\}^{\mathbb{N}})$ is a G_δ set. Let π be the map given by $(F,y) \mapsto y$. As $U(f) = \{0,1\}^{\mathbb{N}}$, we have $\pi[S] = \{0,1\}^{\mathbb{N}}$. By Lemma A.47 there is a compact subset K of S such that $\pi[K]$ is uncountable. Let

$$V = \pi[K] \text{ and } W = \{P : (P,y) \in K \text{ for some } y\},$$

so $K \subset W \times V$. With π_V, the natural projection of the product onto V, apply Lemma A.48 to find a nonempty, perfect, compact subset V_0 of V and a continuous map $s : V_0 \to W$ such that $(y, s(y)) \in (\pi_V|K)^{-1}[\{y\}]$ for every y in V_0. As $s[V_0]$ is a compact set in $2^{\{0,1\}^{\mathbb{N}}}$ we have that $X^* = \bigcup_{P \in s[V_0]} P$ is a perfect subset of $\{0,1\}^{\mathbb{N}}$. Observe that $\mathrm{graph}(f|X^*) = \mathrm{graph}(f) \cap (X^* \times V_0)$. Hence $(f|X^*)^{-1}[\{y\}] = s(y)$ for each y in V_0. Let $Y^* = V_0$, and let $h_1\colon \{0,1\}^{\mathbb{N}} \to X^*$ and $h_2\colon Y^* \to \{0,1\}^{\mathbb{N}}$ be homeomorphisms. Define $\Theta = h_1$, $\vartheta = {h_2}^{-1}$ and $f^* = h_2 f h_1$. Then $f \succ_B f^*$. Observe that ${h_1}^{-1}$ yields a homeomorphism Φ between 2^{X^*} and $2^{\{0,1\}^{\mathbb{N}}}$. As $h(y) = (fh_1)^{-1}[\{y\}] = {h_1}^{-1}(f|X^*)^{-1}[\{y\}] = {h_1}^{-1}[s(y)] = \Phi(s(y))$, the continuity of h is established. □

A.7.4. Third reduction of the proof. The second reduction of the proof has led us to a continuous map $f\colon \{0,1\}^{\mathbb{N}} \to \{0,1\}^{\mathbb{N}}$ such that $f^{-1}[\{y\}]$ is a nonempty perfect subset of $\{0,1\}^{\mathbb{N}}$ for each y in $\{0,1\}^{\mathbb{N}}$, the map h given by $y \mapsto f^{-1}[\{y\}]$ is continuous, and $U(f) = \{0,1\}^{\mathbb{N}}$.

PROPOSITION A.59. *Let $f\colon \{0,1\}^{\mathbb{N}} \to \{0,1\}^{\mathbb{N}}$ be a continuous map such that $f^{-1}[\{y\}]$ is nonempty perfect subset of $\{0,1\}^{\mathbb{N}}$ for each y in $\{0,1\}^{\mathbb{N}}$, let $h : y \mapsto f^{-1}[\{y\}]$ be continuous, and let $U(f) = \{0,1\}^{\mathbb{N}}$. Then there exists a homeomorphism g of $\{0,1\}^{\mathbb{N}} \times \{0,1\}^{\mathbb{N}}$ onto $\mathrm{graph}(f)$ such that $g(\cdot, y)$ is a homeomorphism of $\{0,1\}^{\mathbb{N}} \times \{y\}$ onto $f^{-1}[\{y\}] \times \{y\}$ whenever $y \in \{0,1\}^{\mathbb{N}}$.*

PROOF. The proof follows immediately from Corollary A.55. □

COROLLARY A.60. *Let f be as in the previous proposition. For the continuous map f^* defined by $f^*(x,y) = y$, $(x,y) \in \{0,1\}^{\mathbb{N}} \times \{0,1\}^{\mathbb{N}}$, there exist homeomorphisms*

$$\Theta\colon \{0,1\}^{\mathbb{N}} \times \{0,1\}^{\mathbb{N}} \to \{0,1\}^{\mathbb{N}} \text{ and } \vartheta\colon \{0,1\}^{\mathbb{N}} \to \{0,1\}^{\mathbb{N}}$$

such that $f \succ_B f^*$.

PROOF. Let $\pi\colon \mathrm{graph}(f) \to \mathrm{domain}(f)$ be the natural projection. Then $\vartheta = \mathrm{id}$ and $\Theta = \pi g$ are the required homeomorphisms. □

As all of the maps Θ's and ϑ's in the propositions that appear in the reductions of the proof are continuous injections we have the final lemma.

LEMMA A.61. *Let X be an absolute Borel space and Y be a separable metrizable space. If $f\colon X \to Y$ is a Borel measurable map and $U(f)$ is uncountable, then there exist continuous injections*

$$\Theta\colon \{0,1\}^{\mathbb{N}} \times \{0,1\}^{\mathbb{N}} \to X \text{ and } \vartheta\colon \{0,1\}^{\mathbb{N}} \to Y$$

such that the map f^* defined on $(x,y) \in \{0,1\}^{\mathbb{N}} \times \{0,1\}^{\mathbb{N}}$ by $f^*(x,y) = y$ satisfies $f \succ_B f^*$.

It is well-known that there is a Borel subset B of $\{0,1\}^{\mathbb{N}} \times \{0,1\}^{\mathbb{N}}$ such that $f^*[B]$ is an analytic set that is not a Borel set, whence f^* is not a $\mathfrak{B}$-map.

A.7.5. The proof of necessity. The proof is now quite straightforward. For if there is a $\mathfrak{B}$-map $f: X \to Y$ with $\mathrm{card}(U(f)) > \aleph_0$, then, by Lemma A.61, there is a Borel measurable map f^* such that f^* is not a $\mathfrak{B}$-map, domain(f^*) is an absolute Borel space and $f \succ_B f^*$. But, by Lemma A.46, this map f^* is a $\mathfrak{B}$-map, thereby a contradiction has been established.

A.8. Comments

Except for the development of Purves's theorem, the material of this chapter is found in standard reference books, for example, Kuratowski [**85**] and Federer [**55**].

In mathematical literature, the name $\mathfrak{B}$-map appears for the first time here. As stated in the footnote to the definition (see page 183), Purves [**129**] calls these maps bimeasurable maps. $\mathfrak{B}$-maps appear in Chapters 2 and 3.

Purves's theorem, proved in 1966, sharpens the well-known theorem of N. Lusin [**93**, pages 237–252] that a countable-to-one Borel measurable map defined on a separable, completely metrizable space preserves Borel sets. The later proof of Purves's theorem by Mauldin [**103**] used the notion of parametrization. Proposition A.59 is an example of a parametrization. See [**102, 27, 29, 147**] for further references to parametrizations. Another proof of Purves's theorem is indicated in the book [**145**] by S. M. Srivastava. The proof of Purves's theorem given here has been modeled after the one in [**129**] but with the following changes. The relation $\succ_B$ introduced here is more suitable for the proof than the equivalence relation defined in [**129**]. Lemma A.54 is an improvement over the one proved in [**129**]. Darst [**37**] was the first one to strengthen Purves's construction to yield a weaker form of Proposition A.58. The final form of this proposition appears here for the first time.

Exercises

A.1 For a σ-finite Borel measure space $\mathrm{M}(X,\mu)$ define $\mathsf{c}\,\mathfrak{M}(X,\mu)$ to be the family of all subsets M of X with the property that there exist Borel subsets A and B of X such that $A \subset M \subset B$ and $\mu(B \setminus A) = 0$. Show that this family is a σ-algebra. Show that the measure μ has an extention to $\mathsf{c}\,\mathfrak{M}(X,\mu)$ and that this extension of μ is a complete Borel measure on X with $\mathfrak{M}(X,\mu) \subset \mathsf{c}\,\mathfrak{M}(X,\mu)$. Show that $\mathrm{M}(X,\mu)$ is complete if and only if $\mathfrak{M}(X,\mu) = \mathsf{c}\,\mathfrak{M}(X,\mu)$.

A.2 Prove Proposition A.45.

Appendix B

Probability theoretic approach

The notions of universally measurable set and universally null set in a space found their way in the 1960s into probability theory. This is not surprising since these notions use Borel measure theory. In a purely *measure theoretic setting*, questions concerning properties of universally null sets in a measure space were studied very early on as witnessed by works of Polish mathematicians in the first half of the twentieth century. The book has included one of the questions of that time, namely, problem P 21 in S. Banach [**7**]. This article contains a commentary by E. Marczewski in which sequences of characteristic functions were applied. In earlier appearing papers Marczewski (= E. Szpilrajn) [**150, 151**] had formalized this use of characteristic functions. Our development of universally measurable sets and universally null sets in a purely measure theoretic sense will use terminology of these earlier works. For probability theory we shall use the terminology from the book *Measure Theory* by D. L. Cohn [**32**, pages 288–296].[1] Another useful reference is the recent book [**145**] by S. M. Srivastava.

B.1. Basic definitions

Let X be a set and let $\mathfrak{A}$ be a σ-algebra of subsets of X. The pair $(X, \mathfrak{A})$ is called a *measurable space* (of course, in the measure theory sense). A nonempty subset F of X is called an *atom*[2] of the measurable space $(X, \mathfrak{A})$ if $F = \bigcap\{ A \in \mathfrak{A} : x \in A \}$ for some x in F. Clearly each x of X is a member of some atom of $(X, \mathfrak{A})$. Observe that $F_1 = F_2$ whenever F_1 and F_2 are atoms of $(X, \mathfrak{A})$ with $F_1 \cap F_2 \neq \emptyset$. For measurable spaces $(X_1, \mathfrak{A}_1)$ and $(X_2, \mathfrak{A}_2)$ a map $f : X_1 \to X_2$ is said to be *measurable* if $f^{-1}[A] \in \mathfrak{A}_1$ whenever $A \in \mathfrak{A}_2$. For a measurable space $(X, \mathfrak{A})$ a map $f : X \to Y$, where Y is a topological space, is said to be *Borel measurable* if $f^{-1}[U] \in \mathfrak{A}$ whenever U is an open set of Y (often we shall simply say measurable in this context). A measurable space $(X_1, \mathfrak{A}_1)$ is said to be *isomorphic* to a measurable space $(X_2, \mathfrak{A}_2)$ if there is a bijection $f : X_1 \to X_2$ such that f is measurable and f^{-1} is measurable. A *probability measure space* is a triple $(X, \mathfrak{A}, P)$ where $(X, \mathfrak{A})$ is a measurable space and P is a

[1] In addition to the cited book, useful resources for this discussion are the Blackwell article [**12**] and the paper [**94**] by G. W. Mackey.

[2] The definition follows that of Szpilrajn [**150, 151**] and not of the book *Measure Theory* by D. L. Cohn [**32**, pages 288–296].

probability measure on $(X, \mathfrak{A})$ (that is, $P(X) = 1$). Of great interest in probability theory is conditional probability, hence it is not assumed that all atoms are singleton sets. In his article [**12**], D. Blackwell pointed out the existence in the literature of three negative examples concerning formulas in probability theory and also the fact that no such examples existed for perfect probability measure spaces, which will be defined next (see [**60**]).

Definition B.1. *A probability measure space $(X, \mathfrak{A}, P)$ is said to be* perfect *if, for each measurable real-valued function f and for each subset E of $\mathbb{R}$ such that $f^{-1}[E] \in \mathfrak{A}$, there is a Borel set B contained in E such that $P(f^{-1}[E]) = f_{\#}P(B)$.*

Notice that this notion is defined by properties of measurable real-valued functions. Actually many properties in probability theory are defined by imposing conditions on real-valued functions, as we shall see later. This permits the use of the existence of many natural classes of subsets of $\mathbb{R}$ that are invariant under Borel isomorphisms. The space $\mathbb{R}$ has a countable subbase $\mathfrak{S}$ for the open sets, which of course generates $\mathfrak{B}(\mathbb{R})$, the collection of all Borel subsets of $\mathbb{R}$. It will be convenient if the measurable space $(X, \mathfrak{A})$ also has this property. For this we have the following simple proposition.

Proposition B.2. *Let $(X, \mathfrak{A})$ be a measurable space and f be a real-valued map. If f is an isomorphism between $(X, \mathfrak{A})$ and $(f[X], \mathfrak{B}(f[X]))$, then there is a countable subcollection $\mathfrak{E}$ of $\mathfrak{A}$ such that the smallest σ-algebra of X that contains $\mathfrak{E}$ is precisely $\mathfrak{A}$, and $\{f^{-1}[\{r\}] : r \in f[X]\}$ is precisely the collection of all atoms of $(X, \mathfrak{A})$.*

The proof is obvious, let $\mathfrak{E} = \{f^{-1}[S] : S \in \mathfrak{S}\}$, where $\mathfrak{S}$ is a countable subbase for the open sets of $f[X]$. A useful fact is that every separable metrizable space is isomorphic to some subspace of $\mathbb{R}$. Clearly the isomorphism is generally not a homeomorphism. A second is that if Y is a separable metrizable space and $f : X \to Y$ is a map defined on a set X then there is a natural smallest topology on X such that f is continuous; moreover, if f is an injection then f is a homeomorphism of X into Y.

With the aid of Exercise B.2 we can prove that surjections induce natural σ-algebras and probability measures.

Lemma B.3. *Let X and X' be sets, let $g : X' \to X$ be a surjection, and let $(X, \mathfrak{A})$ be a measurable space. If $\mathfrak{A}'$ is the smallest σ-algebra of X' that makes g measurable, then the following statements hold.*

(1) *If $(X, \mathfrak{A}, P)$ is a probability measure space, then there is a probability measure space $(X', \mathfrak{A}', g^{\#}P)$ such that $g_{\#}g^{\#}P = P$. Moreover, if $(X, \mathfrak{A}, P)$ is perfect, then $(X', \mathfrak{A}', g^{\#}P)$ is perfect.*
(2) *If $(X', \mathfrak{A}', P')$ is a probability space, then $(X, \mathfrak{A}, g_{\#}P')$ is such that $g^{\#}g_{\#}P' = P'$. Moreover, if $(X', \mathfrak{A}', P')$ is perfect, then $(X, \mathfrak{A}, g_{\#}P')$ is perfect.*

Proof. We leave the proof of the first assertion of each statement to the reader.

Assume that $(X, \mathfrak{A}, P)$ is perfect and let $f' : X' \to \mathbb{R}$ be measurable. As $\mathfrak{A}'$ is the smallest σ-algebra that makes g measurable, the map $f = f'g^{-1}$ is a measurable map of X into $\mathbb{R}$. Let E be a subset of $\mathbb{R}$ such that $f'^{-1}[E] \in \mathfrak{A}'$. Then $f^{-1}[E] = g[f'^{-1}[E]] \in \mathfrak{A}$. There is a Borel set B in $\mathbb{R}$ such that $B \subset E$ and

$P(f^{-1}[B]) = P(f^{-1}[E])$. As $g^{\#}P(f'^{-1}[B]) = P(g[f'^{-1}[B]]) = P(f^{-1}[B])$ and $g^{\#}P(f'^{-1}[E]) = P(g[f'^{-1}[E]]) = P(f^{-1}[E])$ we have shown that $(X', \mathfrak{A}', g^{\#}P)$ is perfect.

Assume that $(X', \mathfrak{A}', P')$ is perfect and let $f: X \to \mathbb{R}$ be measurable. Clearly $f' = fg$ is a measurable map of X' into $\mathbb{R}$. Let E be a subset of $\mathbb{R}$ such that $f^{-1}[E] \in \mathfrak{A}$. Then $f'^{-1}[E] = g^{-1}[f^{-1}[E]] \in \mathfrak{A}'$. There is a Borel subset B of $\mathbb{R}$ such that $B \subset E$ and $P'(f'^{-1}[B]) = P'(f'^{-1}[E])$. It follows that $g_{\#}P'(f^{-1}[B]) = g_{\#}P'(f^{-1}[E])$, whence $(X, \mathfrak{A}, g_{\#}P')$ is perfect. □

We have the following.

Proposition B.4. *Let $(X, \mathfrak{A})$ be isomorphic to an analytic separable metrizable space and let Y be a separable metrizable space. If $f: X \to Y$ is a measurable map, then $f[E]$ is an analytic subspace of Y whenever $E \in \mathfrak{A}$.*

Proof. Let A be an analytic space and let $g: X \to A$ be an isomorphism between $(X, \mathfrak{A})$ and $(A, \mathfrak{B}(A))$. If E is in $\mathfrak{A}$, then $g[E]$ is a Borel subset of A. Hence $g[E]$ is also an analytic subspace of A. Since $g|E: E \to g[E]$ is an isomorphism between the measurable spaces $(E, \mathfrak{A}\,|E)$ and $(g[E], \mathfrak{B}(g[E]))$, where $\mathfrak{A}\,|E = \{A \cap E: A \in \mathfrak{A}\}$, it follows that $(E, \mathfrak{A}\,|E)$ is isomorphic to the analytic space $g[E]$. It is easily seen that $F = f(g|E)^{-1}: g[E] \to Y$ is a Borel measurable map from the analytic space $g[E]$ into Y. Hence $f[E] = F[g[E]]$ is an analytic subspace of Y. □

Corollary B.5. *If $(X, \mathfrak{A})$ is isomorphic to an analytic separable metrizable space and if $E \in \mathfrak{A}$, then $(E, \mathfrak{A}\,|E)$ is isomorphic to an analytic space.*

B.2. Separable metrizability

Let us turn to the question of what condition is necessary and sufficient on $\mathfrak{A}$ to assure that $(X, \mathfrak{A})$ is isomorphic to a separable metrizable space. The solution was known many years ago. A necessary condition is easily found. Indeed, from Proposition B.2 we infer that if $(X, \mathfrak{A})$ is isomorphic to a separable metrizable space Y, then a subbase $\mathfrak{S}$ of Y and the isomorphism will yield a countable subcollection $\mathfrak{E}$ of $\mathfrak{A}$ such that the smallest σ-algebra that contains $\mathfrak{E}$ is precisely $\mathfrak{A}$. (Recall that the collection of all finite intersections of $\mathfrak{S}$ forms a base for the open sets of Y.) Hence there must exist a countable subcollection $\mathfrak{E}$ of $\mathfrak{A}$ such that $\mathfrak{A}$ is the smallest σ-algebra that contains $\mathfrak{E}$. Let us turn to the proof that this condition is also sufficient.

B.2.1. An embedding. A separable metrizable topology can be associated with a measure space $(X, \mathfrak{A})$ by imposing the following natural conditions on the σ-algebra $\mathfrak{A}$.

Definition B.6. *Let $(X, \mathfrak{A})$ be a measurable space. A* countably generated [3] *σ-algebra $\mathfrak{A}$ is a pair $(\mathfrak{A}, \mathfrak{E})$ such that $\mathfrak{E}$ is a countable subset of $\mathfrak{A}$ and $\mathfrak{A}$ is the smallest σ-algebra of X that contains $\mathfrak{E}$. (Often $\mathfrak{E}$ will not be displayed; the collection $\mathfrak{E}$ is*

[3] The usual terminology in the literature for this notion is "separable." Consistent with **[32]** we have used "countably generated" because the word separable is already being used in another context.

said to generate $\mathfrak{A}$.) *A* σ*-algebra* $\mathfrak{A}$ *on* X *is said to be* separating *if the collection of all atoms of* $(X, \mathfrak{A})$ *is* $\{\{x\} : x \in X\}$.

The notion of separating by a σ-algebra $\mathfrak{A}$ concerns the property of disjoint subsets being contained in disjoint members of $\mathfrak{A}$. Obviously distinct atoms as defined earlier are separated by $\mathfrak{A}$. Consequently if every atom of $\mathfrak{A}$ is a singleton then distinct points are separated by $\mathfrak{A}$. An analogous situation occurs for distinguishing pseudo-metric spaces and metric spaces in metric topology theory. We shall see shortly that atoms are members of $\mathfrak{A}$ for countably generated measurable spaces $(X, \mathfrak{A})$.

In the terminology of the above definition a measurable space $(X, \mathfrak{A})$ is isomorphic to a separable metrizable space only if it is a separating, countably generated measurable space. We now turn to the converse. We first show that a countably generated measurable space corresponds to a natural measurable map $f : X \to \{0,1\}^{\mathbb{N}}$. This argument is given by Marczewski in his commentary included in Banach [7] (see also Szpilrajn [**150, 151**]).

Theorem B.7. *Let* $(X, \mathfrak{A})$ *be a countably generated measurable space and let* $\mathfrak{E}$ *generate* $\mathfrak{A}$. *Then there is a measurable map* $f : X \to \{0,1\}^{\mathbb{N}}$ *such that* $f[E]$ *is both closed and open in the subspace* $f[X]$ *of* $\{0,1\}^{\mathbb{N}}$ *whenever* $E \in \mathfrak{E}$ *and such that* $f^{-1}[\{y\}]$ *is an atom in* $\mathfrak{A}$ *whenever* $y \in f[X]$. *Hence there is a pseudo-metric* d *on* X *such that* $\mathfrak{E}$ *is a subbase for the open sets of the topology* τ *given by* d, *each* E *in* $\mathfrak{E}$ *is both closed and open, and* $\mathfrak{A}$ *is the precisely the collection of Borel subsets of* X *for the topology* τ. *Moreover the continuous map* f *is an open map if* X *is endowed with the topology* τ.

Proof. Let $\{E_i : i \in \mathbb{N}\}$ be a well ordering of the countable collection $\mathfrak{E}$. For each i let f_i be the characteristic function of the set E_i, and let $f : X \to \{0,1\}^{\mathbb{N}}$ be the map whose i-th coordinate map is f_i. Note that if $Y_i = \{y \in \{0,1\}^{\mathbb{N}} : y_i = 1\}$, then $f^{-1}[Y_i] = E_i$ and $X \setminus E_i = f^{-1}\big[\{0,1\}^{\mathbb{N}}\big] \setminus f^{-1}[Y_i]$. As $\mathfrak{A}$ is a σ-algebra of X containing $\mathfrak{E}$ it follows that the $f^{-1}[B] \in \mathfrak{A}$ for each Borel set B of $\{0,1\}^{\mathbb{N}}$. Hence $\mathfrak{A} = \{f^{-1}[B] : B \in \mathfrak{B}(\{0,1\}^{\mathbb{N}})\}$. It is now clear that $f^{-1}[\{y\}]$ is an atom of $\mathfrak{A}$ whenever $y \in f[X]$. A pseudo-metric is now induced on X by a metric on $\{0,1\}^{\mathbb{N}}$. The remaining parts of the proof is left to the reader. □

The next theorem is now obvious.

Theorem B.8. *For a measurable space* $(X, \mathfrak{A})$ *the set* X *has a separable pseudo-metrizable topology* τ *such that* $\mathfrak{A}$ *is precisely the* σ*-algebra of Borel sets of* X *if and only if* $(X, \mathfrak{A})$ *is a countably generated measurable space. Moreover, the sharpening of pseudo-metrizability to metrizability is achieved by requiring the added condition of separating.*

Let us now turn to the definitions of standard measurable spaces and analytic measurable spaces.

Definition B.9. *A measurable space* $(X, \mathfrak{A})$ *is called* standard *if there is a separable completely metrizable space* Y *and there is a measurable bijection*[4] $f: X \to Y$ *such that* $\mathfrak{A}$ *is the smallest* σ*-algebra that contains* $\{f^{-1}[B]: B \in \mathfrak{B}(Y)\}$. *A measurable space* $(X, \mathfrak{A})$ *is called* analytic *if there is an analytic space* Y *and there is a measurable bijection* $f: X \to Y$ *such that* $\mathfrak{A}$ *is the smallest* σ*-algebra that contains* $\{f^{-1}[B]: B \in \mathfrak{B}(Y)\}$.

Let us add definitions that correspond to absolute measurable spaces and absolute null spaces as defined in Chapter 1.

Definition B.10. *A measurable space* $(X, \mathfrak{A})$ *is called* absolute measurable *if there is an absolute measurable space* Y *and there is a measurable bijection* $f: X \to Y$ *such that* $\mathfrak{A}$ *is the smallest* σ*-algebra that contains* $\{f^{-1}[B]: B \in \mathfrak{B}(Y)\}$ *and is called* absolute null[5] *if* Y *is an absolute null space.*

Clearly, every standard measurable space is analytic. It is easily seen that a measurable space $(X, \mathfrak{A})$ is standard if and only if, in the definition, the space Y is some subset of $\mathbb{N}$ or the space Y is equal to $\mathbb{R}$. Standard measurable spaces are invariant under isomorphisms. But standard measurable spaces are not preserved by measurable maps. Indeed, surjective measurable maps of standard measurable spaces yield analytic measurable spaces. But it is true that analytic measurable spaces are, in some sense, invariant under measurable surjections (see the exercises).

B.3. Shortt's observation

R. M. Shortt's observation, announced in [**138**] and published in [**139**], concerns separable metrizable measurable spaces $(X, \mathfrak{A})$ – that is, separating, countably generated measurable spaces. For such measurable spaces he observed that the σ-algebra $\mathfrak{A}$ can have many countable subcollections $\mathfrak{E}$ that generate $\mathfrak{A}$. Consequently, many topologically different metrics can result in the same collection of Borel sets if X is an uncountable set. For a separable metrizable space X with topology τ the collection of Borel sets will be denoted by $\mathfrak{B}(X, \tau)$. Shortt further observed

Proposition B.11. *If* (X, τ_1) *and* (X, τ_2) *are separable metrizable spaces with* $\mathfrak{A} = \mathfrak{B}(X, \tau_1) = \mathfrak{B}(X, \tau_2)$ *and if* $(X, \mathfrak{M}(X, P_1), P_1)$ *and* $(X, \mathfrak{M}(X, P_2), P_2)$ *are continuous, complete, Borel probability measure spaces with* $P_1 | \mathfrak{A} = P_2 | \mathfrak{A}$, *then* $\mathfrak{M}(X, P_1) = \mathfrak{M}(X, P_2)$.

Let us state Shortt's observations as a theorem in topological form. For the definition of universally measurable set and universally null set see Chapter 2.

Theorem B.12 (Shortt). *Let* X *be a set and let* τ_1 *and* τ_2 *be topologies on* X *that make* X *into separable metrizable spaces with* $\mathfrak{B}(X, \tau_1) = \mathfrak{B}(X, \tau_2)$. *Then a subset* M *of* X *is a universally measurable set in the topological space* (X, τ_1) *if and only if it is a universally measurable set in the topological space* (X, τ_2). *Also, a subset* M *of* X *is*

[4] In [**32**] f is required to be an isomorphism in both definitions.

[5] For an isomorphism f this notion is called *nonmeasurable* in the discussion of Grzegorek's example given in Chapter 1.

a universally null set in the topological space (X, τ_1) *if and only if it is a universally null set in the topological space* (X, τ_2).

It is now clear that the study of separating, countably generated measurable spaces $(X, \mathfrak{A})$ of measure theory is equivalent to a study of measurable spaces $(Y, \mathfrak{B}(Y))$ associated with subspaces of Y of $\{0,1\}^{\mathbb{N}}$. As $\{0,1\}^{\mathbb{N}}$ is isomorphic (that is, $\mathfrak{B}$-homeomorphic in the terminology of the book) to $\mathbb{R}$ one may as well assume that only subspaces of $\mathbb{R}$ are of interest. But, as the isomorphism f between $(X, \mathfrak{A})$ and $(Y, \mathfrak{B}(Y))$ need not be a homeomorphism in the event that X is a separable metrizable space and $\mathfrak{A} = \mathfrak{B}(X)$, the measure theoretic approach to absolute measurable spaces may not be useful in the investigation of topological and geometric properties of some separable metric spaces.

We have the following simple theorem, where "absolute measurable space" is in the sense of Chapter 2.

THEOREM B.13. *Suppose that* $(X, \mathfrak{A})$ *is a separating, countably generated measurable space. Then every continuous probability measure space* $(X, \mathfrak{A}, P)$ *is perfect if and only if there is a metric* d *on* X *such that* $\mathfrak{A} = \mathfrak{B}((X, \mathrm{d}))$ *and* (X, d) *is homeomorphic to an absolute measurable space.*

PROOF. Suppose that $(X, \mathfrak{A})$ is a separating, countably generated measurable space such that every continuous complete Borel probability measure space $(X, \mathfrak{A}, P)$ is perfect. Let d be a metric on X such that (X, d) has $\mathfrak{A} = \mathfrak{B}((X, \mathrm{d}))$ and such that (X, d) is homeomorphic to a topological subspace of $\{0,1\}^{\mathbb{N}}$. Denote by f the homeomorphism of X onto $Y = f[X] \subset \{0,1\}^{\mathbb{N}}$ and let g be a homeomorphism of $\{0,1\}^{\mathbb{N}}$ onto the classical Cantor ternary set T in $\mathbb{R}$. Clearly $f' = gf$ is a measurable real-valued function, indeed, a homeomorphism of X onto $Z = g[Y]$. Let μ be a continuous, complete, finite Borel measure on T and let B be a Borel subset of T such that $\mu(B) = \mu^*(Z)$. We have $\big(\mu|Z\big)(Z) = \mu(B)$. If $\mu(B) = 0$ then Z is μ-measurable. Suppose $\mu(B) > 0$ and let $P' = c\,(\mu|Z)$, where $c^{-1} = \mu(B)$. The measure $P = {f'^{-1}}_{\#}P'$ is a continuous probability measure on X such that $f'_{\#}P = P'$. As the probability measure $P|\,\mathfrak{A}$ on $(X, \mathfrak{A})$ is perfect, there is Borel subset A of T such that $A \subset Y$ and $P(f'^{-1}[A]) = P(X)$. Since A is an absolute Borel space contained in Z we have $c\,\mu(A) = P'(A) = f'_{\#}P(A) = P(X) = 1$. Hence Z is μ-measurable. Consequently, Z is μ-measurable for every complete, finite Borel measure μ and thereby Z is a universally measurable set in T. So Z is an absolute measurable space.

To prove the converse, let d be a metric on X such that $\mathfrak{A} = \mathfrak{B}((X, \mathrm{d}))$ and (X, d) is an absolute measurable space. Let $f : X \to \mathbb{R}$ be a Borel measurable function and $(X, \mathfrak{A}, P)$ be a continuous probability measure space. Suppose $A \subset \mathbb{R}$ and $f^{-1}[A] \in \mathfrak{A}$. As $f^{-1}[A]$ is a Borel set in the absolute measurable space X we have that it is also an absolute measurable space. Hence there is an absolute Borel space M contained in $f^{-1}[A]$ such that $P(M) = P(f^{-1}[A])$. Denote the Borel measurable map $f|M$ by g and the measure $P|M$ by ν. Now $g[M]$ is an analytic space because f is a Borel measurable map and M is an absolute Borel set in X. So $\mu = g_{\#}\nu$ is a complete, finite Borel measure on $g[M]$. There is an absolute Borel space B contained in $g[M]$ such that $\mu(B) = \mu(g[M])$. Since $g^{-1}[B] \subset M$ and $\mu(E) = \big(g_{\#}\nu\big)(E) \leq \big(f_{\#}P\big)(E)$

for all Borel sets E contained in $g[M]$, we have $P(f^{-1}[A]) = P(M) = \nu(g^{-1}g[M]) = \mu(g[M]) = \mu(B) \le (f_\# P)(B) \le P(f^{-1}[A])$. Hence $(X, \mathfrak{A}, P)$ is perfect. □

As a result of Lemma B.3 we have the following corollary.

Corollary B.14. *Let $(X, \mathfrak{A})$ be a countably generated measurable space. Then every probability measure space $(X, \mathfrak{A}, P)$ is perfect if and only if there is a pseudo-metric* d *on X such that $\mathfrak{A} = \mathfrak{B}((X, \mathrm{d}))$ and the quotient metric space X / d is homeomorphic to an absolute measurable space.*

Recalling the definition of absolute measurable $(X, \mathfrak{A})$ from Definition B.10, we have the following measure theoretic characterization of absolute measurable $(X, \mathfrak{A})$ and absolute null $(X, \mathfrak{A})$.

Theorem B.15. *A countably generated measurable space $(X, \mathfrak{A})$ is absolute measurable if and only if every continuous probability measure space $(X, \mathfrak{A}, P)$ is perfect. A countably generated measurable space $(X, \mathfrak{A})$ is absolute null if and only if no continuous probability measure space $(X, \mathfrak{A}, P)$ exist.*

As every analytic measurable space $(X, \mathfrak{A})$ is absolute measurable we have

Corollary B.16. *If $(X, \mathfrak{A})$ is analytic, then every probability measure space $(X, \mathfrak{A}, P)$ is perfect.*

From a measure theoretic point of view, a subset E of a set X is defined to be *universally measurable* in a measurable space $(X, \mathfrak{A})$ if for each probability measure P on $(X, \mathfrak{A})$ there are sets A and B in $\mathfrak{A}$ such that $A \subset E \subset B$ and $P(B \setminus A) = 0$. For separable metrizable spaces X, notice that this is the same as the definition for universally measurable sets in X as defined earlier if $\mathfrak{A} = \mathfrak{B}(X)$. The measure theoretic definition of universally null set in a measurable space $(X, \mathfrak{A})$ is easily formulated. The reader is left with this task.

B.4. Lusin measurable space

In [**12**] Blackwell gave the following definition of Lusin measurable space.

Definition B.17. *A measurable space $(X, \mathfrak{A})$ is said to be a* Lusin measurable space[6] *if it is a countably generated measurable space such that $f[X]$ is an analytic subset of $\mathbb{R}$ whenever $f : X \to \mathbb{R}$ is a measurable map.*

He also proved

Theorem B.18. *Let $(X, \mathfrak{A})$ be a measurable space. Then $(X, \mathfrak{A})$ is a Lusin space if and only if it is analytic.*

Proof. Suppose $(X, \mathfrak{A})$ is Lusin. We may assume that $(X, \mathfrak{A})$ is also separating. There is an isomorphism h of X into $\{0, 1\}^{\mathbb{N}}$ and there is a homeomorphism g of

[6] This definition is not the same as the one given in [**32**]; the two definitions are completely different. Blackwell's Lusin spaces are the Suslin spaces in [**32**].

$\{0,1\}^{\mathbb{N}}$ into $\mathbb{R}$. As the real-valued function $f = gh$ is measurable $f[X]$ is analytic space. Consequently $(X, \mathfrak{A})$ is analytic. Conversely, suppose $(X, \mathfrak{A})$ is analytic. Then there is an isomorphism h of $(X, \mathfrak{A})$ onto $(Y, \mathfrak{B}(Y))$ for some analytic space Y. Let $f: X \to \mathbb{R}$ be a measurable map. Then $g = fh^{-1}$ is a Borel measurable map of Y into $\mathbb{R}$. As Y is analytic we have $g[Y]$ is analytic, whence $f[X]$ is analytic. □

Blackwell observed that every continuous probability measure space on a Lusin measurable space $(X, \mathfrak{A})$ is perfect (see Corollary B.16). He asked if the converse implication was also true. We have seen that absolute measurable $(X, \mathfrak{A})$ is characterized by the condition that $(X, \mathfrak{A})$ is a countably generated measure space with the property that every continuous probability measure space $(X, \mathfrak{A}, P)$ is perfect. Hence the answer to Blackwell's question is in the negative. The question was answered by R. B. Darst and R. F. Zink [**41**] – they showed that any uncountable absolute null space X contained in $[0, 1]$ with $\mathfrak{A} = \mathfrak{B}(X)$ is clearly not analytic in the measure theoretic sense because the identity map is measurable; there are no continuous probability measure spaces $(X, \mathfrak{A}, P)$, whence every such probability measure space is perfect. A better example is $X = Z \cup [1, 2]$, where Z is an uncountable absolute null space contained in $[0, 1]$; there are many continuous probability measure spaces.

The Blackwell question was refined further. Observe that not every continuous image of an uncountable absolute null space contained in $[0, 1]$ is a universally measurable set in $\mathbb{R}$; witness the Grzegorek example. This leads to the following definition of a D-space $(X, \mathfrak{A})$.

Definition B.19. *A measurable space $(X, \mathfrak{A})$ is a* D-space *if it is countably generated and if $f[X]$ is a universally measurable set in $\mathbb{R}$ whenever f is a measurable real-valued function on X.*

Obviously we have

Proposition B.20. *If $(X, \mathfrak{A})$ is a D-space, then $(X, \mathfrak{A})$ is absolute measurable.*

Darst showed in [**33**] the existence of D-spaces that are not Lusin measurable spaces provided Lusin singular sets in $\mathbb{R}$ exists. (We have seen that Lusin singular sets in $\mathbb{R}$ can be shown to exist by assuming the continuum hypothesis.) Let us give his proof. Darst's example is any Lusin set X contained in $[0, 1]$ with the usual σ-algebra $\mathfrak{A} = \mathfrak{B}(X)$. Clearly $(X, \mathfrak{A})$ is not analytic in the measure theoretic sense and hence is not a Lusin measurable space. It remains to show that $(X, \mathfrak{A})$ is a D-space, that is, the images of Lusin sets in $\mathbb{R}$ under Borel measurable real-valued functions are absolute null spaces. This is proved in Chapter 4. Again, a better example is $X = Z \cup [1, 2]$, where Z is a Lusin singular set contained in $[0, 1]$.

In G. Kallianpur [**80**], D-spaces were characterized – of course, with only measure theoretic notions – as follows.

Theorem B.21 (Kallianpur). *Let $(X, \mathfrak{A})$ be a countably generated measurable space. Then $(X, \mathfrak{A})$ is a D-space if and only if every probability measure space $(X, \mathfrak{A}', P)$ is perfect whenever $(X, \mathfrak{A}')$ is a countably generated measurable space with $\mathfrak{A}' \subset \mathfrak{A}$.*

Proof. There is no loss in assuming $(X,\mathfrak{A})$ is separating. If $(X,\mathfrak{A}')$ is a countably generated measurable space such that $\mathfrak{A}' \subset \mathfrak{A}$, then the identity map g is a measurable bijection of $(X,\mathfrak{A})$ into $(X,\mathfrak{A}')$. Hence a measurable real-valued function f' with respect to $\mathfrak{A}'$ results in a measurable function $f = f'g$ with respect to $\mathfrak{A}$. Also, a measurable real-valued function f with respect to $\mathfrak{A}$ results in a countably generated σ-algebra $\mathfrak{A}' = \{f^{-1}[B]: B \in \mathfrak{B}(\mathbb{R})\}$ contained in $\mathfrak{A}$ and f is measurable on the measurable space $(X,\mathfrak{A}')$. Hence each f that is measurable on $(X,\mathfrak{A})$ can be factored as $f = f'g$ for some f' that is measurable on $(X,\mathfrak{A}')$.

Suppose that $(X,\mathfrak{A})$ is a D-space and let $\mathfrak{A}' \subset \mathfrak{A}$. If f' is a measurable real-valued function with respect to $\mathfrak{A}'$, then $f = f'g$ is also measurable with respect to $\mathfrak{A}$ and hence $f[X]$ is an absolute measurable metrizable space contained in $\mathbb{R}$. So $f'[X]$ is an absolute measurable space. By Corollary B.14 and Lemma B.3 we have that every probability measure space $(X,\mathfrak{A}',P')$ is perfect.

For the converse let f be a measurable real-valued function on X with respect to $\mathfrak{A}$. Then there is an $\mathfrak{A}'$ contained in $\mathfrak{A}$ and a factorization $f = f'g$. As every probability measure space $(X,\mathfrak{A}',P')$ is given to be perfect, by Corollary B.14, $f'[X]$ is an absolute measurable space contained in $\mathbb{R}$. Since $f[X] = f'[X]$, the converse now follows by Lemma B.3. □

B.5. Comments

The literature of the subject matter of this appendix uses measurable mappings into $\mathbb{R}$. This restriction is not really needed since the only property of $\mathbb{R}$ that is used is the metric completeness of the usual metric of $\mathbb{R}$. Actually the needed properties are that $\mathbb{R}$ is an absolute measurable space in the sense of Chapter 1 and that every absolute measurable space is Borel isomorphic to a subspace of $\mathbb{R}$. We have stated Theorem B.13 and its corollary with this comment in mind.

For countably generated measurable spaces $(X,\mathfrak{A})$ the conditions of standard, analytic, D-space, absolute measurable, and countably generated become successively weaker; and none of them are equivalent. It is easily seen that $(X,\mathfrak{A})$ is any one of these if and only if $(E,\mathfrak{A}\,|E)$ is for each E in $\mathfrak{A}$. Darst's example shows that not every D-space is an analytic measurable space; his example relies on the continuum hypothesis. It would be interesting to know if there exists one such in ZFC, the usual axioms of set theory.

Question. In ZFC, are there uncountable absolute null spaces such that every Borel measurable image of them are absolute null spaces?

The example that shows that not every absolute measurable space is a D-space is the example of Grzegorek (see Section 1.4.2 on page 20).

The separating requirement on a countably generated measurable space can be ignored; the only consequence of this is that the atoms may not be singletons, just as for pseudo-metric spaces (see Theorem B.7). Perhaps definitions using the prefix pseudo would have been appropriate; we have not done so.

As mentioned at the start of this appendix, the book *Measure Theory* by Cohn **[32]** is a major source. Its Chapter 8 has a rather self-contained development of the measure theory of completely metrizable spaces. The reader may wish to turn to it as another source for analytic spaces. Cohn's development does not cover the intricacies of absolute null spaces which has been included in our appendix.

The reader is reminded again that the probability theoretic approach to absolute measurable space and to universally measurable set is not very suited to the study of topological properties or of geometric properties of the measurable space $(X, \mathfrak{A})$ because the σ-algebra structure $\mathfrak{A}$ does not necessarily carry any information on such properties.

Exercises

B.1. Use the Bernstein decomposition of the interval $[0, 1]$ to show the existence of a probability measure space $(X, \mathfrak{A}, P)$ that is not perfect.

B.2. Let $g: X' \to X$ be any surjection and let $(X, \mathfrak{A})$ be a measurable space. Prove the following:

(a) $\mathfrak{A}' = \{g^{-1}[A]: A \in \mathfrak{A}\}$ is the smallest σ-algebra of X' such that g is measurable.

(b) $g[A'] \in \mathfrak{A}$ and $g^{-1}\big[g[A']\big] = A'$ for every A' in $\mathfrak{A}'$.

(c) If $(X, \mathfrak{A}, P)$ is a probability measure space, then there is a probability measure space $(X', \mathfrak{A}', g^{\#}P)$ such that $g_{\#}g^{\#}P = P$, where $g^{\#}P(A') = P(g[A'])$, $A' \in \mathfrak{A}'$.

(d) If $(X', \mathfrak{A}', P')$ is a probability space, then $(X, \mathfrak{A}, g_{\#}P')$ is a probability space such that $g^{\#}g_{\#}P' = P'$.

B.3. Prove:

(a) A measurable space $(X, \mathfrak{A})$ is standard if and only if it is isomorphic to $(Z, \mathfrak{B}(Z))$ for some subset Z of $\mathbb{N}$ or for Z equal to $\mathbb{R}$.

(b) Standard measurable spaces are invariant under isomorphisms.

(c) If $(X, \mathfrak{A})$ is a standard measurable space and if $E \in \mathfrak{A}$, then $(E, \mathfrak{A}\,|E)$ is a standard space. Hint: An absolute Borel space is isomorphic to a subspace of $\mathbb{N}$ or to $\mathbb{R}$.

(d) Characterize those separating countably generated measurable spaces $(X, \mathfrak{A})$ with the property that $f[X]$ is an absolute Borel space for every measurable real-valued map on $(X, \mathfrak{A})$.

B.4. Prove: *Let $(X, \mathfrak{A})$ be an analytic measurable space and let $(X', \mathfrak{A}')$ be a separating countably generated measurable space. If $f: X \to X'$ is a measurable surjection, then $(X', \mathfrak{A}')$ is analytic.* Prove that in the above assertion that analytic may be replaced with D-space, but not with standard or absolute measurable. Does the Purves Theorem have any consequences for standard measurable spaces?

B.5. Prove: *Each bijective measurable map between standard measurable spaces is an isomorphism.* Prove: *Each bijective measurable map between analytic measurable spaces is an isomorphism.* The last assertion is Proposition 8.6.2 of [**32**]. Hint: Use Theorem A.8.

Appendix C

Cantor spaces

It is well-known that a compact metrizable space X is homeomorphic to $\{0,1\}^{\mathbb{N}}$ if and only if X is nonempty, perfect and totally disconnected (hence, zero-dimensional). The classical Cantor ternary set in $\mathbb{R}$ is one such, thus the name Cantor spaces. There are many other classical examples. A useful one is the product space $k^{\mathbb{N}}$, where k is a finite space endowed with the discrete topology and with $\operatorname{card}(k) > 1$. It will be necessary that Cantor spaces be investigated not only as topological spaces but also as metric spaces with suitably assigned metrics.

The development presented in this appendix is based on E. Akin [**2**], R. Dougherty, R. D. Mauldin and A. Yingst [**47**], and O. Zindulka [**162, 161**]. There are two goals. The first is to present specific metrics on Cantor spaces which are used in the computations of Hausdorff measure and Hausdorff dimension in Chapter 5. The second is to discuss homeomorphic measures on Cantor spaces. The lack of an analogue of the Oxtoby–Ulam theorem for Cantor spaces motivates this goal.

Topologically characterizing homeomorphic, continuous, complete, finite Borel measures on Cantor spaces is a very complex task which has not been achieved yet. Simple topological invariants do not seem to characterize the homeomorphism classes of such measures. By introducing a linearly ordered topology consistent with the given topology of a Cantor space, which is always possible, a linear topological invariant has been discovered by Akin in [**2**]. In another direction, by restricting the investigation to those measures that are product probability measures, a courser equivalence relation can be introduced (see Mauldin [**105**]). This equivalence relation permits the introduction of algebraic methods. Recent advances in this approach have been made by Dougherty, Mauldin and Yingst [**47**] and by T. D. Austin [**6**].

The appendix begins with a discussion of properties of the simultaneously closed and open sets of $k^{\mathbb{N}}$. The next section concerns metrics on Cantor spaces. There are many topologically compatible metrics on $\{0,1\}^{\mathbb{N}}$ – of special interest are those induced by homeomorphisms between $\{0,1\}^{\mathbb{N}}$ and $k^{\mathbb{N}}$. The remaining sections concern shift invariant product measures on $k^{\mathbb{N}}$ – these are the Bernoulli measures. Akin's development of the uniform Bernoulli measures is presented. For $\operatorname{card}(k) = 2$, the Bernoulli measures are closely tied to the binomial coefficients; such measures will be called binomial Bernoulli measures. The discussion of binomial Bernoulli measures is based on Dougherty, Mauldin and Yingst [**47**], Akin [**3, 2**], and Austin [**6**]. Also presented are the works of Huang [**78**], Mauldin [**105**], Navarro-Bermúdez [**115, 116**], and Navarro-Bermúdez and Oxtoby [**117**].

C.1. Closed and open sets

Let k be a finite set consisting of more than one point and let n be a positive integer. Then the natural projection $\pi_n\colon k^{\mathbb{N}} \to k^n$ defines a natural collection $\mathcal{P}_n$ of simultaneously closed and open sets of $k^{\mathbb{N}}$; namely,

$$\mathcal{P}_n = \{\pi_n{}^{-1}[\{w\}]\colon w \in k^n\}.$$

The members of this collection are called *cylinder sets* of $k^{\mathbb{N}}$. The collection $\mathcal{P}_{n+1}$ refines $\mathcal{P}_n$, and the collection $\bigcup_{n\in\mathbb{N}} P_n$ is a countable base for the open sets of $k^{\mathbb{N}}$.

Let X be a nonempty, compact, perfect, zero-dimensional metrizable space and let $h\colon X \to k^{\mathbb{N}}$ be a continuous bijection. Then $h^{-1}[\mathcal{P}_n]$, $n \in \mathbb{N}$, is a sequence of simultaneously closed and open sets of X, and $\bigcup_{n\in\mathbb{N}} h^{-1}[\mathcal{P}_n]$ is a countable base for the open sets of X. Moreover, for any metric on X, $\lim_{n\to\infty} \operatorname{mesh}(h^{-1}[\mathcal{P}_n]) = 0$.

Notation C.1. *For each n in $\mathbb{N}$ let $\pi_n\colon \{0,1\}^{\mathbb{N}} \to \{0,1\}^n$ be the natural projection. For each subset E of $\{0,1\}^n$ let $\langle E\rangle$ denote the simultaneously closed and open set $\pi_n{}^{-1}[E]$. For singleton sets $\{e\}$ the notation $\langle\{e\}\rangle$ will be abbreviated as $\langle e\rangle$.*

Clearly a subset A of $k^{\mathbb{N}}$ is a cylinder set if and only if there exist an n in $\mathbb{N}$ and a point e in k^n such that $\langle e\rangle = A$. If A is a nonempty simultaneously closed and open set in $k^{\mathbb{N}}$, then there is a unique integer n such that $\langle \pi_n[A]\rangle = A$ and $\langle \pi_m[A]\rangle \neq A$ whenever $m < n$. Indeed, by the compactness of A, there is a large m such that some finite subset E_m, which is unique, of k^m is such that $\pi_m[A] = E_m$ and $\langle E_m\rangle = A$. The least such m is the required n.

Lemma C.2. *Let $U = \{U_1, U_2, \ldots, U_k\}$ and $V = \{V_1, V_2, \ldots, V_l\}$ be finite sets. Then the following statements hold.*

1. *If φ is an injection of U into V, then there exists a continuous injection $\Phi\colon U^{\mathbb{N}} \to V^{\mathbb{N}}$, where the respective i-th coordinates x_i and $\Phi(x)_i$ of x and $\Phi(x)$ satisfy $\Phi(x)_i = \varphi(x_i)$. Moreover, if φ is a bijection, then Φ is a bijection.*
2. *If ψ is a map from U to V, then there exists a continuous injection $\Psi\colon U^{\mathbb{N}} \to V^{\mathbb{N}}$.*

The proof of the first statement is straightforward. For the proof of the second statement observe that there exists a natural homeomorphism between $V^{\mathbb{N}}$ and $(V^k)^{\mathbb{N}}$ and there is a natural bijection from U to the graph of ψ.

There are two natural continuous surjections from $k^{\mathbb{N}}$ to itself. It will be convenient to adopt the notation used in the above lemma to make some formulas less cumbersome. That is, for x in $k^{\mathbb{N}}$, the i-th coordinate of x will be denoted by x_i. Associated with the space $k^{\mathbb{N}}$ is the continuous surjection $\mathrm{s}\colon k^{\mathbb{N}} \to k^{\mathbb{N}}$ defined by $\mathrm{s}(x)_i = x_{i+1}$, $i \in \mathbb{N}$, for x in $k^{\mathbb{N}}$, called the *shift map*. The other continuous surjection is formed by permutations of $\mathbb{N}$. Of course, the resulting map is a homeomorphism.

The following is a lemma which provides a homeomorphism between $\{0,1\}^{\mathbb{N}}$ and $\{0,1,2\}^{\mathbb{N}}$.

LEMMA C.3. *For each e in $\{0,1\} \cup \{0,1\}^2$, let $\langle e \rangle$ be the cylinder set in the space $\{0,1\}^{\mathbb{N}}$. Let M be any one of the subsets of $\{0,1\} \cup \{0,1\}^2$ given by*

$$U = \{(0), (1,0), (1,1)\} \quad \textit{and} \quad V = \{(1), (0,1), (0,0)\}.$$

Then there is a homeomorphism $h\colon \{0,1\}^{\mathbb{N}} \to M^{\mathbb{N}}$ and there is a continuous map $n\colon \{0,1\}^{\mathbb{N}} \times \mathbb{N} \to \mathbb{N}$ such that the i-th coordinate of $y = h(x)$ is the unique member of M determined by the $n(x, i-1)$-th and the $n(x,i)$-th coordinates of x, where $n(x,0) = 1$.

PROOF. The key to the proof is the following property of the collection U: If a and b are in $\{0,1\}$, then

$$I(a,b) = \{(a), (a,b)\} \cap U$$

is a singleton set.

Recall that there is a natural homeomorphism between $U^{\mathbb{N}}$ and ${}^{\mathbb{N}}U$, the collection of all functions f on $\mathbb{N}$ into a set U. The topology on U is the discrete topology, and the topology on ${}^{\mathbb{N}}U$ is the topology of pointwise convergence. The map h will be inductively defined at each x of $\{0,1\}^{\mathbb{N}}$; that is, the value $h(x)$, which is a function on $\mathbb{N}$ into U, will be defined inductively. We will inductively define $n(x, \cdot)$ at the same time. For $k = 1$, let $h(x)\big(1\big)$ be the unique member of U determined by $I(x_1, x_2)$. For convenience define $n(x,0) = 1$. Define $n(x,1) = n(x,0) + 1$ if $h(x)\big(1\big) = (x_1)$ and $n(x,1) = n(x,0) + 2$ if $h(x)\big(1\big) \neq (x_1)$. Suppose that $h(x)\big(i\big)$, $i = 1, 2, \ldots, k$, and $n(x,j)$, $j = 0, 1, \ldots, k$, have been determined so that $h(x)\big(i\big)$ is the unique member of $I(x_{n(x,i-1)}, x_{n(x,i-1)+1})$ whenever $1 \leq i \leq k$ and that $n(x,j) = n(x, j-1) + 1$ if $h(x)\big(j\big) = (x_{n(x,j-1)})$, and $n(x,j) = n(x,j-1)+2$ if $h(x)\big(j\big) \neq (x_{n(x,j-1)})$ whenever $1 \leq j \leq k$. The definition of $h(x)\big(k+1\big)$ and $n(x, k+1)$ follows the same procedure as the first step of the induction. Clearly h is surjective. Let us show that h is injective and that h^{-1} is continuous, whence h is a homeomorphism. Let x and x' be distinct members of $\{0,1\}^{\mathbb{N}}$. If $x_1 \neq x'_1$, then $h(x)\big(1\big) \neq h(x')\big(1\big)$. Also, if $(x_1, x_2) \neq (x'_1, x'_2)$, then $h(x)\big(i\big) \neq h(x')\big(i\big)$ for some i with $i \leq 2$. Let k be such that $x_j = x'_j$ for $j < k$ and $x_k \neq x_k$. We may assume $k > 2$. Suppose that $h(x)\big(i\big) = h(x')\big(i\big)$ for every i such that $i \leq 2(k+1)$. Then $n(x,i) = n(x',i)$ for each i with $i \leq 2k+1$. Consequently, $x_j = x'_j$ whenever $j \leq 2k$. But this denies $x_k \neq x'_k$. Hence $h(x)\big(i\big) \neq h(x')\big(i\big)$ for some i. Thereby h is bijective. Finally, let us turn to the continuity of h^{-1} at y in ${}^{\mathbb{N}}U$. Let $x = h^{-1}(y)$ and $\langle(x_1, \ldots, x_k)\rangle$ be a cylinder set for x. By construction, $h(x) = y$. As $k < n(x, 2k+2)$ we have $h^{-1}[\langle(y_1, \ldots, y_{2k})\rangle] \subset \langle(x_1, \ldots, x_k)\rangle$ and the continuity of h^{-1} at y is verified.

Observe that $n(\cdot\,,\,\cdot)$ is continuous at (x,k). This follows from the fact that h^{-1} is a homeomorphism and $\mathbb{N}$ is a discrete space.

The proof for the case $M = V = \{(1), (0,1), (0,0)\}$ follows by interchanging 0 and 1 in the proof for the case $M = U$. □

The homeomorphism $h\colon \{0,1\}^{\mathbb{N}} \to U^{\mathbb{N}}$ has some nice properties that will be useful later. For $n \in \mathbb{N}$ let $\Pi_n\colon U^{\mathbb{N}} \to U^n$ be the natural projection and let $E_n = \{0,1\}^n \cup \{0,1\}^{n+1} \cup \cdots \cup \{0,1\}^{2n}$. There is a natural map $e_n : u \mapsto e_n(u)$ from U^n into

E_n associated with h such that $\langle e_n(u)\rangle$ is a cylinder set in $\{0,1\}^{\mathbb{N}}$, $h^{-1}\big[\Pi_n{}^{-1}[\{u\}]\big] = \langle e_n(u)\rangle$ and such that if $u = (u', u'')$ is in $U^{n-1} \times U$ then $e_n(u) = (e_{n-1}(u'), e_1(u''))$. Hence we have the following

LEMMA C.4. *Let $h\colon \{0,1\}^{\mathbb{N}} \to U^{\mathbb{N}}$, E_n and e_n be as in the preceding paragraph. There is a natural map $H_n\colon u \mapsto \langle e_n(u)\rangle$ of U^n into the collection of all cylinder sets of $\{0,1\}^{\mathbb{N}}$ such that $H_n(u) = h^{-1}\big[\Pi_n{}^{-1}[\{u\}]\big]$ and such that $H_n(u) = \langle (e_{n-1}(u'), e_1(u''))\rangle \subset \langle e_{n-1}(u')\rangle = H_{n-1}(u')$ whenever $u = (u', u'') \in U^{n-1} \times U$.*

There is a natural subset of $\mathbb{R}$ associated with continuous, complete, finite Borel measures on a separable metrizable space X.

NOTATION C.5. *For a separable metrizable space X, the collection of all simultaneously closed and open sets is denoted by $\mathfrak{CO}(X)$. For a continuous, complete, Borel measure μ on X, the* value set *of μ is the subset $\mathrm{vs}(\mu, X)$ of $\mathbb{R}$ defined by*

$$\mathrm{vs}(\mu, X) = \{\mu(U)\colon U \in \mathfrak{CO}(X)\}.$$

A simple compactness argument yields

PROPOSITION C.6. *If X is a nonempty, compact, perfect, 0-dimensional metrizable space, then $\mathrm{card}(\mathfrak{CO}(X)) = \aleph_0$. Furthermore, if μ is a continuous, complete, finite Borel measure on X, then $\mathrm{vs}(\mu, X)$ is a countable dense subset of the interval $[0, \mu(X)]$ and contains the end points of the interval.*

The inverse image of general Borel measurable maps need not preserve open or closed sets, but inverse image of continuous maps preserve both open and closed sets. Hence the following proposition is an easy exercise.

PROPOSITION C.7. *Let X_1 and X_2 be be topological copies of $\{0,1\}^{\mathbb{N}}$ and let μ be a continuous, complete, finite Borel measure on X_1. If $f\colon X_1 \to X_2$ is a continuous map, then $\mathrm{vs}(f_{\#}\mu, X_2) \subset \mathrm{vs}(\mu, X_1)$ with equality whenever f is a homeomorphism.*

C.2. A metric for $k^{\mathbb{N}}$

Let k be a finite set with $\mathrm{card}(k) > 1$ and consider ${}^{\omega}k$, the set of all functions $f\colon \omega \to k$, where ω is the set of all finite ordinal numbers. For each finite subset A in ω the collection of all functions from A into k is denoted by ${}^{A}k$. There is a natural map of ${}^{\omega}k$ onto ${}^{A}k$. By providing k with the discrete topology, the set ${}^{\omega}k$ becomes a compact, metrizable, topological space that is homeomorphic to $k^{\mathbb{N}}$. Let us define metrics on ${}^{\omega}k$ that are useful in the computation of certain Hausdorff measures on $k^{\mathbb{N}}$. For distinct f and g in ${}^{\omega}k$ define

$$\chi(f, g) = \min\{n \in \omega\colon f(n) \neq g(n)\},$$

that is, $\chi(f,g)$ is the length of the initial segment that is common to f and g. Let $0 < \alpha < 1$ and define the metric

$$\mathrm{d}_{(k,\alpha)}(f,g) = \begin{cases} \alpha^{\chi(f,g)}, & \text{if } f \neq g, \\ 0, & \text{if } f = g. \end{cases}$$

That $\mathrm{d}_{(k,\alpha)}$ is indeed a metric is left as an exercise. The *Cantor cube* $\mathbb{C}(k,\alpha)$ is the topological space ${}^{\omega}k$ endowed with the metric $\mathrm{d}_{(k,\alpha)}$.

Let k and l be finite sets and let $0 < \alpha < 1$. Denote by $f \times g$ the members of the metrizable product space $\mathbb{C}(k,\alpha) \times \mathbb{C}(l,\alpha)$, where $f \in {}^{\omega}k$ and $g \in {}^{\omega}l$. Let us display a natural a metric for $\mathbb{C}(k,\alpha) \times \mathbb{C}(l,\alpha)$ that makes the map $\Phi\colon {}^{\omega}k \times {}^{\omega}l \to {}^{\omega}(k \times l)$ defined by

$$\Phi(f \times g)(n) = \langle f(n), g(n) \rangle, \quad n \in \omega,$$

into an isometry. To this end, we write the value $\Phi(f \times g)$ as $\langle f, g \rangle$. Clearly, Φ is an injection. Next let Π_1 and Π_2 be the respective projections of $k \times l$ onto k and onto l. For each h in ${}^{\omega}(k \times l)$ we have $\Pi_1 h \in {}^{\omega}k$ and $\Pi_2 h \in {}^{\omega}l$. Define $\Psi: {}^{\omega}(k \times l) \to {}^{\omega}k \times {}^{\omega}l$ to be the map given by $\Psi(h) = \Pi_1 h \times \Pi_2 h$. Clearly, Ψ is an injection. Moreover, $\Psi\Phi$ and $\Phi\Psi$ are the identity maps on ${}^{\omega}k \times {}^{\omega}l$ and ${}^{\omega}(k \times l)$, respectively. Define δ to be the maximum metric on $\mathbb{C}(k,\alpha) \times \mathbb{C}(l,\alpha)$, that is,

$$\delta(f \times g, f' \times g') = \max\{\mathrm{d}_{(k,\alpha)}(f,f'),\, \mathrm{d}_{(l,\alpha)}(g,g')\}.$$

That δ is an isometry is left as an exercise.

We have the following proposition.

PROPOSITION C.8. *$\mathbb{C}(k,\alpha) \times \mathbb{C}(l,\alpha)$ and $\mathbb{C}(k \times l,\alpha)$ are isometric spaces. Hence there are metrics on $k^{\mathbb{N}}$, $l^{\mathbb{N}}$ and $(k \times l)^{\mathbb{N}}$ such that the natural bijection of $k^{\mathbb{N}} \times l^{\mathbb{N}}$ to $(k \times l)^{\mathbb{N}}$ is an isometry.*

Let us consider the space ${}^{\omega}({}^{m}k)$ where $m \in \omega$ and $1 \leq m$. Recall that each n in ω is expressed uniquely as $n = qm + r$ where q is the quotient and r is the remainder upon division by m. For $0 \leq r < m$ let $\Pi_r\colon {}^{m}k \to k$ be defined by $\Pi_r(f) = f(r)$ whenever $f \in {}^{m}k$. It is easy to see that, for $h \in {}^{\omega}({}^{m}k)$, the map $\Pi_r h$ is in ${}^{\omega}k$. Define the map $\Psi_m(h)$ on ${}^{\omega}k$ as follows:

$$\Psi_m(h)(n) = \Pi_r h(q) \quad \text{where} \quad n = qm + r \in \omega.$$

It is easily seen that $\Psi_m\colon {}^{\omega}({}^{m}k) \to {}^{\omega}k$ is an injection. Next, for $f \in {}^{\omega}k$, define $\Phi_m(f)$ to be the map in ${}^{\omega}({}^{m}k)$ as follows:

$$\Phi_m(f)(q) = \langle f(qm), f(qm+1), \ldots, f(qm+r-1) \rangle \quad \text{where} \quad q \in \omega.$$

It is easily seen that $\Phi_m\colon {}^{\omega}k \to {}^{\omega}({}^{m}k)$ is an injection. Moreover $\Psi_m\Phi_m$ and $\Phi_m\Psi_m$ are identity maps on ${}^{\omega}k$ and ${}^{\omega}({}^{m}k)$, respectively. We leave the following inequalities as exercises.

$$\alpha^{1-m}\ \mathrm{d}_{(k,\alpha)}(f,f') \geq \mathrm{d}_{(k^m,\alpha^m)}(\Phi_m(f),\Phi_m(f')) \quad \text{whenever} \quad f \in k^\omega \text{ and } f' \in k^\omega. \tag{C.1}$$

$$\mathrm{d}_{(k^m,\alpha^m)}(h,h') \geq \mathrm{d}_{(k,\alpha)}(\Psi_m(h),\Psi_m(h')) \quad \text{whenever} \quad h \in (k^m)^\omega \text{ and } h' \in (k^m)^\omega, \tag{C.2}$$

The inequalities yield the next proposition.

PROPOSITION C.9. *Let k be a finite set with* card$(k) > 1$. *For $m \in \omega$ with $m \geq 1$ there exists a bi-Lipschitzian map Φ_m from $\mathbb{C}(k,\alpha)$ onto $\mathbb{C}(k^m,\alpha^m)$. Consequently, there exist metrics on $k^{\mathbb{N}}$ and $(k^m)^{\mathbb{N}}$ and a bi-Lipschitzian homeomorphism between $k^{\mathbb{N}}$ and $(k^m)^{\mathbb{N}}$ when they are endowed with these metrics.*

C.3. Bernoulli measures

There are natural probability measures P on finite sets k, called *Bernoulli measures on k*, where its σ-algebra is the collection of all subsets of k. Of course, $\sum_{w\in k} P(\{w\}) = 1$. We shall call the measure *uniformly distributed* or *uniform distribution* if $P(\{w\}) = \frac{1}{\mathrm{card}(k)}$ for every w in k. Also, we shall call the measure a *binomial Bernoulli distribution* (*binomial distribution* for short) if card$(k) = 2$.

Let k be a finite set with card$(k) > 1$. If μ_n is a Bernoulli measure on k for each n in $\mathbb{N}$, then there is the usual product probability measure $\mu = \times_{n\in\mathbb{N}} \mu_n$. If every factor probability measure μ_n is the same Bernoulli measure P, then the resulting product measure will be called a *Bernoulli measure* for P and will be denoted by $\beta(P,k)$ (or simply $\beta(P)$ if the finite set k is fixed). Clearly the Bernoulli measures $\beta(P,k)$ are continuous Borel measures on $k^{\mathbb{N}}$ which are shift invariant, that is,

$$\mathrm{s}_{\#}\,\beta(P,k) = \beta(P,k), \tag{C.3}$$

and which satisfy the product measure property

$$\beta(P,k)\big({\pi_n}^{-1}[\{w\}]\big) = \Pi_{i=1}^{n} P(\{w_i\}) \quad \text{whenever} \quad n \in \mathbb{N}, \tag{C.4}$$

where π_n is the natural projection of $k^{\mathbb{N}}$ onto k^n. Moreover, if μ is a probability measure on $k^{\mathbb{N}}$ that is shift invariant and satisfies the above product measure property, then μ is the Bernoulli measure $\beta(P,k)$ on $k^{\mathbb{N}}$ with $P = \pi_{1\#}\mu$.

NOTATION C.10. *Let k be a finite set with* $m =$ card$(k) > 1$. *The Bernoulli measure $\beta(P)$ with the uniform distribution P will be denoted by $\beta(\frac{1}{m})$. The Bernoulli measure $\beta(P)$ with binomial distribution P will be denoted by $\beta(r, 1-r)$, where $P[\{w\}] \in \{r, 1-r\} \subset (0,1) \subset \mathbb{R}$ whenever $w \in \{0,1\}$.*

C.4. Uniform Bernoulli distribution

Implicit in the notation $\beta(\frac{1}{m})$ is a finite set k with $m = \operatorname{card}(k) > 1$ and the uniform Bernoulli measure on k. Clearly this notation has assumed that if k_1 and k_2 are finite sets with the same cardinality then the two respective uniform Bernoulli measures on k_1 and k_2 are the same. This assumption will present no topological difficulties since this identification leads to natural bijections between ${k_1}^{\mathbb{N}}$ and ${k_2}^{\mathbb{N}}$ that are homeomorphisms; hence the respective Bernoulli measures induced by the uniform Bernoulli measures on k_1 and k_2 are homeomorphic measures. The reference to $k^{\mathbb{N}}$ will be dropped from the value set also. It is easily seen that the value set of the measure $\beta(\frac{1}{m})$ is

$$\operatorname{vs}\big(\beta(\tfrac{1}{m})\big) = \big\{\tfrac{i}{m^n} : i = 0, \dots, m^n,\ n \in \mathbb{N}\big\}. \tag{C.5}$$

Let us turn to the case of $\beta(\frac{1}{m_1})$ and $\beta(\frac{1}{m_2})$ with $m_1 \neq m_2$, that is, finite sets k_1 and k_2 with unequal cardinalities. The question is: What are necessary and sufficient conditions for the existence of a homeomorphism $h\colon {k_1}^{\mathbb{N}} \to {k_2}^{\mathbb{N}}$ such that $h_{\#}\beta(\frac{1}{m_1}) = \beta(\frac{1}{m_2})$? To answer this question, we make some preliminary observations. For finite sets k_i, $i = 1, \dots, j$, let $k = \times_{i=1}^{j} k_i$, $m_i = \operatorname{card}(k_i)$ for each i, and $m = m_1 \cdots m_j$. Observe that $\times_{i=1}^{j} \beta(\frac{1}{m_i})$ is a Borel measure on k. From Proposition C.8 we infer the existence of a homeomorphism $h\colon k^{\mathbb{N}} \to \times_{i=1}^{j} {k_i}^{\mathbb{N}}$ such that

$$h_{\#}\beta(\tfrac{1}{m}) = \times_{i=1}^{j} \beta(\tfrac{1}{m_i}). \tag{C.6}$$

Similarly, for a finite set k with cardinality $m > 1$ and for a positive integer i we infer from Proposition C.9 the existence of a homeomorphism $h\colon k^{\mathbb{N}} \to (k^i)^{\mathbb{N}}$ such that

$$h_{\#}\beta(\tfrac{1}{m}) = \beta(\tfrac{1}{m^i}). \tag{C.7}$$

We can now answer the question that prompted the last two displayed formulas.

Proposition C.11. *Suppose m_1 and m_2 are integers larger than 1. In order that $h_{\#}\beta(\frac{1}{m_1}) = \beta(\frac{1}{m_2})$ for some homeomorphism h it is necessary and sufficient that m_1 and m_2 have precisely the same prime divisors.*

Proof. Suppose that such a homeomorphism h exists. Then their value sets are the same; that is, $\operatorname{vs}\big(\beta(\frac{1}{m_1})\big) = \operatorname{vs}\big(\beta(\frac{1}{m_2})\big)$. We infer from Exercise C.6 that m_1 and m_2 have precisely the same prime divisors. Conversely, suppose that m_1 and m_2 have precisely the same prime divisors. Exercise C.7 will complete the proof. □

Observed earlier was that the value set $\operatorname{vs}(\mu, X)$ is a topological invariant, that is, homeomorphic measures have the same value set. Akin showed that non homeomorphic measures can have the same value set. We now present his example which uses the uniform Bernoulli measure.

Theorem C.12. *The value set of the binomial Bernoulli measure $\beta(\frac{1}{3}, \frac{2}{3})$ on $\{0, 1\}^{\mathbb{N}}$ and the value set of the uniform Bernoulli measure $\beta(\frac{1}{3}, \frac{1}{3}, \frac{1}{3})$ on $\{0, 1, 2\}^{\mathbb{N}}$ are equal* *to* $\{\frac{a}{3^n} : 0 \leq a \leq 3^n,\ n \in \mathbb{N}\}$. *Furthermore,* $h_\#\beta(\frac{1}{3}, \frac{2}{3}) \neq \beta(\frac{1}{3}, \frac{1}{3}, \frac{1}{3})$ *for every homeomorphism* h *of* $\{0, 1\}^{\mathbb{N}}$ *onto* $\{0, 1, 2\}^{\mathbb{N}}$.

PROOF. It is already known that $\mathrm{vs}(\beta(\frac{1}{3}, \frac{1}{3}, \frac{1}{3}), \{0, 1, 2\}^{\mathbb{N}})$ is equal to $\{\frac{a}{3^n} : 0 \leq a \leq 3^n,\ n \in \mathbb{N}\}$. Also, that $\mathrm{vs}(\beta(\frac{1}{3}, \frac{2}{3}), \{0, 1\}^{\mathbb{N}})$ is a subset of $\{\frac{a}{3^n} : 0 \leq a \leq 3^n,\ n \in \mathbb{N}\}$ is easily shown. The reverse inclusion will be proved by induction – that is, for every n, there exists for each a with $0 \leq a \leq 3^n$ a simultaneously closed and open set U such that $\beta(\frac{1}{3}, \frac{2}{3})(U) = \frac{a}{3^n}$. The statement is true for $n = 1$. For the inductive step we first write

$$\frac{a}{3^n} = \frac{1}{3} \cdot \frac{a_0}{3^{n-1}} + \frac{2}{3} \cdot \frac{a_1}{3^{n-1}}$$

with $a_0,\ a_1 \leq 3^{n-1}$. To see that this is possible, one observes, for $a \leq 2 \cdot 3^{n-1}$, that a_1 is the integer part of $\frac{a}{2}$ and a_0 is 0 or 1; and, for $2 \cdot 3^{n-1} < a \leq 3^n$, that $a_1 = 3^{n-1}$ and $a_0 = a - 2 \cdot 3^{n-1}$ (note that $a_0 \leq (3 - 2) \cdot 3^{n-1}$). Let U_0 and U_1 be simultaneously closed and open sets such that $\beta(\frac{1}{3}, \frac{2}{3})(U_i) = \frac{a_i}{3^{n-1}}$ for $i = 0, 1$. For $i = 0$, 1, define the one-sided inverses s_i of the shift map to be $(x_1, x_2, \ldots) \mapsto (i, x_1, x_2, \ldots)$. Let $U = \mathrm{s}_0[U_0] \cup \mathrm{s}_1[U_1]$. Then

$$\beta(\tfrac{1}{3}, \tfrac{2}{3})(U) = \tfrac{1}{3} \cdot \beta(\tfrac{1}{3}, \tfrac{2}{3})(U_0) + \tfrac{2}{3} \cdot \beta(\tfrac{1}{3}, \tfrac{2}{3})(U_1) = \tfrac{a}{3^n}.$$

The inductive step has been verified.

The second statement of the theorem will be proved by contradiction. Suppose that $h\colon \{0, 1\}^{\mathbb{N}} \to \{0, 1, 2\}^{\mathbb{N}}$ is a homeomorphism such that $h_\#\beta(\frac{1}{3}, \frac{2}{3}) = \beta(\frac{1}{3}, \frac{1}{3}, \frac{1}{3})$. Let π be the natural projection of $\{0, 1, 2\}^{\mathbb{N}}$ onto its first coordinate space $\{0, 1, 2\}$ and let $\mathcal{A}$ be the collection whose members are $A_i = h^{-1}\pi^{-1}[\{i\}]$, $i = 0, 1, 2$. There is an n such that $E_i = \pi_n[A_i]$ has the property that $\langle E_i \rangle = A_i$ for each i. There is an i such that $0 \in E_i$, where $0 = (0, 0, \ldots, 0)$ is the member of $\{0, 1\}^n$ whose coordinates are all 0. We may assume $i = 0$. Then each member of E_1 has a coordinate whose value is 1. Hence $\frac{1}{3} = \beta(\frac{1}{3}, \frac{2}{3})(\langle E_1 \rangle) = 2(\frac{m}{3^k})$, where $\frac{m}{3^k}$ is in reduced form. This is a contradiction. □

Here is a simple lemma related to the counterexample.

LEMMA C.13. *Let* k *and* l *be finite sets and let* $g\colon k \to l$ *be a nonconstant map. If* P *is Bernoulli measure on* k*, then there is a continuous map* $G\colon k^{\mathbb{N}} \to l^{\mathbb{N}}$ *such that* $G_\#\beta(P, k) = \beta(g_\# P, l)$.

C.5. Binomial Bernoulli distribution

In this section we shall concentrate on the binomial Bernoulli distribution $\beta(r, 1-r)$[1] on the Cantor space $\{0, 1\}^{\mathbb{N}}$. Let us begin with a simple consequence of the properties

[1] The discussion of binomial Bernoulli distributions P will be carried out on the set $\{0, 1\}$ of cardinality 2. Our assignment of the values of r and $1 - r$ will be to 0 and 1, respectively, for P, which is opposite of that in many other papers. (The opposite situation is often called *Bernoulli trials*, where 1 stands for success and 0 stands for failure.) Hence the product measure $\beta(r, 1 - r)$ will also differ. Of course there is a simple homeomorphism of $\{0, 1\}^{\mathbb{N}}$ which passes from our assignment to the others. Hence the translations required in statements of assertions and proofs are not difficult to make.

of the value sets $\mathrm{vs}(\mu)$ of measures on $\{0,1\}^{\mathbb{N}}$; namely, from $\mathrm{card}(\mathrm{vs}(\mu)) \le \aleph_0$ and $r \in \mathrm{vs}(\beta(r, 1-r))$ we have

PROPOSITION C.14. *For each r in $(0,1)$ there are only countably many s in $(0,1)$ such that $\beta(r, 1-r)$ and $\beta(s, 1-s)$ are homeomorphic. Hence, among the binomial Bernoulli measures, the set of homeomorphism equivalence classes has cardinality* $\mathfrak{c}$.

This proposition does not give much information about the equivalence classes. Hence a more refined analysis found in the literature will be presented. In order to simplify the notation in our discussion, the binomial Bernoulli distribution $\beta(r, 1-r)$ will be abbreviated as β_r. Let $\mathsf{B}(\{0,1\}^{\mathbb{N}})$ denote the collection of all measures β_r, $r \in (0,1)$. Then $\beta\colon r \mapsto \beta_r$ is a bijection of $(0,1)$ to $\mathsf{B}(\{0,1\}^{\mathbb{N}})$.

C.5.1. Continuity. We begin with the effect of continuous maps f of $\{0,1\}^{\mathbb{N}}$ on the value $s = f_\# \beta_r(\langle 0 \rangle)$ (here, 0 is in $\{0,1\}$). Of course, only those f for which $\beta_s = f_\# \beta_r$ are of interest; this will not happen for every f. We shall say that s *continuously refines* r (denoted $s \le_{\mathrm{c}} r$) if there is a continuous f such that $\beta_s = f_\# \beta_r$. Clearly the relation $\le_{\mathrm{c}}$ is transitive and reflexive. A trivial example is given by the permutation of $\{0,1\}$ which results in $s = 1 - r$. There is the following characterization of such continuous maps f (see Mauldin [**105**, Theorem 1.1, page 619]). Clearly, f must be surjective.

THEOREM C.15. *There is a continuous map $f\colon \{0,1\}^{\mathbb{N}} \to \{0,1\}^{\mathbb{N}}$ such that $\beta_s = f_\# \beta_r$ if and only if there is a positive integer n and there are integers a_i, $0 \le i \le n$, such that*

$$0 \le a_i \le \binom{n}{i}, \quad 0 \le i \le n, \tag{C.8}$$

and

$$s = \textstyle\sum_{i=0}^{n} a_i r^i (1-r)^{n-i}. \tag{C.9}$$

PROOF. Suppose f is a continuous map such that $\beta_s = f_\# \beta_r$. Then $U = f^{-1}[\langle 0 \rangle]$ is a simultaneously closed and open set. There is an integer n and a subset $\mathcal{E}$ of $\{0,1\}^n$ such that $U = \bigcup \{\langle e \rangle : e \in \mathcal{E}\}$. For each i with $0 \le i \le n$ let

$$a_i = \mathrm{card}\big(\{(q_1, \ldots, q_n) \in \mathcal{E} \colon \textstyle\sum_{p=1}^{n} q_p = n - i\}\big).$$

One easily verifies that $0 \le a_i \le \binom{n}{i}$ and that $\beta_r\big({\pi_n}^{-1}[\{(q_1, \ldots, q_n)\}]\big) = r^i (1-r)^{n-i}$ whenever $\sum_{p=1}^{n} q_p = n - i$. Hence

$$s = \beta_s(\langle 0 \rangle) = \beta_r\big(U\big) = \textstyle\sum_{i=0}^{n} a_i r^i (1-r)^{n-i}.$$

To prove the converse assume that the two conditions of the theorem hold and consider the set $\{0,1\}^n$. As $0 \le a_i \le \binom{n}{i}$ for each i, there is a subset $\mathcal{E}$ of $\{0,1\}^n$ such that it has exactly a_i members $(e_1, \ldots, e_n)$ with $\sum_{p=0}^{n} e_p = n - i$ for every i. Let P be the binomial Bernoulli measure on $\{0,1\}$ such that $\beta_r = \beta(P, \{0,1\})$. Let $\varphi\colon \{0,1\}^{\mathbb{N}} \to (\{0,1\}^n)^{\mathbb{N}}$ be the obvious homeomorphism. Then $\varphi_\# \beta(P, \{0,1\}) = \beta(P^n, \{0,1\}^n)$. Obviously $P^n(\mathcal{E}) = \sum_{i=0}^{n} a_i r^i (1-r)^{n-1} = s$. Let

$g\colon \{0,1\}^n \to \{0,1\}$ be such that $g^{-1}[\{0\}] = \mathcal{E}$. Then, by Lemma C.13, there is a continuous map $G\colon (\{0,1\}^n)^{\mathbb{N}} \to \{0,1\}^{\mathbb{N}}$ such that $G_{\#}\beta(P^n, \{0,1\}^n) = \beta_s$. The continuous map $f = G\varphi$ completes the proof. □

It is the above theorem that encourages us to use the name binomial Bernoulli measure for β_r. The following example is given in [**105**]: There is a continuous map f such that $f_{\#}\beta_r = \beta_s$ if $r = \frac{1}{\sqrt{2}}$ and $s = \frac{1}{2}$. Indeed, letting $n = 2$, $a_0 = a_1 = 0$ and $a_2 = 1$ in the theorem, we have the existence of such an f. The theorem also implies that there are no continuous maps g such that $g_{\#}\beta_{\frac{1}{2}} = \beta_{\frac{1}{\sqrt{2}}}$; hence f is not a homeomorphism. We shall return to this example in the comment section at the end of the appendix.

Observe that a permutation of $\mathbb{N}$ induces a homeomorphism h of $\{0,1\}^{\mathbb{N}}$ and $h_{\#}\beta_r = \beta_r$. Consequently, $f_{\#}\beta_r = (fh)_{\#}\beta_r$; and so there are many continuous maps that satisfy the conditions of the theorem.

C.5.2. *Partition polynomials.* The equation $\beta_s = f_{\#}\beta_r$ results in a polynomial

$$p(x) = \textstyle\sum_{i=0}^{n} a_i x^i (1-x)^{n-i}, \tag{C.10}$$

in $\mathbb{Z}[x]$, where $0 \le a_i \le \binom{n}{i}$, $i = 0, 1, \dots, n$. Such polynomials are called *partition polynomials*.[2] The collection of all partition polynomials is denoted by $\mathcal{P}$. Note that the partition polynomial in equation (C.10) satisfies

$$0 \le p(x) \le 1 = \textstyle\sum_{i=1}^{n} \binom{n}{i} x^i (1-x)^{n-i}, \quad x \in [0,1].$$

It is well-known that the collection $\{x^i(1-x)^{n-i} : i = 0, 1, \dots, n\}$ forms a basis for the vector space of polynomials of degree not exceeding n. We shall call this basis the *partition basis* for the polynomials of degree not exceeding n. There is a minimal n for which a partition polynomial $p(x)$ can be expressed in the form (C.10), which will be denoted by part-deg($p(x)$) and called the *partition degree* of $p(x)$. It is easily seen that for each partition polynomial $p(x)$ there is polynomial in the form (C.10) whenever $n \ge$ part-deg($p(x)$). Hence, if $p_k(x)$, $k = 1, 2, \dots, m$, is a finite collection of partition polynomials, then there is an n such that $p_k(x) = \sum_{i=0}^{n} a_{k,i} x^i (1-x)^{n-i}$ is of the form (C.10) for each k. It may happen that part-deg($p(x)$) > deg($p(x)$) for a partition polynomial $p(x)$. For example, the cubic $p(x) = 6x^2(1-x)$ is a partition polynomial with part-deg($p(x)$) > 3.

The following is a useful characterization of partition polynomials (see Dougherty, Mauldin and Yingst [**47**]).

THEOREM C.16 (Dougherty–Mauldin–Yingst). *If $p(x)$ is a polynomial with integer coefficients, then $p(x)$ is a partition polynomial if and only if $p(x)$ maps $(0,1)$ into $(0,1)$, or $p(x)$ is the 0 or 1 polynomial.*

[2] This definition is due to Austin [**6**].

Proof. If $p(x) = \sum_{i=0}^{n} a_i x^i (1-x)^{n-i}$ is a partition polynomial, then either p is the constant polynomial 0, or one of its coefficients is positive, in which case $p(x) > 0$ whenever $x \in (0,1)$. The same is true for $(1-p)(x) = \sum_{i=0}^{n} \left(\binom{n}{i} - a_i\right) x^i (1-x)^{n-i}$, so that $p(x) < 1$ whenever $x \in (0,1)$, or $p(x)$ is the constant polynomial 1. Thereby one implication is proved.

To prove the other implication, let $p(x) = \sum_{j=0}^{k} c_j x^j$ be a polynomial with integer coefficients such that $0 < p(x) < 1$ whenever $x \in (0,1)$. We infer from a theorem of Hausdorff[3] that there are nonnegative integers a_i, $i = 0, 1, \ldots, n$, such that $p(x) = \sum_{i=0}^{n} a_i x^i (1-x)^{n-i}$. Similarly, there are nonnegative integers b_i, $i = 0, 1, \ldots, m$, such that $(1-p)(x) = \sum_{i=0}^{m} b_i x^i (1-x)^{m-i}$. We may assume $n = m$. As $p(x) = 1-(1-p)(x)$ for every x, we have $\sum_{i=0}^{n} a_i x^i (1-x)^{n-i} = \sum_{i=0}^{n} \left(\binom{n}{i} - b_i\right) x^i (1-x)^{n-i}$ and hence, by linear independence, $a_i = \binom{n}{i} - b_i$. Consequently $0 \le a_i \le \binom{n}{i}$. Thereby $p(x)$ is a partition polynomial. □

The following corollary is obvious.

Corollary C.17. *Let $p_1(x)$ and $p_2(x)$ be partition polynomials. Then*

(1) *$p(x) = p_1(x)p_2(x)$ is a partition polynomial,*
(2) *$p(x) = p_2(p_1(x))$ is a partition polynomial,*
(3) *if $p_1(x) < p_2(x)$ whenever $x \in (0,1)$, then $p(x) = p_2(x) - p_1(x)$ is a partition polynomial.*

Let us observe a connection between partition polynomials $p(x)$ and simultaneously closed and open subsets C of $\{0,1\}^{\mathbb{N}}$ that is implicit in the proof of Theorem C.15. As in the proof, note that each nontrivial such set C corresponds to an n such that the cylinder sets of the form $\langle e \rangle$, $e \in \{0,1\}^n$, partitions $\{0,1\}^{\mathbb{N}}$ and refines the binary partition $\{C, \{0,1\}^{\mathbb{N}} \setminus C\}$ of $\{0,1\}^{\mathbb{N}}$. Each measure β_r assigns the value $r^i(1-r)^{n-i}$ to $\langle e \rangle$ for each $e = (e_1, e_2, \ldots, e_n)$ in $\{0,1\}^n$ where i is $n - \sum_{j=1}^{n} e_j$. Hence, for every r in $(0,1)$, $p(x)$ satisfies

$$\beta_r(C) = p(r) = \beta_{p(r)}(\langle 0 \rangle).$$

Conversely, note that every nonconstant partition polynomial $p(x)$ corresponds to a simultaneously closed and open subset C of $\{0,1\}^{\mathbb{N}}$ that satisfies the above equation. Indeed, let $f\colon \{0,1\}^{\mathbb{N}} \to \{0,1\}^{\mathbb{N}}$ be the continuous map such that $\beta_s = f_\# \beta_r$, where $s = p(r)$. Then C is $f^{-1}[\langle 0 \rangle]$, which depends only on $p(x)$. We say that C is a *closed and open set associated with* $p(x)$. For the constant partition polynomial 1 we associate the set $\{0,1\}^{\mathbb{N}}$, and for the constant partition polynomial 0 we associate the empty set. For $r \in (0,1)$, define $\mathcal{P}(r) = \{p(r)\colon p \in \mathcal{P}\}$. So $\mathcal{P}(r)$ is the value set $\mathrm{vs}(\beta_r) = \{\beta_r(C)\colon\ C \text{ and } \{0,1\}^{\mathbb{N}} \setminus C \text{ are open}\}$.

Implicit in the above discussion is the following.

Theorem C.18. *If C_2 is a simultaneously closed and open set in $\{0,1\}^{\mathbb{N}}$ whose associated polynomial is $p_2(x)$ and if $p_1(x)$ is a polynomial with integer coefficients such*

[3] See Exercise C.9 for details.

that $0 < p_1(x) < p_2(x)$ *whenever* $x \in (0,1)$, *then there is a simultaneously closed and open set* C_1 *contained in* C_2 *such that* $p_1(x)$ *is its associated polynomial.*

PROOF. Let n be large enough that $p_1(x) = \sum_{i=0}^{n} a_i x^i (1-x)^{n-i}$, $p_2(x) = \sum_{i=0}^{n} b_i x^i (1-x)^{n-i}$ and $(p_2 - p_1)(x) = \sum_{i=0}^{n} c_i x^i (1-x)^{n-i}$ satisfy $0 \leq a_i + c_i = b_i \leq \binom{n}{i}$. Let n also be large enough to yield the simultaneously closed and open set C_2, which is determined by the coefficients b_i, $i = 0, 1, \ldots, n$. The remaining step is obvious. □

Connected to the last theorem is the following observation: If C_1 and C_2 are simultaneously closed and open sets of $\{0,1\}^{\mathbb{N}}$ associated with the partition polynomials $p_1(x)$ and $p_2(x)$, respectively, and if $C_1 \subset C_2$, then $p_1(x) \leq p_2(x)$ whenever $x \in (0,1)$. This is obvious since $p_1(r) = \beta_r(C_1) \leq \beta_r(C_2) = p_2(r)$ for every r in $(0,1)$.

We have found many properties of partition polynomials that do not rely on the existence of a continuous function f such that $f_{\#}\beta_r = \beta_s$, that is, $s \leq_{\mathrm{c}} r$. Thus the following definition using only partition polynomials seems justified.

DEFINITION C.19. *Let* r *and* s *be in* $(0,1)$. *Then* s *is said to be* binomially reducible *to* r *if there is a partition polynomial* $p(x)$ *such that* $s = p(r)$ *(denoted* $s \leq_{\mathrm{b}} r$*).*

Clearly, $s \leq_{\mathrm{b}} r$ if and only if $s \leq_{\mathrm{c}} r$.

C.5.3. Binomial equivalence. Let us turn the above relation $\leq_{\mathrm{b}}$ into a symmetric one.

DEFINITION C.20. *A pair of numbers* r *and* s *in* $(0,1)$ *is said to be* binomially related *(denoted* $r \sim_{\mathrm{b}} s$*) if the conditions* $r \leq_{\mathrm{b}} s$ *and* $s \leq_{\mathrm{b}} r$ *are satisfied. That is, there are partition polynomials* $p_1(x)$ *and* $p_2(x)$ *such that* $r = p_1(s)$ *and* $s = p_2(r)$.

It is easily shown that $\sim_{\mathrm{b}}$ is an equivalence relation. Clearly, if either f or g is a bijection then they are homeomorphisms. This leads to the next definition.

DEFINITION C.21. *A pair of numbers* r *and* s *in* $(0,1)$ *is said to be* homeomorphicly related *if there is a homeomorphism* h *such that* $h_{\#}\beta_r = \beta_s$*; this relation will be denoted by* $r \sim_{\mathrm{h}} s$.

Notice that $\sim_{\mathrm{h}}$ has been defined only for binomial Bernoulli measures, hence it is more restrictive than the notion of homeomorphic measures which applies to any pair of Borel measures.

From the equivalence of the binary relations $\leq_{\mathrm{c}}$ and $\leq_{\mathrm{b}}$ we have the following theorem in the form as stated in [**105**].

THEOREM C.22. *Each of* β_s *and* β_r *are continuous image of the other if and only if there are positive integers* n *and* m *and there are finite collections of integers* a_i,

$0 \le i \le n$, and b_j, $0 \le j \le m$, such that

$$0 \le a_i \le \binom{n}{i} \quad \text{and} \quad 0 \le b_j \le \binom{m}{j}, \tag{C.11}$$

$$r = \sum_{i=0}^{n} a_i s^i (1-s)^{n-i} \quad \text{and} \quad s = \sum_{j=0}^{m} b_j r^j (1-r)^{m-j}. \tag{C.12}$$

Mauldin observed in [**105**] the following existence statement.

PROPOSITION C.23. *For a given pair n and m of nonnegative integers and collections a_i, $i = 0, 1, \ldots, n$, and b_j, $j = 0, 1, \ldots, m$, that satisfy the condition* (C.11) *there exists a pair r and s in* $[0, 1]$ *such that condition* (C.12) *is satisfied.*

PROOF. It is easily shown that the map given by

$$(r,s) = \varphi(x,y) = \left(\sum_{j=0}^{m} b_j y^j (1-y)^{m-j}, \sum_{i=0}^{n} a_i x^i (1-x)^{n-i}\right),$$

where $(x,y) \in [0,1] \times [0,1]$, is continuous and has values in $[0,1] \times [0,1]$. By the Brouwer fixed point theorem there exists a pair (r,s) such that $(r,s) = \varphi(r,s)$. □

In the same article Mauldin asked: Does $r \sim_{\mathrm{b}} s$ imply $r \sim_{\mathrm{h}} s$? This question was answered in the negative by Austin [**6**].

C.5.4. Austin's solution. Consider the partition polynomials

$$p_1(x) = 2x(1-x) \quad \text{and} \quad p_2(x) = 3x^2(1-x) + 3x(1-x)^2.$$

There is a pair (r,s) in $(0,1) \times (0,1)$ such that $s = p_1(r)$ and $r = p_2(s)$. So there are continuous maps f and g such that $f_\# \beta_r = \beta_s$ and $g_\# \beta_r = \beta_s$ and hence $r \sim_{\mathrm{b}} s$.

It will be shown that if there is an h in $\mathsf{HOMEO}(\{0,1\}^{\mathbb{N}})$ such that $\beta_s = h_\# \beta_r$ then $\frac{1}{r}$ is an algebraic integer. But the above polynomials imply that that $\frac{1}{r}$ is not an algebraic integer, thereby achieving a contradiction.

Let us show the first of the above assertions about $\frac{1}{r}$. Let

$$\bar{x} = (0,1,1,\ldots) \quad \text{and} \quad \bar{\bar{x}} = (1,0,1,1\ldots)$$

be the two points of $\{0,1\}^{\mathbb{N}}$ with only one 0 entry as indicated. For each i in $\mathbb{N}$ let $A_i = \{x \in \{0,1\}^{\mathbb{N}} : x_i = 0\}$. Clearly the collection $B_j = h^{-1}[A_j], j \in \mathbb{N}$, will generate a subbase of the topology. Consequently there is a finite intersection $U = \bigcap_{i=1}^{k} U_i$ such that $\bar{x} \in U$ and $\bar{\bar{x}} \notin U$, where each U_i is B_j or $\{0,1\}^{\mathbb{N}} \setminus B_j$ for some j. Hence there exists an i in $\mathbb{N}$ such that $\bar{x} \in B_i$ or $\bar{\bar{x}} \in B_i$. Let $\tilde{x}$ denote that one of the two points $\bar{x}$ and $\bar{\bar{x}}$ that is in B_i. There is an m such that $E = \langle \pi_{m+2}(\tilde{x}) \rangle \subset B_i$, where π_{m+2} is the usual projection of $\{0,1\}^{\mathbb{N}}$ onto $\{0,1\}^{m+2}$.

Define the homeomorphism $\varphi \colon A_i \to \{0,1\}^{\mathbb{N}}$ by the formula

$$\varphi\big((x_1, x_2, \ldots, x_{i-1}, 0, x_{i+1}, \ldots)\big) = (x_1, x_2, \ldots, x_{i-1}, x_{i+1}, \ldots).$$

Then $\nu = \varphi_\#(\beta_s | A_i)$ is a measure on $\{0,1\}^{\mathbb{N}}$ such that $\nu(C) = s\beta_s(C)$ for each simultaneously closed and open set C of $\{0,1\}^{\mathbb{N}}$. Hence $\frac{1}{s}\nu = \beta_s$. With $C = \varphi h[E]$, where E is the cylinder set defined in the previous paragraph, there is a partition polynomial $p_0(x)$ such that $\frac{1}{s}\nu(C) = \beta_s(C) = p_0(s)$. As $h[E] \subset A_i$ we have $\varphi^{-1}[C] = h[E]$ and $\varphi^{-1}[C] \subset A_i$. Consequently, $p_0(s) = \frac{1}{s}\nu(C) = \frac{1}{s}(\beta_s|A_i)(\varphi^{-1}[C]) = \frac{1}{s}h_{\#}\beta_r(h[E]) = \frac{1}{s}\beta_r(E) = \frac{r(1-r)^{m+1}}{s}$. Recalling that $s = p_1(r) = 2r(1-r)$, we have $\frac{r(1-r)^{m+1}}{2r(1-r)} = p_0(p_1(r)) = p(r)$, where $p(x)$ is a partition polynomial. Writing $p(x)$ in the form $\sum_{j=0}^{n} c_j x^j(1-x)^{n-j}$ with $m < n$, we have

$$(1-r)^m \sum_{j=0}^{n-m} \binom{n-m}{j} r^j (1-r)^{n-m-j} = 2\sum_{j=0}^{n} c_j r^j (1-r)^{n-j}.$$

Dividing by r^m and collecting terms, we have

$$(2c_0 - 1)(\tfrac{1}{r} - 1)^n + \sum_{k=0}^{n-1} d_k(\tfrac{1}{r} - 1)^k = 0,$$

where $d_k \in \mathbb{Z}$, $k = 0, 1, \ldots, n-1$, and $0 \le c_0 \le \binom{n}{0}$. As $\frac{1}{r} - 1$ is a root of a monic polynomial in $\mathbb{Z}[x]$, we have that $\frac{1}{r}$ is an algebraic integer, thereby the first assertion is proved. It remains to be prove that $\frac{1}{r}$ is not an algebraic integer to arrive at a contradiction.

Notice that the equation $r = p_2(s) = 3s(1-s)^2 + 3s(1-s)^2$ was not used in the above argument. Substituting $s = p_1(r) = 2r(1-r)$ into the last equation, we derive the equation $5 - 18r + 24r^2 - 12r^3 = 0$. Dividing by r^3, we have that $\frac{1}{r}$ is a root of the polynomial $5x^3 - 18x^2 + 24x - 12$ in $\mathbb{Z}[x]$. By Eisenstein's criterion, using the prime 3, this polynomial is irreducible in $\mathbb{Z}[x]$ and not monic. Hence $\frac{1}{r}$ is not an algebraic integer, thereby establishing the contradiction.

The above Austin example shows that $r \sim_{\mathrm{b}} s$ and $r \not\sim_{\mathrm{h}} s$.

C.5.5. Some early results. The equivalence relations $\sim_{\mathrm{b}}$ and $\sim_{\mathrm{h}}$ on $(0,1)$ induces equivalence relations on $\mathsf{B}(\{0,1\}^{\mathbb{N}})$, the collection of all binomial Bernoulli measures, because $\beta : r \mapsto \beta_r$ is a bijection. We shall write $\beta_r \sim_{\mathrm{b}} \beta_s$ and $\beta_r \sim_{\mathrm{h}} \beta_s$, respectively, for these equivalence relations on $\mathsf{B}(\{0,1\}^{\mathbb{N}})$. These equivalence classes of $\mathsf{B}(\{0,1\}^{\mathbb{N}})$ determined by $\sim_{\mathrm{b}}$ and $\sim_{\mathrm{h}}$ are now known to be different due to Austin's example. We now turn to the question of the cardinality of these equivalence classes. It is obvious that the cardinalities cannot be smaller than 2.

A simple observation, which follows from the partition polynomials, is that if r is a rational number and $\beta_r \sim_{\mathrm{b}} \beta_s$ then s is a rational number. Also, if r is a transcendental number and $\beta_r \sim_{\mathrm{b}} \beta_s$, then s is a transcendental number. As $r \sim_{\mathrm{b}} s$ is easier to write than $\beta_r \sim_{\mathrm{b}} \beta_s$ we shall revert back to it. The first result on the cardinality of the equivalence classes for rational numbers r in $(0,1)$ is due to Navarro-Bermúdez [**115, 114**].

THEOREM C.24 (Navarro-Bermúdez). *If r is a rational number in $(0,1)$ and $r \sim_{\mathrm{b}} s$, then either $s = r$ or $s = 1-r$; hence, $r \sim_{\mathrm{h}} s$ if and only if $s = r$ or $s = 1-r$.*

We shall not give his proof, a straightforward one can be found in his article [**115**, Theorem 3.3]. He also proved in the same article

THEOREM C.25 (Navarro-Bermúdez). *If r is a transcendental number in $(0, 1)$ and $r \sim_b s$, then either $s = r$ or $s = 1-r$; hence, $r \sim_h s$ if and only if $s = r$ or $s = 1-r$.*

PROOF. The equations (C.12) hold for r and s. So we have polynomial expressions with integer coefficients

$$r = r_0 + r_1 s + \cdots + r_j s^j,$$

$$s = s_0 + s_1 r + \cdots + s_k r^k,$$

such that $r_0 = a_0$ and $s_0 = b_0$ (which are equal to 0 or 1) and $r_j s_k \neq 0$, and such that $1 \leq j \leq m$, $1 \leq k \leq n$. Eliminating s from the two equations, we result in a polynomial $p(x)$ of degree jk. Since the coefficients of $p(x)$ are all integers and r is a transcendental number with $p(r) = 0$, we have that all the coefficients of $p(x)$ are zero. If $jk > 1$, then the jk-th coefficient of $p(x)$ is $r_j(s_k)^j$ which cannot be zero. Hence $jk \leq 1$. It follows that $0 = r_0 + r_1 s_0 + (r_1 s_1 - 1)r$. Hence $r_0 + r_1 s_0 = 0$ and $r_1 s_1 - 1 = 0$. Recall that r_0 is either 0 or 1, whence $r_1 s_0 = 0$ or $1 + r_1 s_0 = 0$, respectively, from the first equality. The second yields $r_1 = s_1 = 1$ or $r_1 = s_1 = -1$. If $r_0 = 0$ and $r_1 = 1$, then either $s_0 = 0$ or $1 + s_0 = 0$, whence $s_0 = 0$ because $1 + s_0 \neq 0$. Hence $r = s$ if $r_0 = 0$ and $r_1 = 1$. If $r_0 = 0$ and $r_1 = -1$, then $-s_0 = 0$ or $-s_1 - 1 = 0$, whence $s_0 = 0$ because $-s_1 - 1 \neq 0$. Hence $r = s$ if $r_0 = 0$ and $r_1 = -1$. Following the same procedure, we have that the two remaining cases of $r_0 = 1$ and $r_1 = 1$, and $r_0 = 1$ and $r_1 = -1$ will result in $r = 1 - s$. □

The next theorem, due to Huang **[78]**, is similar to the last one.

THEOREM C.26 (Huang). *If an algebraic integer r of degree 2 and s in $(0, 1)$ are such that $r \sim_b s$, then $s = r$ or $s = 1 - r$; hence $r \sim_h s$ if and only if $s = r$ or $s = 1 - r$.*

The proof relies on the following nice lemma concerning the fractional part of a real number. Recall that the integer part of a real number x is the largest integer, denoted by $[x]$, not exceeding x, and the fractional part of x is $\langle x \rangle = x - [x]$. Of course, $0 \leq \langle x \rangle < 1$. The proof of this lemma is left as an exercise.

LEMMA C.27. *Let t be an irrational number and m be an integer with $|m| > 1$. If $r = \langle mt \rangle$, then t can never be $\langle kr \rangle$ for any integer k.*

PROOF OF THEOREM C.26. Let r be an algebraic integer in $(0, 1)$. That is, there is an irrational root ρ of a monic quadratic polynomial with integer coefficients such that 1 and ρ generate an integral domain. Every member of this integral domain is of the form $c + d\rho$ for unique integers c and d. Hence every polynomial expression $p(\rho)$ with integer coefficient is equal to $c + d\rho$ for some unique integers c and d. Moreover $s = c + d\rho$ is also an algebraic integer.

Since r is an algebraic integer in $(0, 1)$, we have that r is an irrational number. As $r \sim_b s$ the numbers r and s satisfy conditions (C.11) and (C.12). Hence $r = c + ds$ for some integers c and d, and $s = c' + d'r$ for some integers c' and d'. We have $r = c + [ds] + \langle ds \rangle$, $0 < r < 1$ and $0 < \langle ds \rangle < 1$, whence $c + [ds] = 0$. Consequently,

$r = \langle ds \rangle$. Analogously, $s = \langle d'r \rangle$. The lemma gives $d = \pm 1$. As $0 < s < 1$ we have that $\langle ds \rangle = s$ for $d = 1$, and that $\langle ds \rangle = 1 - s$ for $d = -1$. Thereby the theorem is proved. □

Huang also proved that, for each degree n larger than 2, some algebraic integer r of degree n fails to have the property that the cardinality of its $\sim_{\mathrm{b}}$ equivalence class is 2. He proved

THEOREM C.28 (Huang). *For each integer n with $n > 2$ there is an algebraic integer r of degree n such that $r \sim_{\mathrm{b}} r^2$ and $r^2 \neq 1 - r$.*

An improvement of this theorem, where $\sim_{\mathrm{b}}$ is replaced by $\sim_{\mathrm{h}}$, results from a theorem in [**47**], which will be proved in a later section.

For the case $n = 3$, Navarro-Bermúdez and Oxtoby [**117**] proved that the finer condition $r \sim_{\mathrm{h}} r^2$ holds by elementary means. That is,

THEOREM C.29 (Navarro-Bermúdez–Oxtoby). *Let r be the algebraic integer solution of the irreducible polynomial $x^3 + x^2 - 1$ in $(0, 1)$ and let $s = r^2$. Then there is a homeomorphism h of the Cantor space $\{0,1\}^{\mathbb{N}}$ such that $h_{\#}\beta(r, 1-r) = \beta(s, 1-s)$.*

Though later results have generalized this result, we present their proof for the case $n = 3$. We first prove their lemma.

LEMMA C.30. *Let U and V be subsets of $\{0,1\} \cup \{0,1\}^2$ given by*

$$U = \{(0), (1,0), (1,1)\} \quad \textit{and} \quad V = \{(1), (0,1), (0,0)\}.$$

For p and q in $(0,1)$, there exist homeomorphisms $h\colon \{0,1\}^{\mathbb{N}} \to U^{\mathbb{N}}$ and $g\colon \{0,1\}^{\mathbb{N}} \to V^{\mathbb{N}}$ such that $h_{\#}\beta_p = \beta(P)$ and $g_{\#}\beta_q = \beta(Q)$, where $P = (P_1, P_2, P_3)$ and $Q = (Q_1, Q_2, Q_3)$ are Bernoulli measures on U and V, respectively, with $P_1 = \beta_p(\langle(0)\rangle)$, $P_2 = \beta_p(\langle(1,0)\rangle)$, $P_3 = \beta_p(\langle(1,1)\rangle)$ and $Q_1 = \beta_q(\langle(1)\rangle)$, $Q_2 = \beta_q(\langle(0,1)\rangle)$, $Q_3 = \beta_q(\langle(0,0)\rangle)$. If $\psi\colon U \to V$ is a bijection, then the homeomorphism $\Psi\colon U^{\mathbb{N}} \to V^{\mathbb{N}}$ induced by ψ satisfies $\Psi_{\#}\beta(P) = \beta(Q)$ if and only if

$$\beta_p\big(\langle \psi^{-1}(v) \rangle\big) = \beta_q\big(\langle v \rangle\big) \quad \textit{whenever} \quad v \in V. \tag{C.13}$$

The homeomorphism $f = g^{-1}\Psi h$ of $\{0,1\}^{\mathbb{N}}$ yields $\beta_q = f_{\#}\beta_p$ if and only if p and q satisfy the constraints consisting of the simultaneous equations produced by the above requirements.

PROOF. By Lemmas C.2 and C.3 there are homeomorphism h, g, and Ψ such that the following diagram commutes.

$$\begin{array}{ccc}
\{0,1\}^{\mathbb{N}} & \xrightarrow{f} & \{0,1\}^{\mathbb{N}} \\
h \downarrow & & \downarrow g \\
U^{\mathbb{N}} & \xrightarrow{\Psi} & V^{\mathbb{N}}
\end{array}$$

We infer from Lemma C.4 that $h_{\#}\beta_p$ is the Bernoulli measure $\beta(P)$ on $U^{\mathbb{N}}$ and $g^{-1}{}_{\#}\beta_q$ is the Bernoulli measure $\beta(Q)$ on $V^{\mathbb{N}}$. Using the homeomorphism Ψ induced by ψ, we get $\Psi_{\#}\beta(P) = \beta(\psi_{\#}P)$. Hence $f_{\#}\beta_p = \beta_q$ if and only if $\psi_{\#}P = Q$. The condition (C.13) determines a constraint on p and q. □

We apply the lemma in the following proof of the Navarro-Bermúdez–Oxtoby theorem (Theorem C.29 above).

PROOF OF THEOREM C.29. In the lemma, let ψ be given by

$$\psi((0)) = (0,0), \ \psi((1,0)) = (1), \ \psi((1,1)) = (0,1).$$

Then condition (C.13) becomes

$$q = p^2, \qquad q(1-q) = 1-p, \qquad (1-q)^2 = p(1-p).$$

The first two equations $q = p^2$ and $q(1-q) = 1-p$ together with the identity $1 - q(1-q) = q^2 + q(1-q) + (1-q)^2$ show that $p \sim_b q$. Such p and q exist by Proposition C.23. On eliminating q from all three equations, we find that p must be the unique root of the irreducible polynomial $x^3 + x^2 - 1$ in the interval $(0, 1)$. Clearly, $p > p^2 > p^3 = 1 - p^2$. The theorem is proved. □

There are several other possible maps ψ available in Lemma C.30. Consider the map $\psi_1((0)) = (1)$, $\psi_1((1,0)) = (0,1)$, $\psi_1((1,1)) = (0,0)$. Here, p and q must satisfy $p = 1-q$, with any choice of q in $(0, 1)$. Consider $\psi_2((0)) = (1)$, $\psi_2((1,0)) = (0,0)$, $\psi_2((1,1)) = (0,1)$. Here, $p = q = \frac{1}{2}$. The other cases will be left for the reader to consider.

C.6. Linear ordering of $\{0,1\}^{\mathbb{N}}$ and good measures

At the end of Section C.4 it was shown that $\text{vs}(\mu, \{0,1\}^{\mathbb{N}})$, the value set of the measure μ on $\{0,1\}^{\mathbb{N}}$, does not characterize equivalence classes of homeomorphic measures on the Cantor space $\{0,1\}^{\mathbb{N}}$. We shall turn to a characterization due to Akin that uses a finer structure on Cantor spaces by the introduction of a linear order. There are many linear orders that induce the Cantor space topology. Hence a continuous, complete, finite Borel measure μ on $\{0,1\}^{\mathbb{N}}$ can be very different if one considers two different linear orders which give the usual topology on $\{0,1\}^{\mathbb{N}}$.

It is not difficult to show that $\{0,1\}^{\mathbb{N}}$ is homogeneous; that is, if x_1 and x_2 are points of $\{0,1\}^{\mathbb{N}}$, then there is an h in $\mathsf{HOMEO}(\{0,1\}^{\mathbb{N}})$ such that $h(x_1) = x_2$ and $h(x_2) = x_1$. Hence it follows that, for distinct points x_0 and x_1 of the classical Cantor ternary set, there is a self-homeomorphism of the Cantor ternary set such that $h(x_0) = 0$ and $h(x_1) = 1$. Consequently we have

PROPOSITION C.31. *If X is a Cantor space and if x_0 and x_1 are distinct points of X, then there is a linear order $\leq$ on X such that*

(1) $x_0 \leq x \leq x_1$ *whenever* $x \in X$,
(2) *the order topology induced by* $\leq$ *on* X *is precisely the topology of* X.

Akin has made an extensive study of Cantor spaces with linear orders that satisfy the two conditions enumerated above. A separable metrizable space X with a linear order $\leq$ that satisfies the above two conditions will be denoted by $(X,\leq)$ and will be called a *linearly ordered topological space* (or, more briefly, an *ordered space*). We need two definitions.

DEFINITION C.32. *Let $(X,\leq)$ be an ordered space and let μ be a complete, finite Borel measure on X. The function $F_\mu\colon X \to [0,\mu(X)]$ defined by*

$$F_\mu(x) = \mu([x_0,x]), \quad x \in X,$$

where x_0 is the minimal element of X in the order $\leq$, is called the cumulative distribution function *of μ. (Recall that the collection of all simultaneously closed and open sets of X is denoted by $\mathfrak{CO}(X)$.) Define $\widetilde{\mathrm{vs}}(\mu,X,\leq)$ to be the set of values*

$$\widetilde{\mathrm{vs}}(\mu,X,\leq) = \{\mu([x_0,x])\colon [x_0,x] \in \mathfrak{CO}(X)\} \cup \{0\},$$

called the special value set.

DEFINITION C.33. *Let $(X_1,\leq_1)$ and $(X_2,\leq_2)$ be ordered spaces. $\varphi\colon X_1 \to X_2$ is said to be an* order preserving map *if $\varphi(a) \leq_2 \varphi(b)$ whenever $a \leq_1 b$. Such a map that is also bijective is called an* order isomorphism.

To prove his main theorems, Akin [**2**, Lemma 2.9] established the following "lifting lemma."

LEMMA C.34 (Akin). *For ordered spaces $(X_1,\leq_1)$ and $(X_2,\leq_2)$, let μ_1 and μ_2 be positive, continuous, complete Borel probability measures on X_1 and X_2, respectively. An order isomorphism $\varphi\colon (I,\leq) \to (I,\leq)$ satisfies the condition*

$$\varphi\big[\widetilde{\mathrm{vs}}(\mu_1,X_1,\leq_1)\big] \supset \widetilde{\mathrm{vs}}(\mu_2,X_2,\leq_2)$$

if and only if there exists a continuous map $h\colon X_1 \to X_2$ such that the following diagram commutes.

$$\begin{array}{ccc} X_1 & \xrightarrow{\ h\ } & X_2 \\ {\scriptstyle F_{\mu_1}}\downarrow & & \downarrow{\scriptstyle F_{\mu_2}} \\ I & \xrightarrow{\ \varphi\ } & I \end{array}$$

If such a lifting map h exists, then it is unique and is a surjective order preserving map of $(X_1 \leq_1)$ to $(X_2,\leq_2)$. Furthermore, h is an order isomorphism if and only if $\varphi\big[\widetilde{\mathrm{vs}}(\mu_1,X_1,\leq_1)\big] = \widetilde{\mathrm{vs}}(\mu_2,X_2,\leq_2)$.

Here is a useful proposition.

PROPOSITION C.35 (Akin). *Let $(X,\leq)$ be an ordered space. If F is an order preserving map of X into $[0,1]$ which is right continuous and satisfies $F(x_1) = 1$, where x_1 is the maximal member of X, then there is a unique probability measure μ on X such that $F = F_\mu$.*

Proof. As usual we define μ on the collection of all half open intervals $(a,b]$ of the space $(X,\leq)$ to be $\mu\big((a,b]\big) = F(b) - F(a)$, and $\mu(\{x_0\}) = F(x_0)$ where x_0 is the minimal member of X. Since this collection of subsets of X generate the Borel sets of X, the measure μ is defined. The uniqueness is easily proved. □

It is not difficult to construct a continuous, complete Borel probability measure μ on the Cantor ternary set X such that its special value set $\widetilde{\text{vs}}(\mu,X,\leq)$ for the usual order $\leq$ on X is a proper subset of the value set $\text{vs}(\mu,X)$ (see Exercise C.11). Also, it is quite clear that there are some homeomorphisms of a linearly ordered X that are not order isomorphisms. A consequence of the lifting lemma is Akin's theorem [**2**, Theorem 2.10] which asserts that $\widetilde{\text{vs}}(\mu,X,\leq)$ classifies the order isomorphism equivalence classes of positive, continuous, complete Borel probability measures on ordered spaces.

Theorem C.36 (Akin). *Suppose that $(X_1,\leq_1)$ and $(X_2,\leq_2)$ are ordered spaces and that μ_1 and μ_2 are positive, continuous, complete probability measures on X_1 and X_2, respectively. Then there exists an order isomorphism $h\colon (X_1,\leq_1) \to (X_2,\leq_2)$ such that $h_\#\mu_1 = \mu_2$ if and only if*

$$\widetilde{\text{vs}}(\mu_1,X_1,\leq_1) = \widetilde{\text{vs}}(\mu_2,X_2,\leq_2).$$

For Cantor spaces X, Akin in [**2**] showed that every linear order $\leq$ on X which results in an ordered space $(X,\leq)$ is order isomorphic to the classical Cantor ternary set endowed with the usual order.

Theorem C.37. *Suppose that $(X_1,\leq_1)$ and $(X_2,\leq_2)$ are ordered Cantor spaces. Then there exists an order isomorphism h of $(X_1,\leq_1)$ onto $(X_2,\leq_2)$.*

Proof. We may assume X_1 is the Cantor ternary set. Let μ_1 be the measure induce on X_1 by the Cantor function on X_1. Let $\beta(\frac{1}{2})$ be the uniform Bernoulli measure on $\{0,1\}^{\mathbb{N}}$ and let $\varphi\colon \{0,1\}^{\mathbb{N}} \to X_2$ be a homeomorphism. Then $\nu = \varphi_\#\beta(\frac{1}{2})$ is a measure on X_2 and $\widetilde{\text{vs}}(\nu,X_2,\leq_2)$ is a countable dense subset of $[0,1]$. From a well-known fact of dimension theory[4] we infer that there is an order preserving homeomorphism $g\colon [0,1] \to [0,1]$ such that $g\big[\widetilde{\text{vs}}(\nu,X_2,\leq_2)\big] = \widetilde{\text{vs}}(\mu_1,X_1,\leq_1)$. Hence there is a measure μ_2 on X_2 such that $gF_\nu = F_{\mu_2}$ and $\widetilde{\text{vs}}(X_2,\mu_2,\leq_2) = \widetilde{\text{vs}}(X_1,\mu_1,\leq_1)$. Theorem C.36 completes the proof. □

Before we introduce the next notion due to Akin [**3**] let us analyze a particular example. Let X be the usual Cantor ternary set with the usual order and let μ be the measure induced on X by the well-known Cantor function F on X. Clearly, the value set $\text{vs}(\mu,X)$ is $\{\frac{k}{2^n} : k = 0,1,\dots,2^n,\ n = 1,2,\dots\}$ and the cumulative distribution function of μ is F. Moreover, $\widetilde{\text{vs}}(\mu,X,\leq) = \text{vs}(\mu,X)$.

Observe that the Cantor ternary set X has the following nice property: to each nonempty simultaneously closed and open set U there corresponds a positive integer k_0 such that every point x of U has the property that $J \cap X \subset U$ whenever x is in

[4] If D_1 and D_2 are countable dense subsets of $\mathbb{R}$, then there is an order preserving homeomorphism h of $\mathbb{R}$ such that $h[D_2] = D_1$. See [**79**].

J and J is an interval of the k-th step of the Cantor set construction with $k \geq k_0$. An immediate consequence of the observation is that the Cantor ternary set and the corresponding measure μ determined by the Cantor function satisfies the property which Akin calls the "subset condition."

Definition C.38. *Let X be a Cantor space. A Borel measure μ on X is said to satisfy the* subset condition *if, for simultaneously closed and open sets U and V with $\mu(U) \leq \mu(V)$, there exists a simultaneously closed and open set W such that*

$$W \subset V \quad and \quad \mu(W) = \mu(U).$$

A positive, continuous, complete, Borel probability measure μ on X is called good *if it satisfies the subset condition. The collection of all good measures on X will be denoted by* $\mathrm{MEAS}^{\mathrm{good}}$.

Our earlier observation that the measure μ associated with the Cantor function has the property that the two value sets $\mathrm{vs}(\mu, X)$ and $\widetilde{\mathrm{vs}}(\mu, X, \leq)$ coincide, where $\leq$ is the usual order. Akin has given the name "adapted" for this phenomenon.

Definition C.39. *Let X be a Cantor space. A measure μ and an order $\leq$ on X are said to be* adapted *if* $\mathrm{vs}(\mu, X) = \widetilde{\mathrm{vs}}(\mu, X, \leq)$.

In [3] Akin proved the following characterization.

Theorem C.40 (Akin). *Let X be a Cantor space and let μ be a positive, continuous, finite Borel measure on X. Then the following two statements hold.*

(1) *If $\leq$ is a linear order such that $(X, \leq)$ is an ordered space and if μ and $\leq$ are adapted, then μ is a good measure on X.*
(2) *If μ is a good measure on X and if x_0 is a point of X, then there exists a linear order $\leq$ on X such that $(X, \leq)$ is an ordered space with $x_0 \leq x$ whenever $x \in X$ and such that μ and $\leq$ are adapted.*

Consequently, μ is a good measure on X if and only if there exists a linear order on X whose order topology on X is the topology of X and $\mathrm{vs}(\mu, X) = \widetilde{\mathrm{vs}}(\mu, X, \leq)$.

We conclude the section with a theorem that connects the collection $\mathrm{MEAS}^{\mathrm{good}}(X)$ and the group $\mathsf{HOMEO}(X)$.

Theorem C.41 (Akin). *Let X be a Cantor space. Then positive, continuous, finite Borel measures μ_1 and μ_2 are in $\mathrm{MEAS}^{\mathrm{good}}(X)$ and satisfy $h_{\#}\mu_1 = \mu_2$ for some h in $\mathsf{HOMEO}(X)$ if and only if $\mathrm{vs}(\mu_1, X) = \widetilde{\mathrm{vs}}(\mu_2, X, \leq)$ for some order $\leq$ on X.*

C.7. Refinable numbers

Austin's example shows that, for binomial Bernoulli measures, the binomial equivalence classes determined by $\sim_{\mathrm{b}}$ and the homeomorphic equivalence classes determined by $\sim_{\mathrm{h}}$ are not the same. This section is devoted to a discussion of a sufficient condition that assures that they are the same. We have already seen that they are the same for s in $(0, 1)$ whenever s is a rational number or a transcendental number.

Hence sufficient conditions for algebraic numbers are needed. The recent results of Dougherty, Mauldin and Yingst [**47**] will be discussed.

C.7.1. Definitions and properties. Let us begin with comments on the collections $\mathcal{P}$ and $\mathcal{P}(r)$. Recall that $\mathcal{P}$ is the collection of all partition polynomials, which is a collection of functions, and, for an r in $(0, 1)$, that $\mathcal{P}(r)$ is $\{p(r) : p \in \mathcal{P}\}$, a collection of real numbers. From properties of partition polynomials, we know that each polynomial of the form $x^m(1-x)^n$, where m and n are nonnegative integers, are partition polynomials, and that each partition polynomial is a finite sum of such polynomials. A polynomial of the form $x^m(1-x)^n$ will be called a *partition monomial*. It is obvious that a finite sum $f(x) = \sum_{i=1}^k g_i(x)$ of nontrivial partition polynomials $g_i(x)$ is a partition polynomial if and only if $f \equiv 1$ or $f(x) < 1$ whenever $x \in (0, 1)$. But there is no such nice characterization for finite sums $\sum_{i=1}^k g_i(r)$ from the collection $\mathcal{P}(r)$ even if the sum is a number $f(r)$ from the collection $\mathcal{P}(r)$. Indeed, there is no assurance that the corresponding polynomial $p(x) = \sum_{i=1}^k g_i(x)$ is a partition polynomial and that $p(x)$ is the partition polynomial $f(x)$. Hence the following definition is meaningful.

Definition C.42. *A number r in* $(0, 1)$ *is said to be* refinable *if $f(x), g_1(x), \ldots, g_k(x)$ in $\mathcal{P}$ are such that $f(r) = \sum_{i=1}^k g_i(r)$, then there is a collection $h_1(x), \ldots, h_k(x)$ in $\mathcal{P}$ such that $h_i(r) = g_i(r)$ for each i and $f(x) = \sum_{i=1}^k h_i(x)$.*

It would seem reasonable that a weaker condition would suffice, that is, only partition monomials $x^m(1-x)^n$ need be considered. Hence the definition

Definition C.43. *A number r in* $(0, 1)$ *is said to be* weakly refinable *if $f(x)$, $g_1(x), \ldots, g_k(x)$ are partition monomials such that $f(r) = \sum_{i=1}^k g_i(r)$, then there is a collection $h_1(x), \ldots, h_k(x)$ in $\mathcal{P}$ such that $h_i(r) = g_i(r)$ for each i and $f(x) = \sum_{i=1}^k h_i(x)$.*

The two definitions are somewhat hard to verify by the very nature of the requirements imposed by them. The sole exception is the case where r is transcendental.[5] This observation is implicit in Theorem C.25 by Navarro-Bermúdez. But fortunately there is a characterization with a much simpler test for algebraic numbers. The proof of this characterization is not easy – it very cleverly uses the Bernstein polynomial approximation of continuous functions f on $[0, 1]$; that is, the formula $B_nf(x) = \sum_{i=0}^n f(\frac{1}{n})\binom{n}{i}x^i(1-x)^{n-i}$ which looks like a partition polynomial and is precisely the binomial expansion of the constant 1 function.

The characterization [**47**, Theorem 11] will be stated without proof. Also, a crucial lemma used in the proof of the characterization will be stated without proof since the lemma will be used later.

Lemma C.44 (Dougherty–Mauldin–Yingst). *Suppose $p(x)$ is a partition polynomial that is not the* 0 *polynomial and suppose $f(x)$ is a polynomial with real coefficients such that $f(x) > 0$ whenever $x \in (0, 1)$. If $R(x)$ is in $\mathbb{Z}[x]$ with $|R(0)| = 1$ and $|R(1)| = 1$ and if $p(r) < f(r)$ whenever $R(r) = 0$ and $r \in (0, 1)$, then there is a*

[5] See Exercise C.13.

polynomial $Q(x)$ *in* $\mathbb{Z}[x]$ *such that* $0 < p(x) + Q(x)R(x) < f(x)$ *whenever* $x \in (0,1)$. *Moreover,* $q(x) = p(x) + Q(x)R(x)$ *is a partition polynomial such that* $q(x) < f(x)$ *whenever* $x \in (0,1)$ *and such that* $q(r) = p(r)$ *whenever* $R(r) = 0$ *and* $r \in (0,1)$.

THEOREM C.45 (Dougherty–Mauldin–Yingst). *Let* r *be an algebraic number in* $(0,1)$. *The following are equivalent.*

(1) r *is refinable.*
(2) r *is weakly refinable.*
(3) *There is a polynomial* $R(x)$ *in* $\mathbb{Z}[x]$ *such that* $|R(0)| = 1$, $|R(1)| = 1$, *and* $R(r) = 0$. *Moreover,* $R(x)$ *may be assumed to be the unique, up to sign, irreducible polynomial solved by* r *such that the* gcd *of the coefficients is* 1.

Note that the only rational number that is refinable is $\frac{1}{2}$. Hence the method which requires r to be refinable is only sufficient for determining $\sim_{\mathrm{h}}$ equivalence from $\sim_{\mathrm{b}}$ equivalence.

Note that if $g(r) = h(r)$ for partition polynomials $g(x)$ and $h(x)$, then $g(r) = \sum_{j=1}^{n} a_j r^j (1-r)^{n-j}$ where $h(x) = \sum_{j=1}^{n} a_j x^j (1-x)^{n-j}$. Here is a second characterization of weakly refinable numbers.

THEOREM C.46. *Let* r *be in* $(0,1)$ *and let* $f(x), g_1(x), \ldots, g_k(x)$ *be partition monomials. Then there is a positive integer* m *such that, for each* n *larger than* m, $f(x) = \sum_{j=0}^{n} b(n,j) x^j (1-x)^{n-j}$, *and* $g_i(r) = \sum_{j=0}^{n} a(i,n,j) r^j (1-r)^{n-j}$, $i = 1, \ldots, k$. *Moreover,* r *is weakly refinable if and only if there is an* n_0 *such that* $n_0 > m$ *and* $b(n,j) = \sum_{i=1}^{k} a(i,n,j)$ *for each* j *and* $n \geq n_0$.

PROOF. The first statement is easily shown. If r is weakly refinable, then the equations connecting the coefficients are straightforward computations using the definition of weakly refinable numbers in $(0,1)$. For the converse, the proof follows from the inequality $0 \leq b(n,j) \leq \binom{n}{j}$. □

Here is a useful theorem which follows immediately from the first characterization.

THEOREM C.47. *For* r *and* s *in* $(0,1)$, *if* $r \leq_{\mathrm{b}} s$ *(that is, there is a partition polynomial* $p(x)$ *such that* $r = p(s)$*), and* r *is refinable, then* s *is refinable.*

The following is the main theorem of [**47**].

THEOREM C.48 (Dougherty–Mauldin–Yingst). *Suppose* r *and* s *are numbers in* $(0,1)$ *that are refinable. If* r *and* s *are binomially equivalent (that is,* $r \sim_{\mathrm{b}} s$*), then the measures* β_r *and* β_s *are homeomorphic (that is,* $\beta_r \sim_{\mathrm{h}} \beta_s$*).*

Our proof will use limits of inverse systems which will be discussed next.

C.7.2. Inverse systems. Let P be the collection of all finite partitions of $\{0,1\}^{\mathbb{N}}$ by open sets. There is a partial order $\leqslant$ on P defined by refinement, that is $P \leqslant Q$ if Q refines P. There is a natural map π_Q^P, called the *bonding map*, from the finite set Q to the finite set P defined by inclusion. The collection P is a directed set that has an obvious cofinal sequence P_n, $n = 1, 2, \ldots$. Indeed, the cylinder sets

$P_n = \{\varphi_n{}^{-1}[\langle e\rangle] \colon e \in \{0,1\}^n\}$, $n \in \mathbb{N}$, is cofinal in $\{\mathsf{P}, \leqslant\}$, where φ_n is the natural projection of $\{0,1\}^{\mathbb{N}}$ onto $\{0,1\}^n$. The limit of the inverse system $\{\mathsf{P}, \leqslant\}$, denoted $\varprojlim\{\mathsf{P}, \leqslant\}$, is a closed subset of the product space $\mathsf{X}\{P \colon P \in \mathsf{P}\}$ consisting of all threads of the product space.[6] The inverse limit is homeomorphic to $\{0,1\}^{\mathbb{N}}$. If Q is a cofinal directed subset of $\{\mathsf{P}, \leqslant\}$, then the limit of the inverse system $\{\mathsf{Q}, \leqslant\}$ is also homeomorphic to $\{0,1\}^{\mathbb{N}}$. As an aside, let us mention that P corresponds to uniform continuity and the inverse system $\{\mathsf{P}, \leqslant\}$ corresponds to uniform convergence.

Consider a continuous map $f \colon \{0,1\}^{\mathbb{N}} \to \{0,1\}^{\mathbb{N}}$. As we are interested in the equation $f_\# \beta_r = \beta_s$, the map will be assumed to be surjective. For each P in P we have $f^{-1}[P] = \{f^{-1}[u] \colon u \in P\} \in \mathsf{P}$, and that $P_1 \leqslant P_2$ if and only if $f^{-1}[P_1] \leqslant f^{-1}[P_2]$. For each Q and each P in P such that $f^{-1}[P] \leqslant Q$ there is a natural surjective map $f_Q^P \colon v \mapsto u$, where $v \in Q$, $u \in P$ and $v \subset f^{-1}[u]$.

Observe that $f^{-1}[\mathsf{P}]$, $P \in \mathsf{P}$ is a cofinal directed set in $\{\mathsf{P}, \leqslant\}$ if and only if f is a homeomorphism (indeed, $f[\mathsf{P}] = \mathsf{P}$ if and only if f is a homeomorphism). Let us define a map $\psi \colon \mathsf{P} \to \mathsf{P}$ as follows. If f is a homeomorphism, then $\psi(P) = f^{-1}[P]$ for every P. If f is not a homeomorphism, then, for each P in P, $\psi(P)$ is the cylinder set $P_{n(P)}$, where $n(P)$ is the least n such that P_n refines $f^{-1}[P]$. The maps f_Q^P satisfy the commutative diagram

$$\begin{array}{ccc} P & \xrightarrow{\pi_P^{P'}} & P' \\ {\scriptstyle f^P_{\psi(P)}}\uparrow & & \uparrow{\scriptstyle f^{P'}_{\psi(P')}} \\ \psi(P) & \xrightarrow[\pi_{\psi(P)}^{\psi(P')}]{} & \psi(P') \end{array} \tag{C.14}$$

where P and P' are members of P such that $P' \leqslant P$. The above map ψ satisfies the following property: $\psi \colon \mathsf{P} \to \mathsf{P}$ *is such that* $\psi(P') \leqslant \psi(P'')$ *whenever* $P' \leqslant P''$, *and is such that* $\psi(P)$, $P \in \mathsf{P}$, *is cofinal in* $\{\mathsf{P}, \leqslant\}$. This property is called the *monotone-cofinal property* for ψ. For the above defined ψ the collection $\psi(P)$, $P \in \mathsf{P}$, is cofinal because f is surjective, hence a monotone-cofinal ψ always exists. If f is a homeomorphism, then $f[\psi(P)] = P$ for each P in P. Hence $f^P_{\psi(P)}$ is injective for every P in P. If f is not a homeomorphism, let x' and x'' be distinct points such that $f(x') = f(x'')$ and let $\psi(P')$ be a partition that separates x' and x''. Then $f^P_{\psi(P)}$ is not injective for every P in P with $P' \leqslant P$. Hence f is a homeomorphism if and only if $f^P_{\psi(P)}$ is bijective for every P in P. Finally, if $f_\#\mu = \nu$, then f_Q^P satisfies

$$\nu(u) = \mu\big((f^P_{\psi(P)})^{-1}[\{u\}]\big) = \sum\{\mu(v) \colon v \in (f^P_{\psi(P)})^{-1}[\{u\}]\} \tag{C.15}$$

whenever $u \in P$ and $P \in \mathsf{P}$. Such measures μ and ν are said to be *compatible* with $\{f_Q^P, \mathsf{P}, \psi\}$.

Conversely, let $\{f_Q^P, \mathsf{P}, \psi\}$ be a system of maps $f_Q^P \colon Q \to P$, where Q and P are members of P, $\psi \colon \mathsf{P} \to \mathsf{P}$, and μ and ν are measures such that

[6] For *inverse system*, *limit* of an inverse system, *thread* of an inverse system, see R. Engelking [**51**, page 135].

(1) ψ has the monotone-cofinal property,
(2) f_Q^P are surjections that satisfy the diagram (C.14),
(3) μ and ν are compatible with $\{f_Q^P, \mathsf{P}, \psi\}$, that is, satisfies condition (C.15).

With $Q = \psi(P)$, to each f_Q^P let f_P be a map defined on $\{0,1\}^{\mathbb{N}}$ such that it is constant on each v in Q and such that $f_P(x) \in f_Q^P(v)$ whenever $x \in v \in Q$ and $v \subset (f_Q^P)^{-1}[\{u\}]$. Obviously f_P is continuous. The net of maps f_P, $P \in \mathsf{P}$, is uniformly convergent to a continuous surjection f. Indeed, let $P_0 \in \mathsf{P}$. If $P \in \mathsf{P}$ and $P_0 \leqslant P$, then $\psi(P_0) \leqslant \psi(P)$. Hence $f_P(x) \in u \in P_0$ whenever $x \in v \in \psi(P_0)$ and $P_0 \leqslant P$. Moreover, $f^{-1}[u] = (f_{\psi(P)}^P)^{-1}[\{u\}]$ whenever $u \in P$ and $P_0 \leqslant P$. Let us show $f_\#\mu = \nu$. For a nonempty simultaneously closed and open set w such that $w \neq \{0,1\}^{\mathbb{N}}$ let P_0 be $\{w, \{0,1\}^{\mathbb{N}} \setminus w\}$. Let $P \in \mathsf{P}$ be such that $P_0 \leqslant P$. Then $f^{-1}[w] = f^{-1}[\bigcup\{u \in P\colon u \subset w\}]$. As $f_\#\mu(u) = \mu((f_{\psi(P)}^P)^{-1}[\{u\}]) = \nu(u)$ whenever $u \in P$ we have $f_\#\mu(w) = \mu(f^{-1}[w]) = \sum\{\nu(u)\colon u \subset w, u \in P\} = \nu(w)$. Finally, f is a homeomorphism if and only if $f_{\psi(P)}^P$ is bijective for every P in P for some monotone-cofinal map ψ.

Let us summarize the above discussion as a theorem.

THEOREM C.49. *Let μ and ν be continuous, complete, finite Borel measures on* $\{0,1\}^{\mathbb{N}}$.

If f is a continuous surjection such that $f_\#\mu = \nu$, then there are maps f_Q^P, $(Q,P) \in \mathsf{P} \times \mathsf{P}$, *and there is a monotone-cofinal map ψ such that $f_{\psi(P)}^P$ is surjective for each P, the system $\{f_Q^P, \mathsf{P}, \psi\}$ satisfies the commutative diagram* (C.14), *and μ and ν are compatible with this system. Moreover, f is a homeomorphism if and only if $f_{\psi(P)}^P$ is a bijection for each P in* P.

Conversely, let f_Q^P be maps for each $(Q,P) \in \mathsf{P} \times \mathsf{P}$ *and let ψ be a monotone-cofinal map such that $f_{\psi(P)}^P$ is surjective for each P. If the system $\{f_Q^P, \mathsf{P}, \psi\}$ satisfies the commutative diagram* (C.14) *and if μ and ν are compatible with this system, then there is a continuous surjection f such that $f_\#\mu = \nu$. Moreover, f is a homeomorphism if and only if there is a ψ such that $f_{\psi(P)}^P$ is a bijection for each P in* P.

See Exercise C.17 for another characterization for bijection.

Compositions of maps and compositions of inverse systems are easily described due to the requirements of the monotone-cofinal property of the map ψ. Indeed, for a composition gf, there are maps ψ_f, ψ_g and ψ_{gf} such that $\psi_{gf}(P) \leqslant \psi_f(\psi_g(P))$ for each P and such that $\psi_f(\psi_g(P)) \leqslant \psi_f(\psi_g(P'))$ whenever $P \leqslant P'$. Let $\psi(P) = \psi_f(\psi_g(P))$. Then $(gf)_{\psi(P)}^P$ can be defined from $(gf)_{\psi_{gf}(P)}^P$, and $g_{\psi_g(P)}^P f_{\psi(P)}^{\psi_g(P)} = (gf)_{\psi(P)}^P$. The composition of inverse systems has a natural definition.

Let us consider pairs of continuous surjections f and g such that $f_\#\mu = \nu$ and $g_\#\nu = \mu$. Observe that there is a homeomorphism f with this property if and only if there is a homeomorphism g with this property. Note that fg and gf are measure preserving, that is, $(fg)_\#\nu = \nu$ and $(gf)_\#\mu = \mu$. As f and g are surjective, f and g are homeomorphisms if and only if $h = fg$ is a homeomorphism. Thus we find that only continuous measure preserving surjections need to be considered. For the Austin example, both fg and gf are not homeomorphisms. Hence there are measures μ for which continuous measure preserving surjections h that are not injective exist.

DEFINITION C.50. *Let μ be a continuous, complete, finite Borel measure on $\{0,1\}^{\mathbb{N}}$. A continuous surjection h of $\{0,1\}^{\mathbb{N}}$ such that $h_{\#}\mu = \mu$ is said to be* μ-refinable *if for each simultaneously closed and open set u of $\{0,1\}^{\mathbb{N}}$ and for each finite partition $Q = \{v_1, v_2, \ldots, v_n\}$ of $h^{-1}[u]$ by open sets there is a partition $P = \{u_1, u_2, \ldots, u_n\}$ of u by open sets such that $\mu(u_i) = \mu(v_i)$ for every i.*

THEOREM C.51. *Let μ be a continuous, complete finite Borel measure on $\{0,1\}^{\mathbb{N}}$. A continuous surjection h of $\{0,1\}^{\mathbb{N}}$ such that $h_{\#}\mu = \mu$ is μ-refinable if and only if h is a homeomorphism.*

PROOF. Suppose that h is μ-refinable. Then h defines the system $\{h_Q^P, \mathsf{P}, \psi\}$ for which μ and μ are compatible. For P in P consider the map h_Q^P. By definition, Q is a refinement of the partition $h^{-1}[P]$. Let $Q = \psi(P)$. As h is μ-refinable, there is a measure preserving bijection $\zeta\colon \{v \in \psi(P) \colon v \subset h^{-1}[u]\} \longrightarrow \{u' \in P_u \colon u' \subset u\}$ for each u in P, where P_u is a partition of u. Extend ζ to all of $\psi(P)$ in the natural way. Let $\eta(P) = \bigcup\{P_u \colon u \in P\}$, a partition of $\{0,1\}^{\mathbb{N}}$ that refines P. This defines a continuous map g_P on $\{0,1\}^{\mathbb{N}}$ that is constant on each v in $\psi(P)$ and $g_P(x) \in \zeta(v)$ whenever $x \in v$. As $\eta(P)$ refines P we have that $\{\eta(P) \colon P \in \mathsf{P}\}$ is a cofinal net. Hence g_P, $P \in \mathsf{P}$ is a uniformly convergent net of continuous maps whose limit g is a homeomorphism. The continuous map h satisfies $h(x) \in u \in P$ whenever $x \in v \in \psi(P)$ with $\zeta(v) \subset u$. Hence $g = h$.

The converse is obvious. □

PROOF OF MAIN THEOREM. Suppose there are continuous maps f and g such that $\beta_s = f_{\#}\beta_r$ and $\beta_r = g_{\#}\beta_s$. Then $h = fg$ is such that $\beta_s = h_{\#}\beta_s$. Let us show that s being refinable implies h is β_s-refinable. To this end let u be a simultaneously closed and open set and let $\{v_1, v_2, \ldots, v_n\}$ be a partition of $h^{-1}[u]$. Then $\beta_s(u) = h_{\#}\beta(u) = \beta_s(h^{-1}[u]) = \sum_{i=1}^{n} \beta_s(v_i)$. Hence there are polynomial numbers $p_i(s) = \beta_s(v_i)$, $i = 1, 2, \ldots, n$, and a polynomial number $p_u(s) = \beta_s(u)$. As s is refinable, we may assume that $p_u(x) = \sum_{i=1}^{n} p_i(x)$. We infer from Theorem C.18 that there is a partition $\{u_1, u_2, \ldots, u_n\}$ of u such that $\beta_s(u_i) = \beta_s(v_i)$ for every i. Hence h is β_s-refinable. By Theorem C.51, h is a homeomorphism. Thereby Theorem C.48 is proved. □

THEOREM C.52. *Let $r \sim_{\mathrm{b}} s$. Then $r \sim_{\mathrm{h}} s$ if and only if there exist continuous maps f and g such that $f_{\#}\beta_r = \beta_s$ and $g_{\#}\beta_s = \beta_r$ with the property that fg is β_s-refinable (equivalently, gf is β_r-refinable).*

C.7.3. An application. Let us consider the polynomial $R(x) = x^n + x - 1$ in $\mathbb{Z}[x]$. As $R(0) = -1$ and $R(1) = 1$, there is an algebraic number r in $(0,1)$ that is a root of $R(x)$. Implicit in R. G. E. Pinch [**126**] is the fact that if d is an integer factor of n with $1 < d < n$, then r^d and $1 - r^d$ are binomially equivalent to r. This follows immediately from Theorem C.16. Hence the Dougherty–Mauldin–Yingst theorem yields that β_r and β_{r^d} are topologically equivalent measures. Clearly r is not $\frac{1}{2}$ if $n > 2$, and $r > r^d > r^n = 1 - r$.

Some results due to E. S. Selmer [**136**] on irreducible polynomials in $\mathbb{Z}[x]$ will be used in applications. We quote his theorem.

Theorem C.53 (Selmer). *Consider polynomials in $\mathbb{Z}[x]$. The polynomials $p_1(x) = x^n - x - 1$ are irreducible for all n. The polynomials $p_2(x) = x^n + x + 1$ are irreducible if $n \equiv 2$ fails* (mod 3), *but have a factor $x^2 + x + 1$ for $n \equiv 2$* (mod 3). *In the latter case, the second factor is irreducible.*

Observe that Selmer's theorem implies, for even n, that the polynomial $p(x) = x^n + x - 1$ is irreducible, and that the root r that is in $(0, 1)$ is an algebraic integer with degree n. Hence the vector space over $\mathbb{Q}$, the field of rational numbers, generated by $r^k, k = 0, 1, 2, \dots, n-1$, has dimension n.[7]

Theorem C.54. *For $k \geq 0$ let $n = 2^{k+1}$. Then there are $2k$ algebraic integers which are topologically equivalent to each other.*

Proof. Let $R(x) = x^n + x - 1$, an irreducible monic polynomial. As $R(0) = -1$ and $R(1) = 1$, there is a root r in $(0, 1)$, which is an algebraic integer. We have already mentioned that Pinch has observed that r^{2^i}, $i = 1, 2, \dots, k$, and $1 - r^{2^i}$, $i = 1, 2, \dots, k$, form an *indexed* set of binomially equivalent algebraic integers, no two distinct members of which are the same numbers. Theorem C.48 completes the proof. □

C.8. Refinable numbers and good measures

Here is an application of the Dougherty–Mauldin–Yingst theorem to good Bernoulli probability measures on $\{0, 1\}^{\mathbb{N}}$. Remember that not all good measures are good Bernoulli measures since the discussion of Section C.6 concerned general Borel probability measures; the discussion in this section concerns measures that are homeomorphic to some Bernoulli probability measure. Observe that if β_r is a good measure then r is necessarily an algebraic number.[8] Dougherty, Mauldin and Yingst proved

Theorem C.55 (Dougherty–Mauldin–Yingst). *Let r be an algebraic number in* $(0, 1)$. *A necessary and sufficient condition that β_r be good is that r be refinable and r be the only root of its minimal polynomial in* $(0, 1)$.

Proof. To prove necessity, it is easily seen that β_r being good implies r is refinable. Indeed, suppose $f(r) = \sum_{i=1}^{k} g_i(r)$ with $f(r), g_1(r), \dots, g_k(r)$ in $\mathcal{P}(r)$. Let C and C'_i, $i = 1, 2, \dots, k$, be simultaneously closed and open sets associated with $f(r)$ and $g_i(r)$ respectively. As β_r is good, there are mutually disjoint, simultaneously closed and open sets C_i contained in C such that $\beta_r(C_i) = \beta_r(C'_i)$. Clearly $\beta_r(C) = \sum_{i=1}^{k} \beta_r(C_i)$. Let $p_k(x) = f(x)$, and for $i < k$ let p_i be a partition polynomial associated with $\bigcup_{j \leq i} C_j$. Obviously $p_{i-1}(x) \leq p_i(x)$ for $x \in (0, 1)$. Let $h_i(x) = p_i(x) - p_{i-1}(x)$,

[7] See Exercise C.10.
[8] See Exercise C.15.

$i = 1, 2, \ldots, k$. Hence r is refinable. Now suppose that r' is a root in $(0, 1)$ of a minimal polynomial $M(x)$ in $\mathbb{Z}[x]$ of r, that is, $p(x) = q(x)M(x)$ with $q(x)$ in $\mathbb{Z}[x]$ whenever $p(x) \in \mathbb{Z}[x]$ with $p(r) = 0$. Suppose that A and B are simultaneously closed and open sets with $\beta_r(A) = \beta_r(B)$. Then we assert that $\beta_{r'}(A) = \beta_{r'}(B)$. Indeed, let $p_A(x)$ and $p_B(x)$ be partition polynomials associated with A and B, respectively. Then $(p_A - p_B)(r) = 0$. As $(p_A - p_B)(x) = q(x)M(x)$ for some $q(x)$ in $\mathbb{Z}[x]$ and $M(r') = 0$, we have $\beta_{r'}(A) = p_A(r') = p_B(r') = \beta_{r'}(B)$. Now suppose that U and V are simultaneously closed and open sets such that $\beta_r(U) \leq \beta_r(V)$. As β_r is a good measure there is a simultaneously closed and open set W contained in V such that $\beta_r(W) = \beta_r(U)$. So $\beta_{r'}(U) = \beta_{r'}(W) \leq \beta_{r'}(V)$. That is, if $\beta_r(U) \leq \beta_r(V)$, then $\beta_{r'}(U) \leq \beta_{r'}(V)$. It is a simple exercise[9] to show that there exist simultaneously closed and open sets U and V such that $\beta_r(U) < \beta_r(V)$ and $\beta_{r'}(U) > \beta_{r'}(V)$ whenever $r \neq r'$. Consequently, $r = r'$.

To prove sufficiency, let U and V be simultaneously open and closed sets such that $\beta_r(U) < \beta_r(V)$. Then there are partition polynomials $p_U(x)$ and $p_V(x)$ such that $p_U(r) < p_V(r)$. We infer from Lemma C.44 the existence of a partition polynomial $\hat{p}(x)$ such that $p_U(r) = \hat{p}(r)$ and $\hat{p}(x) < p_U(x)$ whenever $x \in (0, 1)$. Theorem C.18 completes the proof. □

For rational numbers r in $(0, 1)$, it is now clear that β_r is a good measure if and only if $r = \frac{1}{2}$. We close the discussion with

THEOREM C.56 (Dougherty–Mauldin–Yingst). *There exist refinable numbers r and s in $(0, 1)$ such that $r \sim_{\mathrm{b}} s$, $r \neq s$ and $r \neq 1 - s$, and such that both β_r and β_s are not good measures.*

PROOF. The polynomial $R(x) = -14x^6 + 21x^4 - 8x^2 - x + 1$ is irreducible and has three roots in $(0, 1)$. It is also minimal in $\mathbb{Z}[x]$ for each root r in $(0, 1)$. So r is refinable and β_r is not a good measure. Let $s = r^2$. Then $r = p(s) = -14s^3 + 21s^2 - 8s + 1 = (s - 1)(-14s^2 + 7s - 1)$. As $0 < p(x) < 1$ whenever $x \in (0, 1)$, $p(x)$ is a partition polynomial. Hence $r \sim_{\mathrm{b}} s$. β_s is not a good measure since $r \sim_{\mathrm{h}} s$. As r is an algebraic number of degree 6, the vector space over the field $\mathbb{Q}$ generated by r^i, $i = 0, 1, \ldots, 5$, has dimension 6; hence $s \neq r$ and $1 - s \neq r$. □

C.9. Comments

This appendix has been unusually long. But its length is justified by the inclusion of the recent activity and advances.

The relation $\sim_{\mathrm{b}}$ defined on $\mathsf{B}(\{0, 1\}^{\mathbb{N}})$, the collection of all binomial Bernoulli measures on $\{0, 1\}^{\mathbb{N}}$, involves polynomial conditions. These conditions were discovered by Navarro-Bermúdez [**114, 115**] as necessary conditions for $\sim_{\mathrm{h}}$. The transitivity of $\sim_{\mathrm{b}}$ was proved by Huang in [77]. As observed by Mauldin [**105**], the transitivity is easily proved by using the relation $\leq_{\mathrm{c}}$ and its equivalent $\leq_{\mathrm{b}}$. Later, using his notion

[9] See Exercise C.16.

of a-convexity [**125**], Pinch gave a second proof for the fact that relation $\sim_{\mathrm{b}}$ is an equivalence relation [**126**, Proposition 1].

There are many early results cited in Section C.5.5. They were based on algebraic numbers arising from irreducible polynomials $R(x)$, in particular, $x^{2m} + x - 1$, $2x^{2m} + 2x^{2m-1} - x^2 - x - 1$, $2x^{2m+1} - 2x^{2m} + 2x^m - 1$ and $x^3 + x^2 - 1$. Each of these polynomials satisfy $|R(0)| = 1$ and $|R(1)| = 1$. Hence their roots are refinable, whence $r \sim_{\mathrm{b}} s$ implies $r \sim_{\mathrm{h}} s$ for roots r of $R(x)$ in $(0, 1)$.

The refinable condition is one that permits the existence of a homeomorphism f such that $f_\# \beta_r = \beta_s$ by means of a factorization using limits of inverse systems of measures whose factor measures "commute." That is, if $f \sim_{\mathrm{b}} g$ then fg and gf result in measure preserving transformations. It is seen that f is a homeomorphism if and only if fg is a homeomorphism. Our inverse limits are closed subsets of products of discrete spaces. Factorizations using product spaces were used in our version of the proof of Theorem C.29 by Navarro-Bermúdez and Oxtoby. The use of factorization was foreshadowed in the original Navarro-Bermúdez and Oxtoby proof. A combination of this factorization method and Theorem C.52 has the potential for a characterization of $r \sim_{\mathrm{h}} s$.[10]

The only rational number that is refinable is $\frac{1}{2}$. In Austin's example, r and s are algebraic numbers that are not rational numbers since $r \sim_{\mathrm{b}} s$ and $r \nsim_{\mathrm{h}} s$. As neither of them is refinable, again since $r \sim_{\mathrm{b}} s$ and $r \nsim_{\mathrm{h}} s$, there exist irrational algebraic numbers that are not refinable.

On page 223 it was shown by a direct polynomial computation that $r = \frac{1}{\sqrt{2}}$ and $s = \frac{1}{2}$ are such that $r \preceq_{\mathrm{b}} s$ fails and $s \preceq_{\mathrm{b}} r$ holds. Observe that r is a root of $2x^2 - 1$ and s is a root of $2x - 1$. Hence this will give a second proof, albeit a more complicated one, that $r \nsim_{\mathrm{b}} s$.

Let us turn to good measures among the collection $\mathsf{B}(\{0,1\}^{\mathbb{N}})$. These have been characterized by Theorem C.55. Among the rational numbers r in $(0, 1)$ only $r = \frac{1}{2}$ is good. Is there a simple way to find simultaneously closed and open sets U and V with $\beta_{\frac{1}{9}}(U) < \beta_{\frac{1}{9}}(V)$ that fails the subset condition? Let us turn to a more serious question. Not all Borel probability measures are binomial Bernoulli measures.

QUESTION. Is there a good Borel probability measure μ on $\{0,1\}^{\mathbb{N}}$ such that $\mu \neq h_\# \beta_r$ for any h in $\mathsf{HOMEO}(\{0,1\}^{\mathbb{N}})$ and any good binomial Bernoulli measure β_r?

We conclude with two other questions.

QUESTION. [**105**, Problem 1067] Is there an infinite $\sim_{\mathrm{h}}$ equivalence class in $(0, 1)$? Is there an infinite $\sim_{\mathrm{b}}$ class in $(0, 1)$?

QUESTION. [**47**] Is every number in a nontrivial $\sim_{\mathrm{h}}$ equivalence class of $(0, 1)$ refinable?

[10] A very recent preprint by Yingst [**157**] results in necessary and sufficient conditions for a partition polynomial that corresponds to a measure preserving surjection to be injective.

Exercises

C.1. Prove Proposition C.6 on page 217.

C.2. Prove Proposition C.7 on page 217.

C.3. Prove $d_{(k,\alpha)}$, defined on page 218, is a metric for the space ${}^{\omega}k$.

C.4. Establish that δ as defined on page 218 is an isometry by proving the following.

(a) Prove $\chi(\langle f,f'\rangle,\langle g,g'\rangle) = \min\{\chi(f,f'),\chi(g,g')\}$ whenever $\langle f,f'\rangle$ and $\langle g,g'\rangle$ are in $(k \times l)^{\omega}$.

(b) Prove that Φ and Ψ are Lipschitzian bijections with respect to the metrics defined above and that the Lipschitz constants are equal to 1.

C.5. Prove the following:

(a) Prove

$$\chi(f,f') \leq m{\cdot}\chi(\Phi_m(f),\Phi_m(f')) + m - 1$$
$$\text{whenever} \quad f \in k^{\omega} \text{ and } f' \in k^{\omega}.$$

Verify equation (C.1) on page 219.

(b) Prove

$$m{\cdot}\chi(h,h') \leq \chi(\Psi_m(h),\Psi_m(h'))$$
$$\text{whenever} \quad h \in (k^m)^{\omega} \text{ and } h' \in (k^m)^{\omega}.$$

Verify equation (C.2) on page 219.

C.6. Prove: p is a prime factor of a positive integer m if and only if $\frac{1}{p} \in \mathrm{vs}\big(\beta(\frac{1}{m})\big)$.

C.7. Prove: If k is a finite set with $\mathrm{card}(k) = m$ and if p_i, $i = 1,\ldots,j$, are all the prime divisors of a positive integer m, then there is a homeomorphism $h\colon k^{\mathbb{N}} \to \mathsf{X}_{i=1}^{j} {k_i}^{\mathbb{N}}$, where k_i is a finite set with cardinality p_i for each i. Hence $h_{\#}\beta(\frac{1}{m}) = \mathsf{X}_{i=1}^{j} \beta(\frac{1}{p_i})$.

C.8. Let $n = n_1 + n_2$ and $E \subset \{0,1\}^n$. Let π_j be a map of $\{0,1\}^n$ onto $\{0,1\}^{n_j}$, $j = 1,2$, such that

$$\pi_1(e_1,\ldots,e_{n_1},e_{n_1+1},\ldots,e_n) = (e_1,\ldots,e_{n_1})$$

and

$$\pi_2(e_1,\ldots,e_{n_1},e_{n_1+1},\ldots,e_n) = (e_{n_1+1},\ldots,e_n).$$

Prove: If $E_1 \subset \{0,1\}^{n_1}$, $E_2 \subset \{0,1\}^{n_2}$ and

$$E = {\pi_1}^{-1}[E_1] \cap {\pi_2}^{-1}[E_2],$$

then $\beta(t,1-t)\big(E\big) = \beta(t,1-t)\big(E_1\big){\cdot}\beta(t,1-t)\big(E_2\big)$.

C.9. Hausdorff [72] proved the following theorem: *If $p(x)$ is a polynomial with real coefficients is such that $p(x) > 0$ whenever $x \in (-1,1)$, then $p(x)$ is a* *finite sum of the form*

$$\sum_{i,j} c_{i,j}(1+x)^i(1-x)^J \text{ with } c_{i,j} \geq 0.$$

Prove that if $p(x)$ is positive on $(0,1)$ then $p(x)$ is a finite sum of the form $\sum_{i,j} a_{i,j}x^i(1-x)^j$ with $a_{i,j} \geq 0$. Indeed, for some n, $p(x) = \sum_{i=1}^{n} a_i x^i(1-x)^{n-i}$, with $a_i \geq 0$. Moreover, if $p(x)$ has integer coefficients, then $c_i \in \mathbb{Z}$ for each i.

C.10. Prove Lemma C.27 on fractional part of a real number on page 228.

C.11. Let $(X, \leq)$ be the Cantor ternary set with the usual order and let μ_0 be the probability measure on X given by the well-known Cantor function. Show that there is an order preserving homeomorphism $g\colon [0,1] \to [0,1]$ such that the measure μ as provided by Proposition C.35 for the order preserving map $F = gF_{\mu_0}$ yields a special value set $\widetilde{\mathrm{vs}}(\mu, X, \leq)$ that is a proper subset of $\mathrm{vs}(\mu, X)$. Hint: Use a g such that $g(y) = y$ whenever $0 \leq y \leq \frac{1}{2}$ and such that $g(\frac{3}{4}) = \frac{2}{3}$.

C.12. For the Cantor ternary set X show that to each nonempty simultaneously closed and open set U there corresponds a positive integer k_0 such that for every point x of U there is J such that $J \cap X \subset U$ whenever $x \in J$ and J is an interval of the k-th step of the Cantor set construction with $k \geq k_0$.

C.13. Prove that each transcendental number in $(0,1)$ is refinable.

C.14. Let $P_s(x) = x^{2n} + x - 1$ be a Selmer irreducible polynomial and let r be a root of the polynomial in $(0,1)$. (See Theorem C.53 on page 239 for the Selmer polynomials.) Prove: If k is a positive integer that divides n, then $r \sim_{\mathrm{b}} r^k$. Moreover, for each such k, $r^k \neq 1 - r^l$ for every l with $l \neq k$ and $0 \leq l < 2n$. Consequently, there are at least $2m$ distinct numbers t in $(0,1)$ such that $r \sim_{\mathrm{b}} t$ where m is the total number of positive integers less than $2n$ that divide $2n$.

C.15. Prove that if the binomial Bernoulli measure β_r is good then r is an algebraic number. Hint: If U is a nonempty simultaneously closed and open subset of $\langle(0)\rangle$ such that $U \neq \langle(0)\rangle$ and $p(x)$ is a partition polynomial associated with U, then $\deg(p(r)) \geq 2$.

C.16. Let μ_1 and μ_2 be different continuous, complete, finite Borel probability measures on an uncountable absolute Borel space. Prove that there exist disjoint compact sets U and V such that $\mu_1(U) < \mu_1(V)$ and $\mu_2(V) < \mu_2(U)$. Hint: First prove the statement for the space $[0,1]$ with the aid of the cumulative distribution functions of μ_1 and μ_2. For the general case, consider a $\mathfrak{B}$-homeomorphism of the space onto $[0,1]$.

C.17. Let $f\colon \{0,1\}^{\mathbb{N}} \to \{0,1\}^{\mathbb{N}}$ be a continuous surjection. Prove that there exists a Borel measurable injection $g\colon \{0,1\}^{\mathbb{N}} \to \{0,1\}^{\mathbb{N}}$ such that g^{-1} is the map f restricted to the compact set $g\big[\{0,1\}^{\mathbb{N}}\big]$. Hence f is a homeomorphism if and only if $g\big[\{0,1\}^{\mathbb{N}}\big] = \{0,1\}^{\mathbb{N}}$. Interpret these assertions in terms of inverse limit systems. Hint: Devise a selection scheme that reverses the commutative diagram (C.14).

Appendix D

Dimensions and measures

This appendix contains a summary of the needed topological dimension theory, and, for metric spaces, the needed Hausdorff measure theory and the Hausdorff dimension theory.

D.1. Topological dimension

There are three distinct dimension functions in general topology, two of which are inductively defined and the third is defined by means of open coverings. Each definition has its advantages and its disadvantages. Fortunately, the three agree whenever the spaces are separable and metrizable. Let us give their definitions.

DEFINITION D.1. *Let X be a topological space.*

The space X is said to have small inductive dimension -1 *if and only if it is the empty space. For each positive integer n, the space X is said to have* small inductive dimension *not exceeding n if each point of X has arbitrarily small neighborhoods whose boundaries have small inductive dimension not exceeding $n-1$. These conditions are denoted by* $\operatorname{ind} X \leq n$. *The definition of* $\operatorname{ind} X = n$ *is made in the obvious manner for $n = -1, 0, 1, \ldots, +\infty$.*

The space X is said to have large inductive dimension -1 *if and only if it is the empty space. For each positive integer n, the space X is said to have* large inductive dimension *not exceeding n if each closed subset of X has arbitrarily small neighborhoods whose boundaries have large inductive dimension not exceeding $n-1$. These conditions are denoted by* $\operatorname{Ind} X \leq n$. *The definition of* $\operatorname{Ind} X = n$ *is made in the obvious manner for $n = -1, 0, 1, \ldots, +\infty$.*

If the space X is nonempty, then it is said to have covering dimension *not exceeding n if each finite open cover of X has a finite open cover refinement whose order*[1] *does not exceed $n+1$. This is denoted by* $\dim X \leq n$. *(For convenience, the modifier "covering" will be deleted.)* $\dim X = n$ *is defined in the obvious manner, and* $\dim X = -1$ *if and only if $X = \emptyset$.*

Here follows the important properties of topological dimension for separable metrizable spaces.

[1] For a collection $\{ U_\alpha : \alpha \in A \}$ of subsets of a set X, the *order at a point* x is the cardinal number of $\{ \alpha : x \in U_\alpha \}$ (finite or ∞) and the *order* is the supremum of the orders at the points of X.

THEOREM D.2. *If X is a separable metrizable space, then* $\operatorname{ind} X = \operatorname{Ind} X = \dim X$.

THEOREM D.3. *Let n be a nonnegative integer. If X is a separable metrizable space, then $\dim X \leq n$ if and only if there exists a base $\mathcal{B}$ for the open sets such that $\dim \operatorname{Bd}_X(U) \leq n - 1$ for every U in $\mathcal{B}$.*

THEOREM D.4. *Let n be a nonnegative integer. If X is a separable metrizable space and A is a subset of X with $\dim A \leq n$, then there is a base $\mathcal{B}$ for the open sets of X such that $\dim(A \cap \operatorname{Bd}_X(U)) \leq n - 1$ for every U in $\mathcal{B}$.*

THEOREM D.5. *Let X be a separable metrizable space and n be a nonnegative integer. If $\dim X = n$, then X can be topologically embedded into $[0, 1]^{2n+1}$.*

THEOREM D.6. *Let X be a separable metrizable space. Then the following statements hold.*

(1) **Normed.** *If $X = \emptyset$ then $\dim X = -1$; if $\operatorname{card}(X) = 1$ then $\dim X = 0$; if $X = [0, 1]^n$, then $\dim X = n$.*
(2) **Monotone.** *If $A \subset B \subset X$, then $\dim A \leq \dim B$.*
(3) **Sum.** *If A_i, $i = 1, 2, \ldots$, are closed subsets of X, then*

$$\dim \left(\bigcup_{i=1}^{\infty} A_i\right) \leq \sup \{ \dim A_i : i = 1, 2, \ldots \}.$$

(4) **Decomposition.** *If $0 \leq \dim X = n < \infty$, then there are subsets A_i, $i = 0, 1, \ldots, n$, of X with $\dim A_i = 0$ such that $X = \bigcup_{i=0}^{n} A_i$.*
(5) **G_δ Hull.** *If $A \subset X$, then there is a G_δ set B such that $A \subset B \subset X$ and $\dim A = \dim B$.*
(6) **Compactification.** *There exists a metrizable compactification Y of X such that $\dim X = \dim Y$.*
(7) **Addition.** *If $A \cup B \subset X$, then $\dim(A \cup B) \leq \dim A + \dim B + 1$.*
(8) **Product.** *If A and B are subsets of X, then $\dim(A \times B) \leq \dim A + \dim B$.*

These theorems can be found in any standard book on dimension theory. See, for example, W. Hurewicz and H. Wallman [**79**], R. Engelking [**50, 51**] and J-I. Nagata [**113**].

Infinite dimensional spaces are of two types. Clearly one can construct infinite dimensional spaces by taking unions of countably many finite dimensional ones whose dimensions are not bounded above. These will be the union of countably many zero-dimensional subsets. Not every infinite dimensional space is like this. A space is said to be *strongly infinitely dimensional* if it cannot be written as the union of countably many zero-dimensional subsets.

D.1.1. Borel measures. Let us take a small digression. It is well-known that every finite Borel measure μ on a separable metrizable space X and every Borel set M of X has the property that there exists an F_σ set K contained in M with $\mu(M) = \mu(K)$. A strengthening of this property is easily shown.

PROPOSITION D.7. *Let X be a separable metrizable space and let μ be a finite Borel measure on X. If M is a Borel set, then there is an F_σ set K such that $K \subset M$,*

$\dim K \leq 0$, *and* $\mu(M) = \mu(K)$. *Moreover, if* M *is an absolute Borel space, then* K *may be selected to be also a* σ*-compact set.*

Proof. Since $\mu(X) < \infty$ there is a countable base $\mathcal{B}$ for the open sets of X such that $\mu(\mathrm{Bd}_X(U)) = 0$ for each U in $\mathcal{B}$. Obviously the set $A = X \setminus \bigcup_{U \in \mathcal{B}} \mathrm{Bd}_X(U)$ satisfies $\dim A \leq 0$ and $\mu(M) = \mu(M \cap A)$. As $M \cap A$ is a Borel set, there is an F_σ set K contained in it such that $\mu(M) = \mu(K)$. Obviously, $\dim K \leq 0$. The final assertion follows easily upon embedding X into the Hilbert cube and extending the measure μ to a Borel measure on the Hilbert cube. □

The following corollary is related to Corollary 4.31. For the definition of a Zahorski set, see Section A.6.

Corollary D.8. *Let* X *be a separable metrizable space and let* μ *be a continuous, complete, finite Borel measure on* X. *If* A *is an absolute Borel space contained in* X, *then there exists a Zahorski set* Z *contained in* A *such that* $\mu(Z) = \mu(A)$. *Moreover,* Z *may be chosen to be a set of the first category of Baire.*

Proof. As $\mu(A)$ is finite there is an absolute Borel space B contained in A with $\mu(B) = \mu(A)$. The theorem yields a σ-compact set E' contained in B such that $\mu(E') = \mu(B)$. Let $E' = \bigcup_{n=1}^{\infty} E'_n$, where E'_n is a compact zero-dimensional set for each n. As μ is continuous there is a topological copy E_n of $\{0,1\}^{\mathbb{N}}$ contained in E'_n such that $\mu(E_n) = \mu(E'_n)$ for each n. Hence $\bigcup_{n=1}^{\infty} E_n$ is a Zahorski set contained in A whose μ-measure is equal to $\mu(A)$. Repeating the argument, for each countable dense subset D_n of E_n, we have a Zahorski set Z_n of $E_n \setminus D_n$ with $\mu(Z_n) = \mu(E_n)$. Clearly Z_n is a set of the first category; hence $Z = \bigcup_{n=1}^{\infty} Z_n$ is a Zahorski set with the required property. □

D.2. Measure theoretical dimension

Let us quickly review the definitions and properties of Hausdorff p-dimensional measures on separable metric spaces that were developed in Chapter 5.

D.2.1. Hausdorff measures. We repeat the definition given in Chapter 5.

Definition D.9. *Let* E *be a subset of* X *and let* p *be a real number with* $0 \leq p$. *For* $\delta > 0$, *define* $\mathsf{H}_p^\delta(E)$ *to be the infimum of the set of numbers* $\sum_{S \in G} (\mathrm{diam}(S))^p$ *corresponding to all countable families* G *of subsets* S *of* X *such that* $\mathrm{diam}(S) \leq \delta$ *and* $E \subset \bigcup_{S \in G} S$.[2] *The* p-dimensional Hausdorff outer measure *on* X *is*

$$\mathsf{H}_p(E) = \sup \{ \mathsf{H}_p^\delta(E) : \delta > 0 \};$$

or equivalently,

$$\mathsf{H}_p(E) = \lim_{\delta \to 0} \mathsf{H}_p^\delta(E)$$

[2] We use the conventions that $\mathrm{diam}(\emptyset) = 0$ and $0^0 = 1$.

since the limit always exists as a nonnegative extended real number. A set E is said to be H_p-measurable *if* $\mathsf{H}_p(T) = \mathsf{H}_p(T \cap E) + \mathsf{H}_p(T \setminus E)$ *whenever* $T \subset X$.

The above definition of H_p is called an "outer measure" since it is defined for all subsets of X. In general, not every subset is H_p-measurable. With $\mathfrak{M}(X, \mathsf{H}_p)$ denoting the collection of all H_p-measurable sets, the measure space $\mathrm{M}(X, \mathsf{H}_p)$ is the triple $\big(X, \mathsf{H}_p, \mathfrak{M}(X, \mathsf{H}_p)\big)$. It is known that $\mathrm{M}(X, \mathsf{H}_p)$ is a complete Borel measure space on the topological space X. If $X = \mathbb{R}^n$, then $\mathsf{H}_n = \alpha_n \lambda_n$, where λ_n is the usual Lebesgue measure on $\mathbb{R}^n$ and α_n is a normalizing constant given by $\mathsf{H}_n([0,1]^n) = \alpha_n \lambda_n([0,1]^n)$. The zero-dimensional Hausdorff measure is often called the counting measure on X. For $p > 0$, $\mathsf{H}_p(E) = 0$ for every singleton set E. In general, the complete Borel measure space $\mathrm{M}(X, \mathsf{H}_p)$ is not σ-finite. But, if $E \subset X$ and $\mathsf{H}_p(E) < \infty$, then the restricted measure space $\mathrm{M}(E, \mathsf{H}_p \,|E)$ of the measure space $\mathrm{M}(X, \mathsf{H}_p)$ is a continuous, complete, finite Borel measure on E – the set E need not be H_p-measurable, recall that $\mathrm{M}(E, \mathsf{H}_p \,|E)$ is defined by employing the outer measure on E induced by the measure space $\mathrm{M}(X, \mathsf{H}_p)$. More precisely, if Y is a subset of X, then Y has a natural metric induced by the restriction of the metric of X, hence it has a p-dimensional Hausdorff outer measure on the metric space Y using the restricted metric. The corresponding p-dimensional Hausdorff outer measure on this metric subspace Y agrees with the outer measure induced by the super measure space $\mathrm{M}(X, \mathsf{H}_p)$ for subsets of Y.

Since $\dim \emptyset = -1$ and since measures take values that are at least equal to 0, the empty set may, at times, require special treatment. The reader should keep in mind the statements in the following proposition in the course of the development of Appendix D.

Proposition D.10. *If X is a separable metric space, then* $\dim \emptyset = -1 < 0 = \mathsf{H}_p(\emptyset)$ *for every p. If $0 \le p$ and if E is a subset of a separable metric space X with* $\dim E \le 0$, *then* $\dim E \le \mathsf{H}_p(E)$.

The reader was asked in Chapter 5 to prove the following theorem concerning Lipschitzian maps.

Theorem D.11. *For separable metric spaces X and Y, let f be a Lipschitzian map from X into Y with Lipschitz constant L. If $0 \le p$ and $E \subset X$, then* $\mathsf{H}_p(f[E]) \le L^p\, \mathsf{H}_p(E)$.

It will be convenient to define at this point the notion of a bi-Lipschitzian embedding.

Definition D.12. *Let X and Y be separable metric spaces with respective metrics* d_X *and* d_Y. *An injection $\varphi \colon X \to Y$ is called a* bi-Lipschitzian embedding *of X onto $M = \varphi[X]$ if φ is a Lipschitzian map and $(\varphi|M)^{-1} \colon M \to X$ is a Lipschitzian map.*

Clearly the Lipschitz constants of the maps φ and $(\varphi|M)^{-1}$ in the above definition are positive whenever $\mathrm{card}(X) > 1$. Hence we have

Theorem D.13. *Let φ be a bi-Lipschitzian embedding of X into Y. If $0 \le p$ and $E \subset X$, then*

$$\mathsf{H}_p(E) < \infty \text{ if and only if } \mathsf{H}_p(\varphi[E]) < \infty$$

and

$$0 < \mathsf{H}_p(E) \text{ if and only if } 0 < \mathsf{H}_p(\varphi[E]).$$

D.2.2. Hausdorff dimension. We will continue to assume that X is a separable metric space. It is easily seen that if $0 \leq p < q$ then $\mathsf{H}_p(E) \geq \mathsf{H}_q(E)$ for subsets E of X. Also, if $\mathsf{H}_p(E) < \infty$ then $\mathsf{H}_q(E) = 0$ whenever $p < q$. This leads to the following definition.

DEFINITION D.14. *For subsets E of X, the* Hausdorff dimension *of E is the extended real number* $\dim_{\mathsf{H}} E = \sup\{p\colon \mathsf{H}_p(E) > 0\}$.

Of course the Hausdorff dimension is dependent on the metric d of X. Unlike the topological dimension function dim which is integer-valued, the Hausdorff dimension function $\dim_{\mathsf{H}}$ is nonnegative extended real-valued. The next theorem is easily proved.

THEOREM D.15. *Let X be a separable metric space.*

(1) $-1 = \dim \emptyset < \dim_{\mathsf{H}} \emptyset = 0$.
(2) *If $A \subset B \subset X$, then* $\dim_{\mathsf{H}} A \leq \dim_{\mathsf{H}} B$.
(3) *If A_i, $i = 1, 2, \ldots$, is a countable collection of subsets of X, then* $\dim_{\mathsf{H}} \bigcup_{i=1}^{\infty} A_i = \sup\{\dim_{\mathsf{H}} A_i\colon i = 1, 2, \ldots\}$.

An important fact to remember is that if $s = \dim_{\mathsf{H}} E$ for a subset E of X then the inequalities $0 < \mathsf{H}_s(E) < \infty$ may fail.

We have the following theorem that can be summarized as "the Hausdorff dimension of a set is a bi-Lipschitzian invariant." It is a consequence of Theorem D.11.

THEOREM D.16. *For bi-Lipschitzian embeddings $\varphi\colon X \to Y$ of X onto $M = \varphi[X]$,*

$$\dim_{\mathsf{H}} E = \dim_{\mathsf{H}} \varphi[E] \text{ whenever } E \subset X.$$

There is an interesting connection between the topological dimension and the Hausdorff dimension of a separable metric space. It is the classical Theorem 5.1 stated in Chapter 5. As the classical proof is not difficult, we shall give it here. A second very different proof will be given in the next section. The proof is a consequence of the following lemma concerning the Lebesgue measure on $\mathbb{R}$.

LEMMA D.17. *Let $0 \leq p < \infty$ and suppose that X is a separable metric space with $\mathsf{H}_{p+1}(X) = 0$. For an arbitrary point x_0 of X and for each nonnegative real number r let $S(r) = \{x \in X\colon \mathrm{d}(x, x_0) = r\}$. If $0 < \delta < \infty$, then the Lebesgue measure of the set*

$$\{r \in \mathbb{R}\colon \mathsf{H}_p^{\delta}(S(r)) \neq 0\}$$

is equal to 0, whence $\mathsf{H}_p(S(r)) = 0$ for Lebesgue almost every r.

PROOF.[3] We shall begin with a simple estimate. Let E be any nonempty subset of X with $\operatorname{diam}(E) \le \delta$. Define $f\colon [0,\infty) \to [0,\delta]$ to be the function given by the formula $f(r) = \operatorname{diam}(E \cap S(r))$. Note that the closed convex hull of $\{ r : f(r) \neq 0 \}$ is a bounded interval I, possibly degenerate. The triangle inequality for the metric d yields $\lambda(I) \le \operatorname{diam}(E)$. Our estimate is $\int_0^\infty (\operatorname{diam}(E) \cdot \chi_I)^p \, dr \le (\operatorname{diam}(E))^{p+1}$. Note also the inequality $f(r) \le \operatorname{diam}(E) \cdot \chi_I(r)$ for every r.

Let $X = \bigcup_{i=1}^\infty E_i$ with $\operatorname{diam}(E_i) \le \delta$ for every i. For each nonempty set E_i let f_i and I_i be the function and closed interval from the previous paragraph that corresponds to E_i and let $g_i = \operatorname{diam}(E_i) \cdot \chi_{I_i}$. Let $g_i = 0$ whenever $E_i = \emptyset$. It is easily seen that

$$\mathsf{H}_p^\delta(S(r)) \le \sum_{i=1}^\infty f_i^{\,p}(r) \le \sum_{i=1}^\infty g_i^{\,p}(r)$$

for every r. It now follows that $G = \sum_{i=1}^\infty g_i^p$ is a Lebesgue measurable function such that $\mathsf{H}_p^\delta(S(r)) \le G(r)$ for every r and $\int_0^\infty G \, dr \le \sum_{i=1}^\infty (\operatorname{diam}(E_i))^{p+1}$. Since $\mathsf{H}_{p+1}(X) = 0$ implies $\mathsf{H}_{p+1}^\delta(X) = 0$, we have for each positive number ε the existence of a Lebesgue measurable function G_ε such that $\mathsf{H}_p^\delta(S(r)) \le G_\varepsilon(r)$ for every r and $\int_0^\infty G_\varepsilon \, dr < \varepsilon$. We infer from the classical Fatou lemma that $\mathsf{H}_p^\delta(S(r)) = 0$ for λ-almost every r. As $\mathsf{H}_p(S(r)) = \lim_{\delta \to 0} \mathsf{H}_p^\delta(S(r))$ for every r, we have that $\mathsf{H}_p(S(r)) = 0$ for λ-almost every r. □

With the aid of the implication $\mathsf{H}_0(E) = 0$ implies $E = \emptyset$ one can prove by induction the following theorem.

THEOREM D.18. *If* $0 \le n < \infty$ *and* $\mathsf{H}_{n+1}(X) = 0$, *then* $\dim X \le n$.

Observe that $\dim X = 0$ implies $0 < \mathsf{H}_0(X)$. Hence the next theorem is equivalent to the previous one.

THEOREM D.19. *If* $0 \le \dim X = m < \infty$, *then* $0 < \mathsf{H}_m(X)$.

Remember that the Hausdorff dimension need not be an integer. The last theorem can be stated in terms of the Hausdorff dimension as follows (which, of course, is Theorem 5.1 of the introduction to Chapter 5).

THEOREM D.20. *For every separable metric space* X,

$$\dim X \le \dim_{\mathsf{H}} X.$$

It is shown in [**79**] that every nonempty, separable, metrizable space X has a metric such that $\dim_{\mathsf{H}} X = \dim X$. For a proof see [**79**, page 106].

D.3. Zindulka's dimension theorem

In order to proceed further we will need Zindulka's dimension theoretic theorem for metric spaces. This theorem concerns general metric spaces. Although the emphasis

[3] The proof given here is the one found in [**113**] which uses properties of the Lebesgue integral on $\mathbb{R}$. An earlier proof given in [**79**], which is due to Szpilrajn [**149**], uses properties of the upper Lebesgue integral on $\mathbb{R}$.

of the book has been on separable metrizable spaces we shall give the proof for general metric spaces since the proof is not any more difficult in the general metric setting. In order to present his results we will need some further results in dimension theory which have not been mentioned earlier (see [**113, 50, 51**]). We give these results now.

THEOREM D.21 (Bing metrization). *Let X be a regular Hausdorff space. In order that X be metrizable it is necessary and sufficient that there is a σ-discrete base for the open sets of X.*

THEOREM D.22 (Morita characterization). *Let X be a metrizable space and n be a nonnegative integer. Then* $\dim X \leq n$ *if and only if there is a σ-locally finite base $\mathcal{B}$ for the open sets such that* $\dim \mathrm{Bd}(U) \leq n-1$ *for every U in $\mathcal{B}$.*

It is well-known that $\dim X = \mathrm{Ind} X$ for every metrizable space. To contrast this with the separable metrizable case, the following theorem has been proved by P. Roy and, subsequently, by J. Kulesza. Observe that every separable subspace X of this counterexample has $\dim X \leq 0$.

THEOREM D.23 (Roy–Kulesza). *There exists a metrizable space X such that* $\dim X = 1$, $\mathrm{ind} X = 0$ *and* $\mathrm{card}(X) = \mathfrak{c}$.

Finally, the generalization of Theorem D.6 to arbitrary metrizable spaces remains true except for the compactification property (6).

We are now ready to present a lemma due to Zindulka. In Chapter 5 we implicitly gave a proof of it with the stronger assumption of separable metric space. It is Bing's metrization theorem that permits the relaxing of the separable condition.

LEMMA D.24. *Let $\mathcal{B} = \bigcup_{n\in\omega} \mathcal{B}_n$ be a σ-discrete base for the open sets of a nonempty metric space X. For each pair $\langle n,j\rangle$ in $\omega \times \omega$, define the Lipschitzian function $g_{\langle n,j\rangle}\colon X \to [0,1]$ given by the formula*

$$g_{\langle n,j\rangle}(x) = 1 \wedge \big(j \,\mathrm{dist}\big(x, X \setminus \bigcup \mathcal{B}_n\big)\big), \quad x \in X;$$

and, for each r in the open interval $(0,1)$, define the set

$$G(r) = \bigcap\nolimits_{\langle n,j\rangle \in \omega\times\omega} g_{\langle n,j\rangle}{}^{-1}\big[\{ s \in [0,1] \colon s \neq r \}\big].$$

Then $G(r)$ is a G_δ set with $\dim G(r) \leq 0$ *for every r. Moreover, if E is an uncountable subset of the open interval $(0,1)$, then $X = \bigcup_{r\in E} G(r)$.*

PROOF. Let $r \in (0,1)$. Clearly $G(r)$ is a G_δ set. Let us show that $\dim G(r) \leq 0$. To this end we first show that the collection

$$\mathcal{D}_r = \bigcup\nolimits_{\langle n,j\rangle\in\omega\times\omega} \{ g_{\langle n,j\rangle}{}^{-1}\big[(r,1]\big] \cap B \colon B \in \mathcal{B}_n \}$$

is a σ-discrete base for the open sets of X. Of course, such a base is also σ-locally finite. As $\mathcal{B}_n$ is discrete, it follows easily that $\mathcal{D}_r$ is a σ-discrete collection. To see that $\mathcal{D}_r$ is a base for the open sets, let $x \in B \in \mathcal{B}_n$. There is a j such that $g_{\langle n,j\rangle}(x) > r$. Consequently, $x \in g_{\langle n,j\rangle}{}^{-1}\big[(r,1]\big] \cap B \subset B$. Hence $\mathcal{D}_r$ is a σ-discrete base for the open sets of X. Next let us show that $D = G(r) \cap g_{\langle n,j\rangle}{}^{-1}\big[(r,1]\big] \cap B$ is closed in the subspace $G(r)$. Using the fact that the distance function $\operatorname{dist}\big(\cdot, X \setminus \bigcup \mathcal{B}_n\big)$ appears in the definition of $g_{\langle n,j\rangle}$, one can easily verify $g_{\langle n,j\rangle}{}^{-1}\big[[r,1]\big] \cap \operatorname{Cl}(B) \subset B$ and $G(r) \cap g_{\langle n,j\rangle}{}^{-1}\big[\{r\}\big] \cap B = \emptyset$. So

$$
\begin{aligned}
\operatorname{Cl}_{G(r)}(D) &\subset G(r) \cap \operatorname{Cl}\big(g_{\langle n,j\rangle}{}^{-1}\big[(r,1]\big] \cap B\big) \\
&\subset G(r) \cap g_{\langle n,j\rangle}{}^{-1}\big[[r,1]\big] \cap \operatorname{Cl}(B) \\
&\subset G(r) \cap g_{\langle n,j\rangle}{}^{-1}\big[(r,1]\big] \cap B = D.
\end{aligned}
$$

Consequently $\dim \operatorname{Bd}_{G(r)}(D) = -1$ and thereby $\dim G(r) \le 0$ follows from the Morita characterization theorem.

Let us show $X = \bigcup_{r\in E} G(r)$ whenever $\operatorname{card}(E) \ge \aleph_1$. Suppose that there is an x in X such that $x \notin G(r)$ for every r in E. From the definition of $G(r)$ there is a pair $\langle n_x, j_x\rangle$ in $\omega \times \omega$ such that x is not in $g_{\langle n_x,j_x\rangle}{}^{-1}\big[\{\, s \in [0,1] \colon s \ne r \,\}\big]$, that is $g_{\langle n_x,j_x\rangle}(x) = r$. This defines a map $\eta \colon r \mapsto \langle n,j\rangle$ of E into $\omega \times \omega$ such that $g_{\eta(r)}(x) = r$. Since E is uncountable and $\omega \times \omega$ is countable, there are two distinct r and r' in E that map to the same $\langle n,j\rangle$. This implies $g_{\langle n,j\rangle}(x) = r$ and $g_{\langle n,j\rangle}(x) = r'$, a contradiction. Thereby the required equality is established. □

An immediate dimension theoretic consequence is the following.

THEOREM D.25 (Zindulka). *If X is a metrizable space, then there is a sequence $G(\alpha)$, $\alpha < \omega_1$, of G_δ sets with $\dim G(\alpha) \le 0$ such that $X = \bigcup_{\alpha<\omega_1} G(\alpha)$.*

This theorem is equivalent to

THEOREM D.26. *If X is a metrizable space, then there is a sequence $B(\alpha)$, $\alpha < \omega_1$, of mutually disjoint Borel sets with $\dim B(\alpha) \le 0$ such that $X = \bigcup_{\alpha<\omega_1} B(\alpha)$.*

PROOF. That the first theorem implies the second is obvious. So assume that the second theorem holds. Let $\alpha \in \omega_1$. As the G_δ hull property (5) of Theorem D.6 holds for all metrizable spaces, there is a G_δ set $G(\alpha)$ such that $B(\alpha) \subset G(\alpha)$ and $\dim G(\alpha) \le 0$. Hence the first theorem holds. □

Here is a restatement of Lemma D.24 that avoids the use of double indexing.

THEOREM D.27 (Zindulka). *If X is a nonempty metric space, then there exists a sequence $h_m \colon X \to [0,1]$, $m = 0, 1, 2, \ldots$, of Lipschitzian functions such that*

$$G(r) = \bigcap\nolimits_{m=0}^{\infty} h_m{}^{-1}\big[\{\, s \in [0,1] \colon s \ne r \,\}\big]$$

is a G_δ set with $\dim G(r) \le 0$ for each r in the open interval $(0,1)$. Moreover, $X = \bigcup_{r\in E} G(r)$ whenever E be an uncountable subset of $(0,1)$.

This theorem leads to the following dimension theoretic result. Notice that $\dim X$ may be infinite in this result. Recall our earlier comment that all the properties listed in Theorem D.6 hold except for the compactification property (6).

THEOREM D.28 (Zindulka). *If X is a metrizable space and if m and n are integers such that $\dim X \geq m \geq n \geq 0$, then to each metric for X there corresponds a countable family $\mathcal{F}$ of Lipschitzian maps of X into $[0,1]^n$ such that for each r in $(0,1)^n$ there is an f in $\mathcal{F}$ with $\dim f^{-1}[\{r\}] \geq m-n$.*

PROOF. We shall use the sequence h_k, $k = 0, 1, 2, \ldots$, of Lipschitzian functions from X into $[0,1]$ as provided by Theorem D.27. For each $\iota = \langle \iota(0), \iota(1), \ldots, \iota(n-1)\rangle$ in ω^n, the function $f_\iota : X \to [0,1]^n$ defined by

$$f_\iota(x) = \langle h_{\iota(0)}(x), h_{\iota(1)}(x), \ldots, h_{\iota(n-1)}(x)\rangle,$$

is Lipschitzian. Define $\mathcal{F}$ to be the countable family $\{f_\iota : \iota \in \omega^n\}$. Let us show that $\mathcal{F}$ satisfies the requirement of the theorem. Let $r \in (0,1)^n$. As $\dim G(r_j) \leq 0$ for $0 \leq j \leq n-1$ we infer from the addition theorem of dimension theory that $\dim \bigcup_{j=0}^{n-1} G(r(j)) \leq n-1$; hence

$$\dim \bigcap_{j=0}^{n-1} F(r(j)) \geq \dim X - \dim \bigcup_{j=0}^{n-1} G(r(j)) - 1 \geq m-n,$$

where $F(r) = X \setminus G(r)$ for every r. On the other hand,

$$\begin{aligned}\bigcap_{j=0}^{n-1} F(r(j)) &= \bigcap_{j=0}^{n-1} \bigcup_{k\in\omega} h_k{}^{-1}[\{r(j)\}] \\ &= \bigcup_{\iota\in\omega^n} \bigcap_{j=0}^{n-1} h_{\iota(j)}{}^{-1}[\{r(j)\}] = \bigcup_{\iota\in\omega^n} f_\iota{}^{-1}[\{r\}].\end{aligned}$$

As $f_\iota{}^{-1}[\{r\}]$ is closed for every ι in ω^n, by the sum theorem of dimension theory, there is an ι in ω^n such that $\dim f_\iota{}^{-1}[\{r\}] \geq m-n$ and the theorem is proved. □

Let us use this theorem to provide a second proof of the classical inequality $\dim X \leq \dim_{\mathsf{H}} X$ for separable metric spaces X.

SECOND PROOF OF THEOREM 5.1. For positive integers m and n such that $\dim X \geq m \geq n \geq 0$, let $\mathcal{F}$ be the countable family provided by Theorem D.28. For each r in $(0,1)^n$ we have $\dim f^{-1}[\{r\}] \geq 0$ for some f in $\mathcal{F}$, whence $f^{-1}[\{r\}] \neq \emptyset$ for this f. Hence $(0,1)^n = \bigcup_{f\in\mathcal{F}} f[X]$. As each f in $\mathcal{F}$ is a Lipschitzian map, we have $\dim_{\mathsf{H}} f[X] \leq \dim_{\mathsf{H}} X$ whenever $f \in \mathcal{F}$. Also $\sup\{\dim_{\mathsf{H}} f[X] : f \in \mathcal{F}\} = \dim_{\mathsf{H}} (0,1)^n = n$. It follows that $n \leq \dim_{\mathsf{H}} X$. □

Notice that this proof does not use Lebesgue integration on $\mathbb{R}$. It uses Zindulka's dimension theorem for metric spaces and the fact that $\dim_{\mathsf{H}} (0,1)^n = n$.

REMARK D.29. There is a simple modification of the above proof. Suppose that F is a nonempty subset of $(0,1)^n$ and let $E_f = f^{-1}[F]$ for each f in $\mathcal{F}$. Define $X' = \bigcup\{E_f : f \in \mathcal{F}\}$. Suppose that E is a subset of X' such that $F \subset \bigcup\{f[E] : f \in \mathcal{F}\}$. Then $\dim_{\mathsf{H}} F \leq \dim_{\mathsf{H}} E \leq \dim_{\mathsf{H}} X$. This modification is used in the proofs of Theorem 5.22 and its corollary in Chapter 5.

D.4. Geometric measure theory

The powerful results on measure theoretic geometric properties of $\mathbb{R}^n$ are used in many areas of mathematics. We also will use some of these results. A discussion of general geometric measure theory will not be presented. We shall narrowly focus our discussion of geometric measure theory to the needs of proving the existence of universally null sets in $\mathbb{R}^n$ that are contained in subsets of $\mathbb{R}^n$.

The geometric measure theory results that are used in the book are nontrivial ones since the Hausdorff p-dimensional measure space $\mathsf{M}(X, \mathsf{H}_p\,|X)$ of compact subsets X of $\mathbb{R}^n$ need not be finitely positive if p is less than n. This makes our investigation of Hausdorff dimension rather delicate for subsets of $\mathbb{R}^n$. We begin with a summary of facts from geometric measure theory that are used.

D.4.1. Geometric measure theory preliminaries. Much of the preliminary discussion is taken from P. Mattila [**99**]. For $n > 1$ we consider m-dimensional linear subspaces V of $\mathbb{R}^n$ and their $(n-m)$-dimensional orthogonal complements $V^{\perp}$, which are linear subspaces. For convenience, we shall call V an m-dimensional plane. The collection of all m-dimensional planes in $\mathbb{R}^n$, denoted by $G(n,m)$, is called the *Grassmannian manifold.* There is a natural metric and a natural Radon probability measure $\gamma_{\mathrm{n,m}}$ on this manifold. The metric is provided by employing the natural orthogonal projections $\pi_{\mathrm{V}}\colon \mathbb{R}^n \to V$, $V \in G(n,m)$. The distance in $G(n,m)$ is given by

$$\mathrm{d}(V,W) = \|\pi_{\mathrm{V}} - \pi_W\|, \quad (V,W) \in G(n,m) \times G(n,m), \tag{D.1}$$

where $\|\cdot\|$ is the usual operator norm. A very nice property of the measures is

$$\gamma_{\mathrm{n,m}}(A) = \gamma_{\mathrm{n,n\text{-}m}}(\{V^{\perp}\colon V \in A\}), \quad A \subset G(n,m). \tag{D.2}$$

For later reference we shall list three theorems without providing their proofs. The first theorem is a projection property (see [**99**, Corollary 9.4]).

THEOREM D.30 (Projection property). *If A is a Borel subset of $\mathbb{R}^n$ such that* $\dim_{\mathsf{H}} A \leq m < n$, *then* $\dim_{\mathsf{H}}(\pi_{\mathrm{V}}[A]) = \dim_{\mathsf{H}} A$ *for $\gamma_{\mathrm{n,m}}$-almost all V in $G(n,m)$.*

The second theorem is a slicing property. The statement given here is equivalent to [**99**, Theorem 10.10].

THEOREM D.31 (Slicing property). *Let $m \leq s \leq n$ and let A be a Borel measurable subset of $\mathbb{R}^n$ such that* $0 < \mathsf{H}_s(A) < \infty$. *Then for all V in $G(n,m)$,*

$$\mathsf{H}_{s-m}\big(A \cap (V^{\perp} + x)\big) < \infty \textit{ for } \mathsf{H}_m\textit{-almost every } x \textit{ in } V,$$

and for $\gamma_{\mathrm{n,m}}$-almost every V in $G(n,m)$,

$$\mathsf{H}_m\big(\{x \in V\colon \dim_{\mathsf{H}}\big(A \cap (V^{\perp} + x)\big) = s - m\}\big) > 0.$$

The third theorem is "the Frostman lemma in $\mathbb{R}^n$." We quote the lemma as it appears in [**99**, Theorem 8.8].

THEOREM D.32 (Frostman lemma). *Let B be a Borel subset of $\mathbb{R}^n$. Then $\mathsf{H}_p(B) > 0$ if and only if there is a finite Radon measure μ on $\mathbb{R}^n$ such that $\emptyset \neq$ support(μ) $\subset B$ and $\mu\big(B(x,r)\big) \leq r^p$ whenever $0 < r$ and $x \in \mathbb{R}^n$.*

Finally, the following Borel measurability lemma is used in the proof of Zindulka's Theorem 5.27.

LEMMA D.33. *Let F be a compact subset of $\mathbb{R}^n$ and V be in $G(n,m)$. Then the function Φ defined by*

$$\Phi(x) = \dim_{\mathsf{H}}\big(F \cap (V^{\perp} + x)\big), \quad x \in \mathbb{R}^n$$

is Borel measurable.

Before beginning the proof, let us define a function f_η. Let F be a nonempty, compact subset of $\mathbb{R}^n$ and let $0 < m < n$. For $\eta > 0$ let $\mathcal{E}(x, V, \eta)$ be the collection of all covers $\{E_i : i \in \mathbb{N}\}$ of $F \cap (V^{\perp} + x)$ with mesh$(\{E_i : i \in \mathbb{N}\}) < \eta$. (Note well the strict inequality is used rather than the weaker mesh$(\{E_i : i \in \mathbb{N}\}) \leq \eta$.) Define f_η to be the extended real-valued function on $(0,\infty) \times \mathbb{R}^n \times G(n,m)$ given by the formula

$$f_\eta(s,x,V) = \inf\big\{ \textstyle\sum_{i=1}^{\infty}\big(\operatorname{diam}(E_i)\big)^s : \{E_i : i \in \mathbb{N}\} \in \mathcal{E}(x,V,\eta) \big\}.$$

Observe that $f_{\eta'}(s,x,V) \leq \mathsf{H}_s^{\eta}\big(F \cap (V^{\perp} + x)\big) \leq f_\eta(s,x,V)$ whenever $\eta < \eta'$.

PROPOSITION D.34. *For each positive η, the function f_η is upper semi-continuous on $(0,\infty) \times \mathbb{R}^n \times G(n,m)$. Moreover, f_η is pointwise decreasing as a function of η.*

The proof will require some preparation. Let F be a nonempty, compact subset of $\mathbb{R}^n$. Clearly there exists a positive number r such that the closed n-ball $B(0,r)$ centered at the origin contains the sets $\pi_{\mathrm{V}}[F]$ for every V in $G(n,m)$. Consider the set

$$M = \{ (x,V) \in \mathbb{R}^n \times G(n,m) : F \cap (V^{\perp} + x) \neq \emptyset \}.$$

Clearly, $M \subset B(0,r) \times G(n,m)$. We assert that M is a closed subset of $B(0,r) \times G(n,m)$. Indeed, let (x_k, V_k), $k = 1,2,\ldots$, be a sequence in M that converges to $\big(\overline{x}, \overline{V}\big)$. Since $B(0,r)$ is compact, we have a sequence $F \cap ({V_k}^{\perp} + x_k)$, $k = 1,2,\ldots$, in the hyperspace $2^{B(0,r)}$ of all nonempty compact subsets of $B(0,r)$. This hyperspace, when endowed with the Hausdorff metric d_H, is a compact space. Hence we may assume that this sequence converges to a nonempty, compact subset E of $B(0,r)$. Let $\overline{y} \in E$. There is a sequence y_k, $k = 1,2,\ldots$, in $B(0,r)$ such that each y_k is in $F \cap ({V_k}^{\perp} + x_k)$ and such that the sequence converges to $\overline{y}$. We infer from equation (D.1) that $\overline{y} \in \overline{V}^{\perp} + \overline{x}$ holds. So E is a subset of $F \cap \big(\overline{V}^{\perp} + \overline{x}\big)$. Hence $\big(\overline{x}, \overline{V}\big) \in M$. We summarize the argument: *If (x_k, V_k), $k = 1,2,\ldots$, converges to $\big(\overline{x}, \overline{V}\big)$ in the metric space $\mathbb{R}^n \times G(n,m)$, then some subsequence (x_{k_i}, V_{k_i}), $i = 1,2,\ldots$, has the property that the sequence $F \cap ({V_{k_i}}^{\perp} + x_{k_i})$, $i = 1,2,\ldots$, converges in the hyperspace $2^{B(0,r)}$ to some nonempty, compact subset E of $F \cap \big(\overline{V}^{\perp} + \overline{x}\big)$ with respect to the Hausdorff metric.*

We turn to the proof of the proposition.

PROOF OF PROPOSITION D.34. Since $A = (0,\infty) \times M$ is a closed subset of $(0,\infty) \times \mathbb{R}^n \times G(n,m)$, we have that f^η is upper semi-continuous at each point in the complement of A. It remains to show that it is upper semi-continuous at each point of A.

Suppose $(\overline{s},\overline{x},\overline{V}) \in A$ and let $\varepsilon > 0$. Select a cover $\{E_i : i \in \mathbb{N}\}$ of $F \cap (\overline{V}^\perp + \overline{x})$ with $\operatorname{mesh}(\{E_i : i \in \mathbb{N}\}) < \eta$, and $\sum_{i=1}^\infty (\operatorname{diam}(E_i))^{\overline{s}} < f^\eta(\overline{s},\overline{x},\overline{V}) + \varepsilon$. For each i there is an open set U_i with $E_i \subset U_i$ such that $\sum_{i=1}^\infty (\operatorname{diam}(U_i))^{\overline{s}} < f^\eta(\overline{s},\overline{x},\overline{V}) + \varepsilon$ and $\operatorname{mesh}(\{U_i : i \in \mathbb{N}\}) < \eta$. As $F \cap (\overline{V}^\perp + \overline{x})$ is compact there exists an integer N such that $\bigcup_{i=1}^N U_i \supset F \cap (\overline{V}^\perp + \overline{x})$. So there is a neighborhood W of $\overline{s}$ such that $\sum_{i=1}^N (\operatorname{diam}(U_i))^s < f^\eta(\overline{s},\overline{x},\overline{V}) + \varepsilon$ whenever $s \in W$. Let (s_k, x_k, V_k), $k = 1,2,\ldots,$ be a sequence in A that converges to $(\overline{s},\overline{x},\overline{V})$ and satisfies $\lim_{k\to\infty} f^\eta(s_k,x_k,V_k) = L$. We may assume that $F \cap ({V_k}^\perp + x_k)$ converges with respect to the Hausdorff metric to a nonempty, compact set E contained in $F \cap (\overline{V}^\perp + \overline{x})$. Hence, for all k sufficiently large, we have $F \cap ({V_k}^\perp + x_k) \subset \bigcup_{i=1}^N U_i$ and $s_k \in W$. Consequently $f^\eta(s_k,x_k,V_k) \leq f^\eta(\overline{s},\overline{x},\overline{V}) + \varepsilon$ for all sufficiently large k. Therefore $L \leq f^\eta(\overline{s},\overline{x},\overline{V})$, and thereby we have shown that f^η is upper semi-continuous. □

We now have the desired Borel measurability assertions.

COROLLARY D.35. *Let F be a nonempty, compact subset of $\mathbb{R}^n$. Then*:

(1) $\mathsf{H}_s(F \cap (V^\perp + x))$, *as a function of* (s, x, V), *is Borel measurable, indeed, a Baire class* 2 *function,*
(2) $\dim_{\mathsf{H}}(F \cap (V^\perp + x))$, *as a function of* (x, V), *is Borel measurable,*
(3) $\{x \in \mathbb{R}^n : \dim_{\mathsf{H}}(F \cap (V^\perp + x)) = s\}$ *is a Borel set of* $\mathbb{R}^n$ *for each* (s, V),
(4) $\{x \in V : \dim_{\mathsf{H}}(F \cap (V^\perp + x)) = s\}$ *is a Borel set of the m-dimensional linear space* V *for each* (s, V).

PROOF. We have $\lim_{\eta\to 0} f^\eta(s,x,V) = \mathsf{H}_s(F \cap (V^\perp + x))$ for each (s,x,V). Hence statement (1) follows.

For the proof of statement (2), let t be a real number. We must show that $E_t = \{(x,V) : \dim_{\mathsf{H}}(F \cap (V^\perp + x)) > t\}$ is a Borel set. To this end, $\dim_{\mathsf{H}}(F \cap (V^\perp + x)) > t$ if and only if there exists an s with $s > t$ and $\mathsf{H}_s(F \cap (V^\perp + x)) = \infty$. By statement (1), the set $F_s = \{(x,V) : \mathsf{H}_s(F \cap (V^\perp + x)) = \infty\}$ is Borel measurable. Let s_i, $i =, 2, \ldots,$ be a sequence that strictly decreases to t. Each of the sets F_{s_i} is Borel measurable and this sequence of sets increases to E_t. Hence E_t is Borel measurable. This proves the Borel measurability of $\dim_{\mathsf{H}}(F \cap (V^\perp + x))$ as a function of (x, V).

Statements (3) and (4) follow easily from statement (2). □

For more on the measurability of dimension functions of geometric measure theory see Mattila and Mauldin [**100**].

D.5. Marstrand's theorem

The theorem of this section concerns lower bounds for Hausdorff measures and Hausdorff dimension of arbitrary subsets of $\mathbb{R}^n$.

THEOREM D.36. *Let* $0 < m < n$, $0 < s \leq m$, $0 < t \leq n - m$ *and* $0 < p$. *For a* V *in* $G(n,m)$ *and an arbitrary subset* F *of* $\mathbb{R}^n$ *let* E *be an arbitrary subset of* $\{x \in V \colon \mathsf{H}_t(F \cap (V^\perp + x)) \geq p\}$. *Then*

$$p\,\mathsf{H}_s(E) \leq \mathsf{H}_{s+t}(F).$$

The planar version of the theorem is stated as Proposition 7.11 in K. J. Falconer [**53**] with the added statement that the higher dimensional one is true. This planar version was proved by J. M. Marstrand in [**98**] with no indication of a higher dimensional version. With the aid of results (namely, R. O. Davies [**45**], J. D. Howroyd [**76**]) that have appeared after [**98**], a proof of the above theorem can be made. We shall give this proof below.[4]

The proof will use the notion of "weighted Hausdorff measures" (see Mattila [**99**, page 59] and Howroyd [**76**, page 585]).

DEFINITION D.37 (Weighted Hausdorff measure). *Let* X *be a separable metric space and* $A \subset X$. *For* $0 \leq s$ *and* $0 < \delta$ *define*

$$\mathrm{w}\,\mathsf{H}_s^\delta(A) = \inf \textstyle\sum_i c_i (\operatorname{diam} E_i)^s,$$

where the infimum is taken over all finite or countable families (*called* weighted δ-covers *of* A) $\{\,(c_i, E_i)\,\}$ *such that* $0 < c_i < \infty$, $E_i \subset X$, $\operatorname{diam} E_i < \delta$ *and*

$$\chi_A \leq \textstyle\sum_i c_i\, \chi_{E_i}\,. \tag{D.3}$$

Then define

$$\mathrm{w}\,\mathsf{H}_s(A) = \lim_{\delta \to 0} \mathrm{w}\,\mathsf{H}_s^\delta(A),$$

which is called the weighted Hausdorff s-measure *of* A.

We infer from [**76**, Theorem 1 and Note 9] and from [**45**, Theorem 8 and Example 1] that

THEOREM D.38. *If* X *is a separable metric space and if* s *is a positive number, then* $\mathrm{w}\,\mathsf{H}_s(A) = \mathsf{H}_s(A)$ *for every subset* A *of* X.

Let us establish some notation that will be used in the proof of Theorem D.36. Let $0 < m < n$. We equate $\mathbb{R}^n = \mathbb{R}^m \times \mathbb{R}^{n-m}$ in the obvious manner. For convenience we write $V = \mathbb{R}^m \times \{0\}$ and $V^\perp = \{0\} \times \mathbb{R}^{n-m}$. Let π_1 and π_2 be the projections of $\mathbb{R}^n$ onto V and $V^\perp$, respectively. Define

$$C_1 = \pi_1[C],\ C_2 = \pi_2[C], \text{ and } \widetilde{C} = C_1 \times C_2 = {\pi_1}^{-1}[C_1] \cap {\pi_2}^{-1}[C_2]$$

for subsets C of $\mathbb{R}^n$. Then $\operatorname{diam} C_1 \leq d$, $\operatorname{diam} C_2 \leq d$, where $d = \operatorname{diam} C$, and

$$\widetilde{C} \cap (V^\perp + z) = C_2 + z \supset C \cap (V^\perp + z) \quad \text{whenever } z \in C_1. \tag{D.4}$$

[4] K. J. Falconer, by e-mail, informed the author of possible sources for a statement and proof of the higher dimensional case. Among them was Falconer and Mauldin [**54**] and Howroyd [**76**]. Though the statement and proof of the higher dimensional case were not found in these sources, the sources did yield the proof that is provided here.

Proof of Theorem D.36.[5] We may assume $V = \mathbb{R}^m \times \{0\}$. The inequality is trivial if $\mathsf{H}_{s+t}(F) = \infty$. So assume $\mathsf{H}_{s+t}(F) < \infty$. Let ε be such that $0 < \varepsilon < p$. For each z in E there is a positive number $\delta(z)$ such that $\mathsf{H}_t^\delta(F \cap (V^\perp + z)) > p - \varepsilon$ whenever $\delta < \delta(z)$. For each positive integer k define $E_k = \{ z \in E \colon 1 \leq k\delta(z) \}$. For $\delta < \frac{1}{k}$ there is a cover $\mathcal{C}$ of F such that $\operatorname{mesh}(\mathcal{C}) < \delta$ and

$$\mathsf{H}_{s+t}^\delta(F) < \textstyle\sum_{C \in \mathcal{C}} (\operatorname{diam} C)^{s+t} < \mathsf{H}_{s+t}(F) + \varepsilon.$$

Let us show that $\mathcal{C}_1 = \left\{ \left(\frac{(\operatorname{diam} C_2)^t}{p-\varepsilon}, C_1 \right) \colon C \in \mathcal{C} \right\}$ is a weighted δ-cover of E_k. To this end we shall verify condition (D.3). Observe that, for each z in E_k, the family $\mathcal{C}_z = \{ \widetilde{C} \cap (V^\perp + z) \colon C \in \mathcal{C} \}$ is a cover of $F \cap (V^\perp + z)$ with $\operatorname{mesh}(\mathcal{C}_z) < \delta$. Hence

$$\textstyle\sum_{C \in \mathcal{C}} (\operatorname{diam} \widetilde{C})^t \geq \mathsf{H}_t^\delta(F \cap (V^\perp + z)) > p - \varepsilon$$

whenever $z \in E_k$. Thereby condition (D.3) results from condition (D.4). We now have

$$\begin{aligned}(p - \varepsilon) \mathrm{w}\, \mathsf{H}_s^\delta(E_k) &\leq \textstyle\sum_{C \in \mathcal{C}} (\operatorname{diam} C_2)^t (\operatorname{diam} C_1)^s \\ &\leq \textstyle\sum_{C \in \mathcal{C}} (\operatorname{diam} C)^{t+s} \leq \mathsf{H}_{t+s}^\delta(F) + \varepsilon,\end{aligned}$$

that is, $(p - \varepsilon) \mathrm{w}\, \mathsf{H}_s^\delta(E_k) \leq \mathsf{H}_{t+s}^\delta(F) + \varepsilon$. Letting $\delta \to 0$, we have, by virtue of Theorem D.38, $(p - \varepsilon)\, \mathsf{H}_s(E_k) \leq \mathsf{H}_{s+t}(F) + \varepsilon$. As $E_k \subset E_{k+1}$ and $E = \bigcup_{k=1} E_k$, we finally have $(p - \varepsilon)\, \mathsf{H}_s(E) \leq \mathsf{H}_{s+t}(F) + \varepsilon$. Theorem D.36 now follows. □

An immediate corollary of this theorem is

Corollary D.39. *Let $0 < m < n$ and $0 < t \leq n - m$. If V is in $G(n, m)$ and F is an arbitrary subset of $\mathbb{R}^n$, then*

$$\dim_{\mathsf{H}} F \geq t + \dim_{\mathsf{H}} \{ x \in V \colon \dim_{\mathsf{H}} \big(F \cap (V^\perp + x)\big) \geq t \}.$$

Exercises

D.1. Prove: *If X is a σ-compact, separable metrizable space, then $X = X_0 \cup X_1$, where X_0 is a countable set and X_1 the union of a countable family of perfect, compact sets.*

[5] P. Mattila pointed out to the author that the theorem is actually implied by a theorem due to Federer [**55**, Theorem 2.10.25 page 158]. The proof provided here is modeled after that of Federer. The cited proof does not use weighted Hausdorff measure, a concept that is motivated by Federer's proof.

Bibliography

1. J. M. Aarts and T. Nishiura, *Dimension and Extensions*, North Holland Mathematical Library, 48, North Holland, Amsterdam, 1993.
2. E. Akin, *Measures on Cantor space*, Topology Proc. **24** (1999) 1–34.
3. ______, *Good measures on Cantor space*, Trans. Amer. Math. Soc. **357** (2005) 2681–2722.
4. P. Alexandroff and P. Urysohn, *Über nulldimensionale Punktmengen*, Math. Ann. **98** (1928) 89–106.
5. S. Alpern and V. S. Prasad, *Typical Dynamics of Volume Preserving Homeomorphisms*, Cambridge Tracts in Mathematics, 139, Cambridge University Press, Cambridge, 2000.
6. T. D. Austin, *A pair of non-homeomorphic measures on the Cantor set*, Math. Proc. Cambridge Phil. Soc., **42** (2007) 103–110.
7. S. Banach, *Sur les suites d'ensembles excluant d'une mesure*, Colloq. Math. **1** (1948) 103–108, Note posthume avec préface et commentaire de E. Marczewski.
8. R. Berlanga and D. B. A. Epstein, *Measures on sigma-compact manifolds and their equivalence under homeomorphisms*, J. London Math. Soc. **27** (1983) 63–74.
9. A. S. Besicovitch, *Concentrated and rarified sets of points*, Acta Math. **62** (1934) 289–300.
10. ______, *On existence of subsets of finite measure of sets of infinite measure*, Indag. Math. **14** (1952) 339–344.
11. M. Bestvina, *Characterizing k-dimensional Universal Menger Compacta*, Memoirs of the American Mathematical Society, **71**, no. 380, American Mathematical Society, Providence, RI, 1988.
12. D. Blackwell, *On a class of probability spaces*, Proceedings of the Third Berkeley Symposium on Mathematical Statistics and Probability, 1954–1955, vol. II, University of California Press, Berkeley and Los Angeles, 1956, pp. 1–6.
13. E. Borel, *Sur la classification des ensembles de mesure nulle*, Bull. Soc. Math. France **47** (1919) 97–125.
14. N. Bourbaki, *Intégration, Livre VI. Chapitre V: Intégration des Mesures*, Hermann, Paris, 1956.

15. S. Braun and E. Szpilrajn, *Annexe*, Fund. Math., nouvelle series **1** (1937), 225–254 en collaboration avec C. Kuratowski.
16. J. B. Brown, *Differentiable restrictions of real functions*, Proc. Amer. Math. Soc. **108** (1990) 391–398.
17. ______, *Continuous-, derivative-, and differentiable-restrictions of measurable functions*, Fund. Math. **141** (1992) 85–95.
18. J. B. Brown and G. V. Cox, *Classical theory of totally imperfect spaces*, Real Anal. Exchange **7** (1981/82) 185–232.
19. ______, *Continuous images of Lusin sets*, Proc. Amer. Math. Soc. **89** (1983) 52–54.
20. J. B. Brown and K. Prikry, *Variations on Lusin's theorem*, Trans. Amer. Math. Soc. **302** (1987) 77–86.
21. M. Brown, *A mapping theorem for untriangulated manifolds*, Topology of 3-Manifolds and Related Topics, Proceedings of the Univ. of Georgia Institute, 1961 (M. K. Fort, ed.), Prentice-Hall, Englewood Cliffs, NJ, 1963, pp. 92–94.
22. A. M. Bruckner, *Differentiation of Real Functions*, Lecture Notes in Mathematics, vol. 659, Springer-Verlag, Berlin, 1978.
23. ______, *Differentiation of Real Functions*, CRM Monogr. Ser., vol. 5, American Mathematical Society, Providence, RI, 1994.
24. A. M. Bruckner, R. O. Davies, and C. Goffman, *Transformations into Baire* 1 *functions*, Proc. Amer. Math. Soc. **67** (1977) 62–66.
25. Z. Buczolich, *Density points and bi-lipschitzian functions in* $\mathbb{R}^m$, Proc. Amer. Math. Soc. **116** (1992) 53–59.
26. C. Cabrelli, U. Molter, V. Paulauskas, and R. Shonkwiler, *Hausdorff measure of p-Cantor sets*, Real Anal. Exchange **30** (2004/2005), no. 2, 413–434.
27. D. Cenzer and R. D. Mauldin, *Measurable parametrizations and selections*, Trans. Amer. Math. Soc. **245** (1978) 399–408.
28. L. Cesari, *Optimization – Theory and Applications, Problems with Ordinary Differential Equations*, Applications of Mathematics, 17, Springer-Verlag, New York, 1983.
29. J. R. Choksi, *Measurable transformations on compact groups*, Trans. Amer. Math. Soc. **184** (1973) 101–124.
30. J. Cichoń, *On Banach numbers*, Bull. Acad. Polon. Sci. Sér. Sci. Math. **29** (1981) 531–534.
31. K. Ciesielski and L. Pawlikowski, *The Covering Property Axiom, CPA, A Combinatorial Core of the Iterated Perfect Set Model*, Cambridge Tracts in Mathematics, 164, Cambridge University Press, Cambridge, 2004.
32. D. L. Cohn, *Measure Theory*, Birkhäuser, Boston, MA, 1980.
33. R. B. Darst, *Properties of Lusin sets with applications to probability*, Proc. Amer. Math. Soc. **20** (1968) 348–350.
34. ______, *A universal null set which is not concentrated*, Fund. Math. **62** (1968) 47–48.
35. ______, *A characterization of universally measurable sets*, Proc. Cambridge Phil. Soc. **65** (1969) 617–618.

36. ______, *A CBV image of a universal null set need not be a universal null set*, Fund. Math. **67** (1970) 219–220.
37. ______, *On bimeasurable images of universally measurable sets*, Fund. Math. **66** (1970) 381–382.
38. ______, *On the* 1-1 *sum of two Borel sets*, Proc. Amer. Math. Soc. **25** (1970) 914.
39. ______, *A characterization of bimeasurable functions in terms of universally measurable sets*, Proc. Amer. Math. Soc. **27** (1971) 566–571.
40. ______, *C^∞-functions need not be bimeasurable*, Proc. Amer. Math. Soc. **27** (1971) 128–132.
41. R. B. Darst and R. E. Zink, *A perfect measurable space that is not a Lusin space*, Ann. Math. Stat. **38** (1967) 1918.
42. R. O. Davies, *Subsets of finite measure in analytic sets*, Indag. Math. **14** (1952) 488–489.
43. ______, *Remarks on measurable sets and functions*, J. Res. Nat. Bur. Standards Sect B **70B** (1966) 83–84.
44. ______, *Some sets with measurable inverses*, Proc. Cambridge Phil. Soc. **65** (1969) 437–438.
45. ______, *Increasing sequence of sets and Hausdorff measure*, Proc. London Math. Soc. **20** (1970) 222–236.
46. P. L. Dordal, *Towers in $[\omega]^\omega$ and $^\omega\omega$*, Ann. Pure Appl. Logic **45** (1989) 247–276.
47. R. Dougherty, R. D. Mauldin, and A. Yingst, *On homeomorphic Bernoulli measures on the Cantor space*, Trans. Amer. Math. Soc., **359** (2007) 6155–6166.
48. H. G. Eggleston, *Concentrated sets*, Proc. Cambridge Phil. Soc. **63** (1967) 931–933.
49. S. Eilenberg and N. Steenrod, *Foundations of Algebraic Topology*, Princeton University Press, Princeton, NJ, 1952.
50. R. Engelking, *Dimension Theory*, North Holland Mathematical Library, 19, North Holland, Amsterdam, 1978.
51. ______, *General Topology*, Mathematical Monographs, vol. 60, PWN–Polish Scientific Publishers, Warsaw, 1978.
52. K. J. Falconer, *The Geometry of Fractal Sets*, Cambridge Tracts in Mathematics, 85, Cambridge University Press, Cambridge, 1986.
53. ______, *Fractal Geometry, Mathematical Foundations and Applications*, John Wiley, Chichester, 1990.
54. K. J. Falconer and R. D. Mauldin, *Fubini-type theorems for general measure constructions*, Mathematica **47** (2000) 251–265.
55. H. Federer, *Geometric Measure Theory*, Springer-Verlag, New York, 1969.
56. D. H. Fremlin, *Chichoń's diagram*, Seminaire Initiation à l'Analyse, no. 23, Univ. Pierre et Marie Curie, Paris, 1983/84.
57. ______, *Consequences of Martin's Axiom*, Cambridge Tracts in Mathematics, 84, Cambridge University Press, Cambridge, 1984.

58. R. J. Gardner and W. F. Pfeffer, *Chapter 22: Borel measures*, Handbook of Set-Theoretic Topology (K. Kunen and J. E. Vaughan, eds.), North Holland, Amsterdam, 1984, pp. 961–1043.
59. B. R. Gelbaum, *Cantor sets in metric measure spaces*, Proc. Amer. Math. Soc. **24** (1970) 341–343.
60. B. V. Gnedenko and A. N. Kolmogorov, *Limit Distributions for Sums of Independent Random Variables*, Addison-Wesley, Cambridge, MA, 1954, Translated by K. L. Chung with an appendix by J. L. Doob.
61. C. Goffman, C. J. Neugebauer, and T. Nishiura, *Density topology and approximate continuity*, Duke Math. J. **28** (1961) 497–505.
62. C. Goffman, T. Nishiura, and D. Waterman, *Homeomorphisms in Analysis*, Mathematical Surveys and Monographs, 54, American Mathematical Society, Providence, RI, 1998.
63. C. Goffman and G. Pedrick, *A proof of the homeomorphism of Lebesgue–Stieltjes measure with Lebesgue measure*, Proc. Amer. Math. Soc. **52** (1975) 196–198.
64. A. J. Goldman, *On measurable sets and functions*, J. Res. Nat. Bur. Standards Sect B **69B** (1965) 99–100.
65. W. J. Gorman III, *The homeomorphic transformation of c-sets into d-sets*, Proc. Amer. Math. Soc. **17** (1966) 825–830.
66. ______, *Lebesgue equivalence to functions of the first Baire class*, Proc. Amer. Math. Soc. **17** (1966) 831–834.
67. E. Grzegorek, *Remarks on σ-fields without continuous measures*, Colloq. Math. **39** (1978) 73–75.
68. ______, *Solution of a problem of Banach in σ-fields without continuous measures*, Bull. Acad. Polon. Sci. Sér. Sci. Math. **28** (1980) 7–10.
69. ______, *On some results of Darst and Sierpiński concerning universal null and universally measurable sets*, Bull. Acad. Polon. Sci. Sér. Sci. Math. **29** (1981) 1–5.
70. E. Grzegorek and C. Ryll-Nardzewski, *A remark on absolutely measurable sets*, Bull. Acad. Polon. Sci. Sér. Sci. Math. **28** (1980) 229–232.
71. ______, *On universal null sets*, Proc. Amer. Math. Soc. **81** (1981) 613–617.
72. F. Hausdorff, *Summationmethod und Momentfolgen. I*, Math. Z. **9** (1921) 74–109.
73. ______, *Summen von $\aleph_1$ Mengen*, Fund. Math. **26** (1936) 241–255.
74. R. Haydon, *On compactness in spaces of measures and measurecompact spaces*, Proc. London Math. Soc. **29** (1974) 1–16.
75. A. Hilgers, *Bemerkung zur Dimensiontheorie*, Fund. Math. **28** (1937) 303–304.
76. J. D. Howroyd, *On dimension and on the existence of sets of finite positive Hausdorff measure*, Proc. London Math. Soc. (3) **70** (1995) 581–604.
77. K. J. Huang, *Algebraic numbers and topologically equivalent measures*, Ph.D. dissertation, North Texas State University, 1983.
78. ______, *Algebraic numbers and topologically equivalent measures in the Cantor set*, Proc. Amer. Math. Soc. **96** (1986) 560–562.
79. W. Hurewicz and H. Wallman, *Dimension Theory*, Princeton Mathematical Series, 4, Princeton University Press, Princeton, NJ, 1948.

80. G. Kallianpur, *A note on perfect probability*, Ann. Math. Stat. **30** (1959) 169–172.
81. H. Kato, K. Kawamura, and E. D. Tymchatyn, *Measures and topological dynamics on Menger manifolds*, Topology Appl. **103** (2000) 249–282.
82. B. Koszela, *Obituary: Tadeusz Świątowski*, Real Anal. Exchange **20** (1994/95) 2–5.
83. K. Kunen, *Set Theory, An introduction to Independence Proofs*, Studies in Logic and Foundations of Mathematics, 102, North-Holland, Amsterdam, 1983.
84. K. Kunen and J. E. Vaughan (eds.), *Handbook of Set-Theoretic Topology*, North-Holland, Amsterdam, 1984.
85. K. Kuratowski, *Topology I*, Academic Press, New York and London, 1966.
86. ———, *Topology II*, Academic Press, New York and London, 1968.
87. M. Laczkovich and D. Preiss, *α-Variation and transformation into C^n functions*, Indiana Univ. Math. J. **34** (1985) 405–424.
88. R. Laver, *On the consistency of Borel's conjecture*, Acta Math. **137** (1976) 151–169.
89. M. Lavrentieff, *Contribution à la théorie des ensembles homémorphes*, Fund. Math. **6** (1924) 149–160.
90. H. Lebesgue, *Leçons sur l'intégration et la recherche des fonctions primitives*, Gauthier-Villars, Paris, 1904.
91. J. Lukeš, J. Malý, and L. Zajíček, *Fine Topology Methods in Real Analysis and Potential Theory*, Lectures Notes in Mathematics, 1189, Springer-Verlag, Berlin, 1986.
92. N. Lusin, *Sur les probleme de M. Baire*, C. R. Paris **158** (1914) 1258–1261.
93. ———, *Leçons sur les Ensembles Analytiques et leur Applications*, Gauthier-Villars, Paris, 1930.
94. G. W. Mackey, *Borel structure in groups and their duals*, Trans. Amer. Math. Soc. **85** (1957) 134–165.
95. P. Mahlo, *Über Teilmengen des Kontinuums von dessen Machtigkeit*, Sitzungberichte der Sächsischen Akademie der Wissenschaften zu Leipzig, Mathematish-naturwissenschaftlische Klasse **65** (1913) 283–315.
96. E. Marczewski, *Remarque sur la measurabilité absolute*, Colloquium Mathematicum, Comptes Rendus **1** (1947) 42–43.
97. ———, *A remark on absolutely measurable sets*, Colloquium Mathematicum, Comptes Rendus **3** (1955) 190–191.
98. J. M. Marstrand, *The dimension of Cartesian product sets*, Proc. Cambridge Phil. Soc. **50** (1954) 198–202.
99. P. Mattila, *Geometry of Sets and Measures in Euclidean Spaces – Fractals and Rectifiability*, Cambridge Studies in Advanced Mathematics, 44, Cambridge University Press, Cambridge, 1995.
100. P. Mattila and R. D. Mauldin, *Measure and dimension functions: measurability and densities*, Math. Proc. Cambridge Philos. Soc. **121** (1997) 81–100.
101. R. D. Mauldin, *Some effects of set-theoretical assumptions in measure theory*, Adv. in Math. **27** (1978) 45–62.
102. ———, *Borel parametrizations*, Trans. Amer. Math. Soc. **250** (1979) 223–234.

103. ______, *Bimeasurable functions*, Proc. Amer. Math. Soc. **83** (1981) 369–370.

104. R. D. Mauldin (ed.), *The Scottish Book: Mathematics from the Scottish café*, Birkhäuser, Boston, MA, 1981.

105. ______, *Chapter 32: Problems in topology arising in analysis*, Open Problems in Topology (J. van Mill and G. M. Reed, eds.), North Holland, Amsterdam, 1990, pp. 618–629.

106. S. Mazurkiewicz and E. Szpilrajn, *Sur la dimension de certains ensembles singuliers*, Fund. Math. **28** (1937) 305–308.

107. C. G. Mendez, *On sigma-ideals of sets*, Proc. Amer. Math. Soc. **60** (1976) 124–128.

108. ______, *On the Sierpiński-Erdös and Oxtoby-Ulam theorems for some new sigma-ideals of sets*, Proc. Amer. Math. Soc. **72** (1978) 182–188.

109. A. W. Miller, *Mapping a set of reals onto the reals*, J. Symbolic Logic **48** (1983) 575–584.

110. ______, *Chapter 5: Special subsets of the real line*, Handbook of Set-Theoretic Topology (K. Kunen and J. E. Vaughan, eds.), North Holland, Amsterdam, 1984, pp. 201–233.

111. ______, *Special sets of reals*, Set theory of the reals (Ramat Gan, 1991) (H. Judah, ed.), Israel Mathematical Conference Proceedings, vol. 6, Bar-Ilan Univ., American Mathematical Society, Providence, RI, 1993, pp. 415–431.

112. Y. N. Moschovakis, *Descriptive Set Theory*, Studies in logic and the foundations of mathematics, 100, North Holland, Amsterdam, 1980.

113. J-I. Nagata, *Modern Dimension Theory*, John Wiley, New York, 1965.

114. F. J. Navarro-Bermúdez, *Topologically equivalent measures in the Cantor space*, Ph.D. dissertation, Bryn Mawr College, 1977.

115. ______, *Topologically equivalent measures in the Cantor space*, Proc. Amer. Math. Soc. **77** (1979) 229–236.

116. ______, *Topologically equivalent measures in the Cantor space II*, Real Anal. Exchange **10** (1984/85) 180–187.

117. F. J. Navarro-Bermúdez and J. C. Oxtoby, *Four topologically equivalent measures in the Cantor space*, Proc. Amer. Math. Soc. **104** (1988) 859–860.

118. T. Nishiura, *Absolutely measurable functions on manifolds*, Real Anal. Exchange **24** (1998/99) 703–728.

119. J. C. Oxtoby, *Homeomorphic measures in metric spaces*, Proc. Amer. Math. Soc. **24** (1970) 419–423.

120. ______, *Measure and Category*, second edn, Graduate Texts in Mathematics, 2, Springer-Verlag, New York, 1980.

121. J. C. Oxtoby and V. S. Prasad, *Homeomorphic measures in the Hilbert cube*, Pacific J. Math. **77** (1978) 483–497.

122. J. C. Oxtoby and S. M. Ulam, *Measure-preserving homeomorphisms and metrical transitivity*, Ann. of Math. (2) **42** (1941) 874–920.

123. E. Pap (ed.), *Handbook of Measure Theory*, vols. I, II, North Holland, Amsterdam, 2002.

124. D. Paunić, *Chapter 1: History of measure theory*, Handbook of Measure Theory (E. Pap, ed.), vol. I, Elsevier, Amsterdam, 2002, pp. 1–26.

125. R. G. E. Pinch, *a-convexity*, Math. Proc. Cambridge Phil. Soc. **97** (1985), 63–68.
126. ______, *Binomial equivalence of algebraic integers*, J. Indian Math. Soc. (N.S.) **58** (1992) 33–37.
127. S. Plewik, *Towers are universally measure zero and always of first category*, Proc. Amer. Math. Soc. **119** (1993) 865–868.
128. K. Prikry, *On images of the Lebesgue measure*, unpublished manuscript dated 20 September 1977.
129. R. Purves, *Bimeasurable functions*, Fund. Math. **58** (1966) 149–157.
130. I. Recław, *On a construction of universally null sets*, Real Anal. Exchange **27** (2001/02) 321–323.
131. C. A. Rogers, *Hausdorff Measures*, Cambridge University Press, Cambridge, 1970.
132. F. Rothberger, *Eine Äquivalenz zwishen der Kontinuumhypothese und der Existenz der Lusinchen und Sierpińskichen Mengen*, Fund. Math. **30** (1938) 215–217.
133. G. E. Sacks, *Forcing with Perfect Closed Sets*, Axiomatic set theory (D. Scott, ed.), Proceeding Symposium in Pure Mathematics (Univ. California Los Angeles, CA, 1967), vol. 13, Part 1, American Mathematical Society, Providence, RI, 1971, pp. 331–355.
134. M. Scheepers, *Gaps in* ω^ω, Set theory of the reals (Ramat Gan, 1991) (H. Judah, ed.), Israel Mathematical Conference Proceedings, vol. 6, Bar-Ilan Univ., American Mathematical Society, Providence, RI, 1993, pp. 439–561.
135. S. Scheinberg, *Topologies which generate a complete measure algebra*, Adv. Math. **7** (1971) 231–239.
136. E. S. Selmer, *On the irreducibility of certain trinomials*, Math. Scand. **4** (1956) 287–302.
137. J. R. Shoenfield, *Martin's axiom*, Amer. Math. Monthly **82** (1975) 610–617.
138. R. M. Shortt, *Universally measurable spaces: an invariance theorem and diverse characterizations*, Abstracts, Amer. Math. Soc. **3** (1982) 792–28–428.
139. ______, *Universally measurable spaces: an invariance theorem and diverse characterizations*, Fund. Math. **121** (1984) 169–176.
140. W. Sierpiński, *Sur l'hypothèse du continu* ($2^{\aleph_0} = \aleph_1$), Fund. Math. **5** (1924) 177–187.
141. ______, *Hypothèse du Continu*, Monografie Matematyczne, vol. 4, Warszawa-Lwów, 1934.
142. W. Sierpiński and E. Szpilrajn, *Remarque sur le problème de la mesure*, Fund. Math. **26** (1936) 256–261.
143. M. Sion, *Topological and measure theoretic properties of analytic sets*, Proc. Amer. Math. Soc. **11** (1960) 769–776.
144. R. M. Solovay, *A model of set-theory in which every set of reals is Lebesgue measurable*, Ann. of Math. **92** (1970) 1–56.
145. S. M. Srivastava, *A Course on Borel Sets*, Graduate Texts in Mathematics, 180, Springer-Verlag, New York, 1998.

146. J. Stern, *Partitions of the real line into* $\aleph_1$ *closed sets*, Higher set theory, Oberwolfach 1977 (G. H. Müller and D. S. Scott, eds.), Lectures Notes in Mathematics, vol. 669, Springer-Verlag, New York, 1978, pp. 455–460.
147. A. H. Stone,*Topology and measure theory*, Measure Theory, Oberwolfach 1975 (A. Bellow and D. Kölzow, eds.) Lecture Notes in Mathematics, vol. 541, Springer-Verlag, Berlin 1976, pp. 43–48.
148. T. Świątowski, *Sur une transformation d'une fonction mesurable en une fonction sommable*, Fund. Math. **52** (1963) 1–12.
149. E. Szpilrajn, *La dimension et la mesure*, Fund. Math. **28** (1937) 81–89.
150. ______, *The characteristic function of a sequence of sets and some of its applications*, Fund. Math. **31** (1938) 207–223.
151. ______, *On the isomorphism and the equivalence of classes and sequences of sets*, Fund. Math. **32** (1939) 133–148.
152. E. Szpilrajn-Marczewski, *Sur les ensembles et les fonctions absolument mesurables*, C. R. de la Soc. Sci. Varsovie **30** (1937) 39–68.
153. F. D. Tall, *Density topology*, Pacific J. Math. **62** (1976) 275–284.
154. E. K. van Douwen, *Chapter 3: The integers and topology*, Handbook of Set-Theoretic Topology (K. Kunen and J. E. Vaughan, eds.), North Holland, Amsterdam, 1984, pp. 111–167.
155. J. von Neumann, *Collected Works, Vol II: Operators, ergodic theory and almost periodic functions in a group*, Pergamon, New York, 1961.
156. W. Weiss, *Chapter 19: Versions of Martin's axiom*, Handbook of Set-Theoretic Topology (K. Kunen and J. E. Vaughan, eds.), North-Holland, Amsterdam, 1984, pp. 827–886.
157. A. Q. Yingst, *A characterization of homeomorphic Bernoulli trial measures*, Trans. Amer. Math. Soc. **360** (2008) 1103–1131.
158. Z. Zahorski, *Über die Menge der Punkte in welchen die Ableitung unendlich ist*, Tôhoku Math. J. **48** (1941) 321–330.
159. ______, *Sur la première dérivée*, Trans. Amer. Math. Soc. **69** (1950) 1–54.
160. O. Zindulka, *Dimension zero vs measure zero*, Proc. Amer. Math. Soc. **128** (2000) 1769–1778.
161. ______, *Universal measure zero sets with full Hausdorff dimension*, (2000), Preprint.
162. ______, *Universal measure zero sets with full Hausdorff dimension*, (2001), Preprint.
163. ______, *Small opaque sets*, Real Anal. Exchange **28** (2002/2003), no. 2, 455–469.

Notation index

Author index

Subject index